Satellite Sensing of a Cloudy Atmosphere:
Observing the Third Planet

Satellite Sensing of a Cloudy Atmosphere: Observing the Third Planet

Edited by
Ann Henderson-Sellers
Department of Geography, University of Liverpool

Taylor & Francis
London and Philadelphia
1984

First published 1984 by Taylor & Francis Ltd
4 John Street, London WC1N 2ET
Published in the USA by Taylor & Francis Inc.
242 Cherry Street, Philadelphia, PA 19106–1906

© 1984 Taylor & Francis Ltd

Typeset by Mathematical Composition Setters, Ltd.
7 Ivy Street, Salisbury, Wiltshire
Printed in the United Kingdom by
Taylor & Francis (Printers) Ltd.
Rankine Road, Basingstoke, Hampshire RG24 0PR

British Library Cataloguing in Publication Data

Henderson-Sellers, Ann
   Satellite sensing of a cloudy atmosphere:
   observing the third planet.
   1. Remote sensing
   I. Title
   551.4'028   G70.4

   ISBN 0-85066-254-0

Library of Congress Cataloging in Publication Data
Main entry under title:

Satellite sensing of a cloudy atmosphere.

   Includes bibliographies and index.
   1. Atmosphere—Remote sensing. 2. Clouds—Remote
sensing. I. Henderson–Sellers, A.
QC871.S28 1984   621.36'78   83–17993

For the guy who's seen clouds all over the world for me

Preface

The aim of this book is to present the principles and methods of remote sensing which are fundamental for the derivation and interpretation of information relating to the planet Earth. Our planet is fundamentally a water-dominated planet. Over two-thirds of its surface is ocean and the biogeochemical cycles between the atmosphere, biosphere and lithosphere all depend on water. The brightest objects are the polar caps and the clouds. Both of these are fundamental features of the Earth's atmosphere – climate system, and hence the environment within which we live. Without water, life on Earth would cease to exist.

The remote senser must accept the ubiquitous nature of water in its three phase states on and around the Earth. He/she must understand the effect that water vapour and sub-grid scale clouds could have upon the data.

This book will satisfy the theoretical and 'application' needs of final year undergraduates and new post-graduate students. Specific chapters will be of interest both to those already using remote sensing technology and to established scientists who wish to explore these new and exciting data. Here we seek to present satellite retrieval methods as part of a new view of the planet Earth — a watery paradise which provides both an hospitable environment for life and a source of frustration for scientists who hope, by the analysis of remotely sensed data, to understand better the home of humanity.

We are grateful to the following agencies: the Remote Sensing Society; NASA, for partial support of Dr A. M. Carleton; the National Science Foundation (USA), for support of both Drs A. Henderson-Sellers and A. M. Carleton; and the Natural Environment Research Council (UK), for support of Dr N. A. Hughes.

There is a short list of references and suggested further reading at the end of each chapter. We have intentionally sought to cite only widely available works although we freely acknowledge that more detailed information about specific satellites, sensors and projects can only be obtained from internal space agency reports and other 'grey' literature. Some copyright material has been drawn from sources other than those cited here. We wish to acknowledge and thank the following for permission to use their material: Plenum Press and J. F. R. Gower (ed.) for tables 1.1, 1.5 and 2.2 from *Oceanography from Space*, 1981; NASA and appropriate authors for figures 1.13, 1.14 and 1.18 and tables 1.6, 1.7, 1.8 and 1.9 from *Meteorological Satellites — Past, Present and Future*, 1982; the Royal Meteorological Society for figures 8.2, 8.3, and 8.4; and Springer Verlag for figures 8.5, 8.6 and 8.7.

I would particularly like to thank my colleagues Drs G. Ohring, K. P. Shine, A. J. Eccleston, R. Crane and J. G. Cogley who read a preliminary version of this book and whose comments are gratefully acknowledged. My friend Kendal McGuffie deserves a very big thank you for his editorial efforts which saved me from, amongst others, the 'Return Bean Vidicon' and the horrors of 'slime'. Many of the diagrams were drafted by my friend and colleague Mrs Sandra Mather. I am, as ever, grateful to the Department of Geography and the University of Liverpool for their continuing support. I am particularly grateful to my husband, Brian, who has, on many occasions during the writing and production of this book, set his own work on one side to help me with mine — thank you.

Ann Henderson-Sellers
Louvain-la-Neuve
April 1983

Contents

Foreword_____

It is appropriate that this book appears at this time. The world meteorological community has just completed perhaps the most magnificent example of international scientific co-operation in the history of mankind: the Global Atmospheric Research Program (GARP), culminating in the highly successful Global Weather Experiment. This same community, joined by scientists of other disciplines, has just embarked upon another major programme of international scientific co-operation: the World Climate Research Program (WCRP). The GARP was mainly a programme for studying those atmospheric processes that control changes of the weather, with a view towards improving the accuracy of weather forecasts for periods of one day to several weeks. The WCRP, which has its seeds in the GARP, has as its two main objectives determinations of: (1) the extent to which climatic fluctuations can be predicted; and (2) the extent of Man's influence on the climate. As its first project the WCRP has selected the International Satellite Cloud Climatology Project (ISCCP). This project will utilize an array of geostationary and polar orbiting satellites to obtain five years of observations of the Earth's clouds for the purpose of preparing a global cloud climatology with particular emphasis on their radiative properties. As its title implies, this book is mainly concerned with the Earth's clouds and their effects on remote radiative sensing of the Earth and its atmosphere.

Clouds modify the streams of solar and terrestrial radiation and thus the net radiative heating of the atmosphere and underlying surface. The net radiative heating influences the dynamics and thermodynamics of the atmosphere, which in turn affect the formation and dissipation of clouds. Uncertainties in estimating the potential feedback effects associated with this cloud–radiation interaction provide one of the greatest sources of uncertainty in determining the sensitivity of the climate to changes in external conditions, such as the solar radiation or the atmospheric carbon dioxide amount. Clouds affect remote observations of the properties of the atmosphere and surface of the Earth. These observations are critical for such diverse activities as weather forecasting, oceanographic research and Earth resources studies. By their effect on the radiation field, clouds themselves can be measured. These topics, and others, are included in this timely book, which has been brought together through the energetic efforts of Ann Henderson-Sellers and includes contributions from a team of active researchers from the United States, Great Britain and FR Germany.

George Ohring
NAS-NRC Senior Resident Research Associate
NASA/Goddard Space Flight Center
Greenbelt, MD 20771
(On leave from Tel Aviv University, Ramat Aviv, Israel)

1

Earth — the water planet

Ann Henderson-Sellers and Nicola A. Hughes
University of Liverpool
Liverpool, UK

1.1 Introduction

Planet Earth is coloured blue and white. Remote sensing of planets in our solar system in the very narrow wavelength region of visible light reveals fundamental characteristics. Venus, the Earth's 'sister' planet, is known as the evening, or morning, 'star' because the dense clouds which completely cover the planet reflect back to the observer over 70% of the incident sunlight. Mars owes it reddish hue to the abundant iron oxide on its surface. It is dark in appearance except for the polar caps, into which is frozen much of the carbon dioxide which would become part of the atmosphere were the surface temperature to rise. The most massive planet in the solar system, Jupiter, exhibits bands of white and red which show respectively the presence of high clouds and complex organic molecules at deeper levels in the atmosphere. The Earth's colour is determined by the presence and phase state of what is, arguably, its most important compound — water.

Over 70% of the Earth's surface is covered by liquid water. Liquid water and ice crystal clouds cover more than half the globe and polar caps dominate the high latitudes. It is clearly incorrect to argue that lithospheric processes control the 'face of the Earth' as viewed from space. Water fulfils an equally important role for the climate as for the natural environment. Two important components of the transport of energy around the globe are the flux of latent heat, which is determined by water phase changes, and the oceanic circulation. Clouds, together with ice and snow covered areas, dominate the reflectivity of the planet, while the height and amount of clouds is pre-eminently important in the control of the infra-red radiation emitted to space. The mean global surface temperature and regional and local scale climates, as well as surface features caused by water erosion, owe their existence and stability to

1

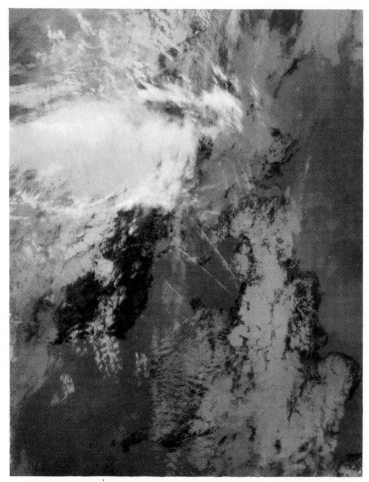

Figure 1.1. TIROS-N (AVHRR) infra-red image of the United Kingdom. The grey scale is proportional to the temperature of the emitting surface. Cold features appear bright while warmer features (such as the land) appear dark. Sea surface temperature is lower than land temperature. This is as would be expected for an early afternoon satellite pass in mid-August (1443 GMT on 18/8/79). The streaked features (such as the 'cirrus cloud' over the Irish Sea and Scotland) are anthropogenerated clouds — aircraft condensation trails (courtesy of P. Baylis, University of Dundee).

water. Any attempt to determine environmental and atmospheric information from remotely derived data is affected by water.

   In this book we present some of the aspects and techniques of remote
sensing which pertain to the troposphere and the surface of the Earth.
The focus is on water in each of its three phase states and in its transitions
between them. The astronaut viewing our planet from space would see

clouds, polar caps and ocean areas. We give each of these topics considerable attention.

The planet Earth is unique in the solar system because of the ubiquity of both water and life. Figure 1.1 illustrates not only the predominance of surface water and water and ice clouds in the environment but also the effect of Man. The streaks or linear cloud features emanating from NW England and Scotland are condensation trails from aircraft. Man may already be modifying the controlling agencies of water on Earth.

Water is one of the most peculiar substances on Earth. An understanding of its extreme or anomalous physical and chemical properties is important for successful interpretation of remotely derived information. Water can be regarded as hydrogen oxide or oxygen hydride. The properties of hydrides of the sixth group of the periodic table change slowly and predictably, with the exception of water. The freezing point of H_2O is higher than both that of H_2S and H_2Se. Its thermal capacity is high and the density variation as a function of temperature is non-linear with a maximum at approximately $4°C$ ($277K$). Many of the properties of water vapour with which remote sensing is concerned are the direct result of the molecular configuration of the hydrogen and oxygen atoms. Figure 1.2 shows how a dipole moment results from a charge displacement within the water molecule. The existence of this dipole moment permits interaction between radiation fields and the molecule causing both vibration and rotation absorption features. Additionally, the strong dipole moment causes aggregation among water molecules to such an extent that the effective molecular weight can be $6-8$ times greater than the theoretical value of 18. This association is the cause of the dimer (or 'n'-mer) absorption in the major 'window'† region in the absorption spectrum of the Earth's atmosphere.

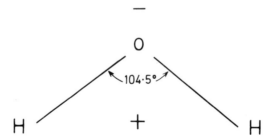

Figure 1.2. Structure of the water molecule.

---

†The widely used analogy between the Earth's atmosphere and the glass of a greenhouse can be extended so that regions of the electromagnetic spectrum where little or no atmospheric absorption occurs may be likened to open windows (see figure 1.6, p. 7).

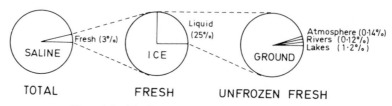

Figure 1.3. Pie diagram of the Earth's water resources.

The vast majority of water on the Earth is saline. Figure 1.3 shows that over 97% of the Earth's water contains salts in solution. Of the small percentage of fresh water most is 'locked' into ice caps. Only $1·3\%$ is in the form of lakes and rivers, and a tiny fraction (approximately $0·14\%$) exists in the vapour phase in the atmosphere. Life almost certainly developed in an aqueous environment. Cells are approximately 80% water and, while it is possible to live for between 70 and 80 days without food, even short-term deprivation of water can be fatal.

Man is interested primarily in the land areas of his planet. Viewing these is difficult because of the atmosphere which, by its very existence, makes them hospitable environments for life. Retrieval of information pertaining to land surface features from satellites is hampered by the presence of clouds and other distortions caused by the atmosphere. For these reasons a section in this book deals particularly with the difficulties encountered by those using remotely sensed data from which atmospheric aberration must be removed. The fundamental feature of our planet as viewed from space is, however, a climate dominated by a dynamic, cloudy atmosphere.

The early and rapid development of satellite systems was focused on the retrieval of meteorological information. The global radiation budget determines the sources and sinks of radiative energy for the Earth–atmosphere system. It is these sources and sinks that drive the general circulation of the atmosphere and thus control the climate. Prior to the satellite era, knowledge of the Earth's radiation budget and climate was based on surface observations (see, for example, the classic study of London, 1957). The importance of radiation budget observations from satellites was recognized very early in the era of spacecraft observations of Earth. Climatological studies were begun as early as 1962, only two years after the launching of the first weather satellite. The resulting estimates of the components of the radiation budget from the visible and infra-red scanning radiometers are reviewed in Chapter 3.

One of the important characteristics of the Earth's climate is its sensitivity to changes in boundary or external conditions. A typical measure of this sensitivity is the response of the mean surface temperature to a change in the solar constant. In turn, this response depends on the sensitivity to changes in surface temperature of the long-wave radiation emitted by the Earth–atmosphere system and of the Earth's planetary albedo. The latter two are components of the radiation budget that are measured from satellites (see, e.g., Chapter 3).

The global hydrological cycle is an important agent in the transport of energy around the globe. It now seems likely that the formation of the hydrosphere followed very closely after the formation of the Earth itself, and that the existence of water vapour and clouds in the early atmosphere compensated for the lowered solar luminosity in the early Precambrian (around four thousand million years ago; Rossow *et al.*, 1982). The total mass of water in the atmosphere is approximately equal to one week's rainfall over the globe. Thus there is a rapid (of the order of 5–6 days) cycling of water vapour from the surface into clouds and subsequently back to the surface as precipitation. Since both precipitation and evaporation vary independently of one another and with latitude, it is clear that to retain a global balance there must be considerable movement of water vapour by the atmosphere. The predominant areas of evaporation are the regions of high surface pressure over the sub-tropical oceans. Here

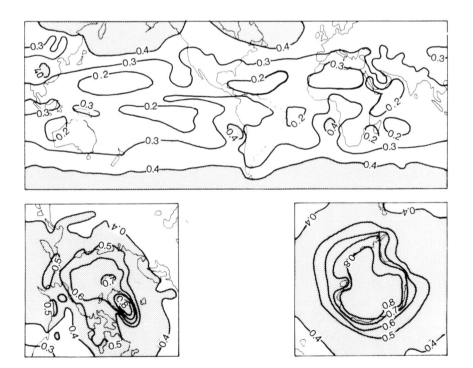

Figure 1.4. Annually averaged planetary albedo map of the world from NOAA satellite data for the period June 1974 to February 1978 (after Winston et al. 1979; see Chapter 3).

descending air inhibits condensation and precipitation while high insolation provides the energy source over a free water surface, leading to
evaporation rates sufficient to provide water vapour for much of the rest
of the globe. Conversely, precipitation considerably exceeds evaporation
both in the Intertropical Convergence Zone (ITCZ) and polewards of
about 40° latitude. Precipitation rates as high as 5000 mm per year have
been observed in the tropical Pacific Ocean. Figure 1.4 is a map of the
Earth's planetary albedo derived from the NOAA polar orbiting
satellites. High albedo areas are the polar caps, the regions which are
persistently cloudy (ITCZ and the eastern edge of many oceans) and the
continental deserts.

The atmosphere of the Earth has been described as a 'heat engine':
terminology which suggests considerable work done in response to input
energy. However, the climate system operates so that there is an almost

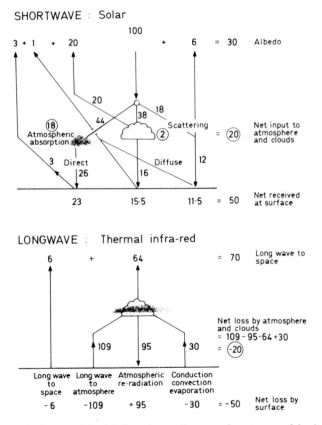

Figure 1.5. Schematic diagram of the globally and annually averaged components of the short-wave and thermal infra-red radiation components of the surface – atmosphere system.

perfect balance between absorbed solar radiation and emitted terrestrial radiation. The frantic motions of the atmosphere: tropical storms, global scale precipitation, droughts and dust devils are the result of the dissipation of approximately $3\,\mathrm{W\,m^{-2}}$, compared to the incident energy at the top of the atmosphere of around $240\,\mathrm{W\,m^{-2}}$. Only a very inefficient engine operates at a conversion rate of the order of 10^{-2}. Atmospheric phenomena are then the result of only a small residual in the overall planetary radiation balance. This balance does however determine the climatic regime, and is itself, in turn, determined by the presence, amount and nature of clouds in the troposphere.

Figure 1.5 shows a schematic representation of the Earth's radiation regime. Only 70 of the 100 incident units of energy are absorbed into the climate system; the majority of the reflection being due to clouds. Longer wavelength thermal radiation is emitted by the Earth's surface but this is absorbed within the atmosphere, particularly by clouds. Re-emission occurs in both upward and downward directions. The final radiation balance on which the surface temperature and hence our ambient climatic environment depends is determined by the amount of radiation intercepted and finally re-emitted by clouds.

The two most important wavelength regions for the Earth's radiation environment are therefore the regions of emission of the Sun and of the

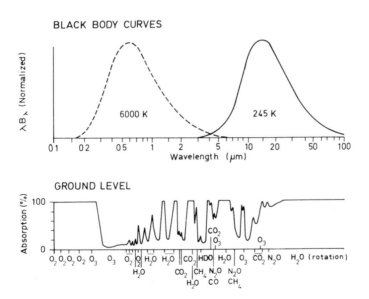

Figure 1.6. Wavelength regions dominated by solar ($0\cdot2-4\ \mu m$) (blackbody temperature 6000K) and terrestrial ($2-50\ \mu m$) (blackbody temperature 245K) radiation. The atmospheric absorption is seen to be closer to zero at short wavelengths than in the region of terrestrial radiation where it varies between 10% and 100% (after Goody, 1964; cf. figure 2.2, p. 50).

Earth itself. Figure 1.6 compares the wavelength regions of the Planck curves of the Sun and the Earth. Solar radiation falls within the region $0 \cdot 2 - 4\mu m$ while terrestrial thermal infra-red radiation extends from around $2\mu m$ to approximately $50\mu m$. In these wavelength regions interaction with water vapour, water droplets and ice crystals is inevitable. Thus, while other wavelength regions (such as microwave/radar) will be considered briefly, the majority of this text will focus on solar and terrestrial wavelength regions. The Planck curves for blackbodies at temperatures of around $6000\,K$ and $300\,K$ are such that only $0 \cdot 4\,\%$ of the emitted solar flux is at wavelengths longer than $5\mu m$ while only $0 \cdot 4\,\%$ of the emitted terrestrial flux occurs at wavelengths shorter than $5\mu m$. The solar and terrestrial streams can thus be treated as independent for most purposes.

The most transient, visible features of the Earth's dynamic atmosphere are clouds. Clouds control the climate by contributing to both the planetary albedo and the emitted terrestrial radiation. The horizontal movement of energy by the atmosphere and the oceans is fundamentally important to the energy budget of all locations. The high latitudes are particularly dependent upon atmospheric input of both energy and precipitation. Water vapour and thus clouds are confined almost entirely to the lowest layer of the Earth's atmosphere: the troposphere (figure 1.7).

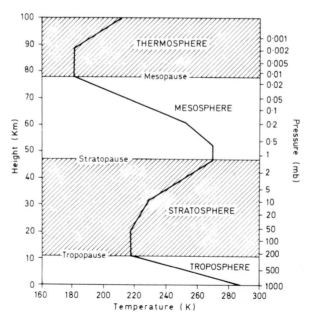

Figure 1.7 Vertical temperature profile of the Earth's atmosphere showing the major levels or 'spheres'.

This is because the tropopause is too cold to allow penetration of water vapour into the stratosphere. The vast majority of the atmosphere (over 80 % by mass) is also within the troposphere. The principles of remote sensing described in this book are confined to the troposphere. Many particular techniques are designed to retrieve information either about the surface or about the gaseous atmosphere. For example, in Chapter 7 the effects of contamination of radiance data being used directly for crop surveillance are described; remote sensing of the ocean surface (Chapter 7) is also hampered by the occurrence and nature of clouds. In Chapter 8 satellite radiation data are used to monitor both the extent and state of the cryosphere and the interactions between cloud, precipitation and sea-ice and snow extent. Satellite retrieval programmes which are designed specifically to establish information about the gaseous state of the troposphere are particularly easily disrupted by the occurrence of clouds. Chapters 4 and 5 are devoted to descriptions of atmospheric data retrieval from satellite-borne sensors. In the case of vertical temperature sounding of the atmosphere, for example, the technique depends on the way in which individual gaseous constituents absorb and re-emit long-wave thermal radiation.

Clouds also interact strongly with terrestrial infra-red radiation in a manner which is strongly frequency dependent. It is only by use of multi-spectral observations on microwave or radar sensors that this effect can be overcome. Multi-spectral radiometers can be used to retrieve vertical temperature profiles in areas that are designated cloud-free or for which 'cloud-cleared' radiances can be established as part of the temperature retrieval. Eyre and Jerrett (1982) describe the use of the High Resolution Infra-red Sounder (HIRS; see table 2.2, p. 70) radiance measurements from the TIROS-N polar orbiting satellites in a programme designed to establish near real-time retrieval of atmospheric layer thicknesses and thermal winds from satellite data. In specific analyses undertaken during June 1981 scientists at the UK Meteorological Office established that an operationally useful meteorological product could be derived from the NOAA-6 and NOAA-7 multi-spectral radiometer measurements (current satellite names and acronyms are listed in table 1.1). Their analysis for a period in June 1981 illustrates the loss of data which is suffered in areas of unbroken cloud cover where only the microwave radiance measurements can be used. The structure of the cloud field is also important since 'cloud clearing' algorithms can successfully 'remove' the effects of only some broken clouds. Although it is clear that satellite-based retrievals of temperature profiles over sparsely populated and in-hospitable regions of the globe can add considerably to the current information available for meteorologists and climatologists, it is also important to note that the quality of data will be affected by the presence and relative persistence of cloud cover.

Table 1.1. Current and planned satellites: names and acronyms.

Acronyms	Full name
COMSS	Coastal Oceans Monitoring Surveillance Satellite. Planned by ESA, now ERS-1.
Cosmos	General name for Russian satellites, continuing.
DMSP	Defense Meteorological Satellite Program.
DOMSAT	Domestic geostationary communication Satellite. Planned, US.
ERS-1	ESA Remote Sensing satellite. Planned; microwave and ocean colour sensors.
ERTS	Earth Resource Technology Satellite, NASA, US, renamed Landsat.
GEOS-3	Geodynamics Experimental Ocean Satellite, NASA, US, 1975–1978. Radar altimeter and tracking systems.
GMS	Geostationary Meteorological Satellite, Japan, continuing. Visible and infra-red full disc imaging.
GOES	Geostationary Operational Environmental Satellite, NOAA, US, continuing. East is located at 75°W and West at 135°W longitude. Visible and infra-red full disc imaging.
GRAVSAT	Gravity Satellite, NASA, US, planned. A satellite pair in very low, drag-corrected orbit.
Intercosmos	Russian satellite with international participation, continuing.
Landsat	Land resources satellite, NASA, US, continuing. High resolution optical sensors; infra-red operational on Landsat-4.
METEOSAT	European geostationary weather satellite, continuing with gaps. Visible, water vapour and infra-red full disc imaging.
Nimbus	A series of experimental weather satellites, NASA, US, continuing. A wide variety of atmosphere, ice and ocean sensors.
NOAA	A series of operational weather satellites, NASA, US, continuing. Visible and infra-red imaging.
SARSAT	Synthetic Aperture Radar Satellite, planned, Canada/US. Imaging radar emphasizing ice mapping.
SEASAT	Experimental ocean satellite, NASA, US, 1978. Microwave sensor experiments.
SMS	Synchronous Meteorological Satellite. Original NASA name for GOES satellites.
SURSAT	Surveillance Satellite, Canada. A project to evaluate satellite capabilities.
TIROS	Television Infra-Red Operational Satellite. NOAA weather satellites starting with world's first weather satellite in 1960. TIROS-N was the prototype for the latest series of NOAA satellites.
TOPEX	Ocean Topography Experiment, planned US. Precision radar altimeter and tracking systems, with supporting microwave instruments.

In all the examples of satellite retrieval of surface and atmospheric information described in this book the occurrence of cloudiness considerably modifies satellite retrieved data. In all the cases considered above, information about cloudiness is not required but inevitably occurs within the total data set. All users of remote sensing techniques are forced to an awareness of cloudiness as a climatological feature. Prior to the existence of satellites, undertaking a description of the Earth's cloud climatology was an almost impossible task. Even with the advent of meteorological satellites in the 1960s, the variable nature of sensors, orbits and cloud retrieval algorithms was such that the composition of a cloud climatology was still very difficult. Since all remotely derived satellite data retrieved in the wavelength regions of solar or terrestrial radiation and pertaining to the troposphere or the surface of our planet are affected by the presence of clouds, it is worthwhile to consider the information available regarding cloud climatologies. The following two sections provide a detailed review of the history and current status of cloud climatological information. The cloud data discussed here will be referred to and drawn upon throughout the rest of this book.

1.2 An historical review of cloud sensing

A cloud climatology can be derived from two basic sources: (*a*) conventional surface observations taken by a trained observer; and (*b*) from data acquired by satellites.

Surface observations of cloud

Surface observations of cloud have been made routinely over many years as part of the regular national meteorological networks. In the United Kingdom cloud observations are taken and archived using an okta scale, while in the United States of America a scale of tenths is used. Other countries use the two scales in approximately equal proportions. Observers are instructed to estimate the cloud amount, type and height from a viewpoint which commands the widest possible view of the sky and to be careful to give equal weights to areas around the zenith and to those at lower angles of elevation. Since perspective 'flattens' the sky in the case of scattered and broken cloud, the observer sees the sides, as well as the bases of clouds; particularly at low elevation angles (figure 1.8). Also, the observer may not see gaps in the cloud field towards the horizon. Therefore, the surface observer frequently overestimates the amount of

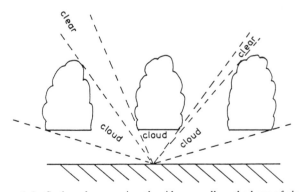

Figure 1.8. Surface observers view the sides as well as the bases of clouds.

cloud. The representativeness of the surface point observation also depends on the height of the clouds and thus the frequency of low cloud situations.

Errors of varying magnitude occur depending on the experience of the observer, the location of the observation site and the method of observation. Surface observations are notoriously poor at night, especially the identification of high, cirrus cloud. Additional errors can result from variations in the national reporting procedures.

A more objective technique for surface observation of cloudiness uses whole-sky cameras. The panoramic photographs are projected onto a linear grid and cloud amount is estimated. For logistic and economic reasons this technique can only be used on a limited basis.

Satellite observations of cloud

Satellite observations are measurements of the spectral distribution and intensity of radiation received by the satellite from the underlying atmosphere and surface. Present satellite data primarily consist of a digital value for each pixel (picture element) of the satellite 'image', representing the amount of radiation received by the sensor for a given spectral band. The meteorological satellites primarily sense in the visible and infra-red wavelength regions. Early researchers, for example, Clapp (1964), manually interpreted the satellite images to produce nephanalyses.† These were then used to estimate cloud amount. The other major technique used is averaging and compositing visible channel brightness data. Such techniques, however, cannot provide cloud observations of the accuracy required, for example by climate modellers, nor fully exploit the potential of the digital satellite data.

The increased volume of cloud amount information available from current satellite systems demands an automated technique for evaluating the cloud parameters. A preliminary review of the automated cloud analysis algorithms, currently being considered for the International Satellite Cloud Climatology Project (ISCCP), is given by Schiffer and Rossow (1983), who list the three groups of cloud retrieval algorithms: threshold techniques; statistical techniques; and radiative transfer techniques. No single version of these algorithms can derive all the necessary cloud properties required. Cloud properties in multi-layered systems are particularly difficult to deduce using any of the cloud analysis schemes. The spatial resolution of the satellite radiance measurements affects the results of all the algorithms. Some of the cloud analysis algorithms calculate cloud cover by assuming each pixel to be either cloud-free or

---

†A nephanalysis is a map of the distribution of cloud amount and cloud information.

cloud-filled. Higher spatial resolution clearly decreases the error in this assumption, but a high enough resolution may result in an impracticably large volume of data. The identification of cirrus cloud, particularly overlying another cloud deck, remains a very difficult problem (see the discussion in Chapter 6).

None of these cloud retrieval techniques has yet been applied to a global and seasonal data set. Each retrieval algorithm requires different data inputs (in addition to the basic satellite data). They are based on different assumptions, utilize varying amounts of computing resources and evaluate different cloud parameters. The products of the cloud analysis algorithms are currently being systematically compared and assessed for the first time, prior to the choice of algorithm for the ISCCP. Since the algorithms differ in accuracy for specific parameters, it is possible that a combination of the currently available algorithms will provide the best cloud estimation technique.

Comparison of surface and satellite cloud observations

Conventional surface observations are point or small area estimates of cloudiness in contrast to the large, areally contiguous coverage of cloud observations from satellites. The distribution of ground stations is uneven. For example, the stations tend to be located on the coast, with few in upland areas. Ship data can contain even greater biases: observations are often restricted to major shipping lanes, while data from island stations are not particularly representative of the surrounding oceans.

Surface-based observations have been shown to systematically overestimate total cloud amount compared to satellite observations. This appears to be a result of the overestimation of cloud amount by surface observers (Figure 1.8). It has been suggested that satellite-derived values of total cloud cover are closer to the estimate of the 'true' cloud cover for use in climate modelling studies, but any underestimate of cloud occurrence can cause serious problems for the remote sensing community (Chapter 7). Cloud observations from the ground are subjective, especially in the estimation of cloud amount, whereas satellite observations can define objectively the cloud amount. However, cloud type can be more accurately evaluated by a surface observer than from analysis of satellite data.

Surface observations provide relatively accurate estimates of low and total cloud coverage; however, middle and high cloud can only be observed in the absence of low or middle level cloud layers respectively. Similarly, satellite observations provide relatively accurate estimates of high and total cloud cover. The two observing techniques provide different, and complementary, perspectives of the cloud amount. These differences

Table 1.2. Characteristics of the global cloud climatologies available at present. Data compiled from the sources referenced in Hughes (1984).

Author	Spatial resolution[a]	Time scale[b]	Time of day[c]	Temporal coverage by maps[d]	Projection[e]	Countour interval[f]
Cloud climatologies compiled from conventional surface observations						
Brooks, 1927	10°	?	40% between 7–10 LST	No maps	NA	NA
Shaw, 1936	10°	?	as Brooks, 1927	Annual, monthly	PS	T
Haurwitz & Austin, 1944	?	?	?	Jan, July	U	T
Landsberg, 1945	?	?	?	Bi-monthly	H/Me	T
Telegadas & London, 1954	5° lat & 10° long	?	1200 GMT	Winter, Summer	PS	T
Seide, 1954	5° lat & 10° long	?	1200 GMT	No maps	NA	NA
London, 1957	5° lat & 10° long	?	1200 GMT	4 seasons	PS	T
Hastenrath & Lamb, 1977	1°	1911–1970	DA	Monthly	SC	T
Cloud climatologies compiled from satellite data						
Arking, 1964	?	12/7/61–10/9/61	Sp day	No maps	NA	NA
Rasool, 1964	?	12/7/61–10/9/61	Sp night	No maps	NA	NA
Taylor & Winston, 1968	5°	1/2/67–1/2/68	Sp day	Seasonal, monthly	Me	11 brightness levels
Kornfield & Hasler, 1969	?	1967	Sp day	Seasonal	PS/Me	brightness levels
Miller & Feddes, 1971	?	1967–1971	Sp day	Annual, seasonal, monthly	PS/Me	5 levels
Bean & Somerville, 1981	2.5°	1974–1977	Sp day	Annual, seasonal	SC	η and γ

patial verage[g]	Zonal averages[h]	Source of cloud data	Comparisons	Motivation
	Annual, monthly (D)	1000 land stations, shiplogs, expedition data	Arrhenius, 1896	To describe world cloud cover
	No	As Brooks, 1927	No	Text book
	Quotes Brooks, 1927	? Conventional surface observations	No	Text book
	No	? Conventional surface observations	No	Text book
H	Winter, Summer (D)	McDonald (1938), plus station data on cloud type/height	No	Climate modelling
H	Spring, Autumn (D)	McDonald (1938), plus station data on cloud type/height	Brooks, 1927	Climate modelling
H	4 seasons (D)	McDonald (1938), plus station data on cloud type/height	No	Climate modelling
tlantic Pacific ceans	No	Ship observations (includes low cloud amount)	No	Compilation of climatological atlas
G	For time available	TIROS-III 1447 photographs: simple threshold technique	London, 1957	Investigating techniques to derive cloud amounts from satellite data
G	For time available	TIROS-III infra-red data	Arking, 1964; Haurwitz & Austin, 1944	
	No	ESSA photographic brightness data	No	
	No	ESSA brightness data	No	
	Compiled by Schutz & Gates 1971 – 1972	ESSA brightness data 'calibrated' against surface observations	No	
	No	Cloud estimated from NOAA SR system albedo – cloud frequency observations modelled with beta distribution	No	

Table 1.2 continued

Author	Spatial resolution[a]	Time scale[b]	Time of day[c]	Temporal coverage by maps[d]	Projection[e]	Contour interval[f]
Chahine, 1982	4° lat & 5° long	1–7 Jan 1975	Sp	7-day average	?	T
Cloud climatologies compiled from satellite-derived nephanalyses						
Clapp, 1964	4°	1/3/62– 1/2/63	Sp day	Annual, seasonal	Me	T
(a) Sadler, 1969 (b) Sadler *et al.* 1976	2.5°	(a) 1/2/65 –1/1/67 (b) 1/1/65 –1/1/72	Sp day	Annual, monthly	Me	Neph-code
Godshall *et al.* 1969 Godshall, 1971	2°	1/8/62– 1/8/68	Sp day	4-monthly	SC	T
Cloud climatologies compiled for climate modelling using a variety of data sources						
Newell, 1970 (Dopplick, 1972)	10°	(a) 1967 –1968 (b) 1964	?	No maps	NA	NA
Van Loon, 1972	?	?	?	Jan, July	PS	20%
Schutz & Gates, 1971, 1972	4° lat & 5° long	(a) 1963 –1968 (b) 1967 –1970	(a) 0000– 1200 (b) 1400 LST	Jan, July	SC	T
Berlyand & Strokina, 1980a & b	5°	30 years	daytime	Monthly (D)	H	T
Regional cloud climatologies						
Sherr *et al.* 1968	NA	Greater than 10 years	DA	No maps	NA	NA
Cloud climatologies 'in preparation'						
Gordon *et al.*, (pc)	?	Jan 1977, July 1979	?	Jan, July	SC	T
London *et al.*,	5°	Since 1902 for land, since 1854 for ocean	DA	?	?	?

tial erage[g]	Zonal averages[h]	Source of cloud data	Comparisons	Motivation
	Yes	VTPR data using a radiative transfer technique	Sadler, 1969	
	Yes	TIROS nephanalyses	Arking, 1964; Vowinckel, 1957; Landsberg, 1954	
North outh	Annual, monthly	TIROS nephanalyses	Landberg, 1954; Clapp, 1964; & rainfall data	
fic an	No	TIROS and ESSA nephanalyses	Surface observations	
	Monthly	Cram; ETAC (TIROS nephanalyses); Gabites, 1960; London, 1957	No	Climate modelling
	Jan	Brooks, 1927; Landsberg, 1945; Vowinckel & Van Loon, 1957; Clapp, 1964; Sadler, 1969; & national climate atlases.	No	Climate modelling
0–15°N QG	Jan, July	(a) ETAC (1971) surface and satellite observations (b) Miller et al. (1970), as Miller & Feddes, 1971	No	Climate model comparison
	Monthly	McDonald, 1938, surface and satellite observations, e.g., Sadler, 1969; Miller & Feddes, 1971	Extensive comparison of zonal averages	Climate modelling
	NA	Used maps of satellite & surface mean cloud to define cloud climatological regions	No	Simulation of global cloud cover
	Jan, July	(a) 3D-nephanalysis data (b) Surface observations	Between themselves	Climate modelling
	?	Ship observations being used to obtain oceanic cloud amount. Total cloud plus cloud types.	?	

Table 1.2 continued

Author	Spatial resolution[a]	Time scale[b]	Time of day[c]	Temporal Coverage by maps[d]	Projection[e]	Conto interva
Rossow et al.	?	1977	Sp	?	?	?
ISCCP	2.5°	5 years 1983–88	3h	?	?	?

[a]Spatial resolution or spatial average of cloud climatology.
[b]Time scale of the observations included in the climatology.
[c]Time of day of the observations: DA = diurnal average, Sp = satellite pass (either day or night if p orbiter).
[d]Temporal average for which maps are available: (D) = numerical data available.
[e]Cartographic projection for which maps are available: PS = polar-stereographic, H = Hamme

between satellite and surface cloud observations need to be considered when comparing cloud climatologies derived from different source data. Cess et al. (1982, p. 58) suggest that "it is by no means clear that a unique definition for cloud amount even exists". Cloud is defined differently by each type of cloud observation or retrieval technique.

A chronological history and description of the published and accepted cloud climatologies is presented by Hughes (1984). The major characteristics of these cloud climatologies are summarized in table 1.2. This review emphasizes the problem of defining a global cloud climatology. The cloud data sets examined by Hughes (1984) are, as far as possible, global in extent. The cloud climatologies can be classified based on the motivation for producing the cloud climatology and the cloud observation technique employed. The following classification includes most published, currently available cloud climatologies.

1. Cloud climatologies compiled from conventional surface observations.
2. Cloud climatologies compiled from satellite data.
3. Cloud climatologies compiled from satellite derived nephanalyses.
4. Cloud climatologies compiled for climate modelling using a variety of data sources.
5. Regional cloud climatologies.
6. Global cloud climatologies 'in preparation'.

Cloud climatologies compiled from conventional surface observations

One of the earliest descriptions of global cloud amount was given in 1927 by Brooks (see table 1.2). Brooks used monthly averages of mean

patial overage[g]	Zonal averages[h]	Source of cloud data	Comparisons	Motivation
;	?	NOAA visible and IR data; using radiative transfer technique	?	
;	?	Polar-orbiter and geostationary satellite data	Extensive comparison anticipated	

Me = Mercator, NA = not applicable, SC = Simple Cylindrical, U = Unknown.
Contour interval used on map: T = tenths or 10 % intervals: NA = not applicable.
Spatial coverage: G = global, QG = quasi-global, NH = Northern Hemisphere, SH = Southern Hemisphere.
Time period for which zonal averages available: (D) = numerical data available, NA = not applicable.

cloudiness from more than 1000 land stations distributed as evenly as possible around the Earth. These data were plotted on large charts of the globe (mercator projection) and isonephs (contour lines of equal cloud amount) were drawn. The charts were ruled into a 10° latitude–longitude grid and the mean cloudiness of each square estimated. More than 40 % of all observations were between 0700 and 1000 LST. Other early surface-based cloud climatologies (see table 1.2) are similarly heterogeneous. Marine observations are concentrated along major shipping routes. Thus, for vast areas, such as parts of the South Pacific Ocean, information is very limited. This is a major problem for any cloud climatologies utilizing marine cloud amount observations.

The most influential published research on global cloud climatologies was undertaken in 1954 by three researchers: Telegadas, London and Seide. Published separately in the same year, these reports jointly provide the data used by London (1957). This climatology is widely used because it provides the only estimates of Northern Hemisphere cloud type and cloud height. It should be noted that researchers using these cloud statistics on a global scale must either seasonally reverse the Northern Hemisphere data for the Southern Hemisphere or use supplementary Southern Hemisphere cloud statistics.

Cloud climatologies compiled from satellite data

The first estimates of global cloud cover from satellite data became available in the early 1960s (table 1.2). Several papers which appeared between 1964 and 1971 used satellite-derived brightness values to evaluate cloud amount distributions. This type of technique gives only an

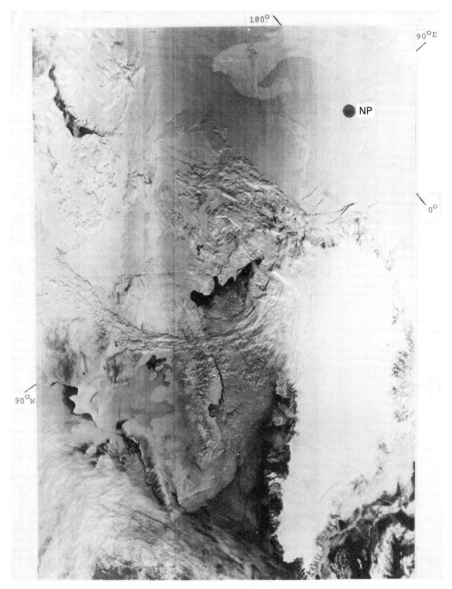

Figure 1.9. DMSP visible image of Greenland and the North Pole (marked to the top right). Open water features in the sea ice (polynyas; see Chapter 8) are easily identified. Discrimination between cloud and the highly reflective surface is more difficult. Additional information, either from sensors in other wavebands (Chapters 6 and 8) or from knowledge of cloud and ice-feature structure (pattern recognition), must be used for interpretation (courtesy of D. Robinson, Lamont Doherty Geological Observatory).

indication of cloud amount as the visible radiation is dependent also on instrumental factors and the contaminating influence of surface albedo.

Multiple exposure techniques have been used where photographic film is exposed to a series of satellite pictures taken daily at the same time and geographical location. The additive property of the photographic process is used to produce a measure of cloud cover for a given time period. In the time-averaging maximum brightness technique, the minimum brightness field is subtracted from the total brightness to provide a value which may be considered as being due to clouds alone. This technique, however, results in relative cloud cover only. Areas of high surface albedo, especially desert and ice/snow covered surfaces, are misinterpreted as areas of high percentage cloud cover. The difficulty of distinguishing cloud from underlying snow and ice-covered snow is well illustrated in figure 1.9. This is a DMSP visible image from May 1979. The shadows cast by the clouds permit differentiation of cloud and ice and identification of structure in cloud. Leads in the ice can be seen near the pole which suggests that cloud cover is thin.

Many of the cloud retrieval algorithms discussed above are still being developed and provisonal results are only now beginning to be published. These are mainly regional studies. Few cloud retrieval algorithms have yet been applied to global data sets.

Cloud climatologies compiled from satellite-derived nephanalyses

Global scale cloud climatologies have also been evaluated from satellite-derived operational nephanalyses (table 1.2). The nephanalyses were mostly prepared by the Data Processing and Analysis Division of the National Environmental Satellite Service (NESS). They were hand-drawn from interpretations of photographs from the vidicon cameras on the early satellites, and since November 1972 (the launch of NOAA-2) from the visible channel imagery of the scanning radiometers. For example, seasonal maps of global cloud amount were produced by Clapp (1964) using TIROS nephanalyses. Major problems resulted from inadequate data coverage and the uncertainty in extracting cloud amount from the nephanalysis codes (see Hughes, 1984). Clapp (1964) also presented a detailed description and critical analysis of the data. This latter summary illustrates the state of research in the mid-1960s. "The meager observational evidence and crude analysis of errors presented [here] suggest that [the total cloud] amounts estimated from the TIROS pictures tend to be too large for large cloud amounts and too small for low cloud amounts, but information is too scanty to justify a quantitative estimate of the average errors" (Clapp, 1964, p. 507).

Cloud climatologies compiled for climate modelling using a variety of data sources

Recently global cloud climatologies have been compiled using both surface and satellite cloud amount observations specifically to provide cloud
amount data for climate modelling (table 1.2). These composite cloud
climatologies are heterogeneous in nature. For example, a Southern
Hemisphere cloud climatology was estimated by Van Loon (1972), which
incorporates a variety of cloud data. These include Brooks' data
(latitudinal cloud amount from surface observations), other surface
observations, whaling-ship observations, TIROS satellite nephanalyses
and various marine climatic atlases (table 1.2). The maps of total cloud
amount for January and July result from "a subjective fusion of these
sources, and probably gives a reasonable picture of total cloudiness" (Van
Loon, 1972, p. 101). The value of such heterogeneous cloud climatologies
is highly questionable, considering the basic differences between satellite
and surface cloud observations.

Several attempts have been made recently to establish a global three-
level cloud cover climatology. For example Becker (see Hughes, 1984)
produced a global cloud climatology for total cloud amount and five cloud
groups: cirrus, nimbostratus (frontal-type cloud), low (stratus and
stratocumulus), cumulus and cumulonimbus. Similarly, Meleshko (see
Chapter 3) has calculated global three-level cloud cover using total cloud
amount data, satellite-derived outgoing infra-red radiation information,
temperature and humidity data.

Regional cloud climatologies

Regional cloud statistics have been compiled by many authors using a
variety of data sources. More interestingly a cloud data set designed
specifically for the remote sensing community, relying on the regionality
of cloudiness characteristics, was produced by Sherr and his colleagues in
1969 (table 1.2). They intended to provide global cloud statistics for the
simulation of Earth-orientated observations from space. Homogeneous
cloud climatological regions were defined based on the climatic zones
proposed by Köppen and the mean cloud amount distributions from
Landsberg, ETAC, Clapp (1964) and early observations by Sadler (see
table 1.2). The objective was to create a master file of global cloud
statistics, including diurnal and monthly variations. Data from a
representative surface station in each region were used to compile the
cloudiness statistics. This represents an alternative approach to compiling
a cloud climatology and is particularly orientated to a specific user community. The basic model of global cloud cover developed by these researchers has been extended by other researchers.

This concept of homogeneous cloud climatological regions may provide a framework within which to compile, collate and compare detailed cloudiness statistics and could facilitate comparisons between regional and global studies and allow the geographical variations in the relationship between satellite-derived and surface observations and between different cloud algorithms to be established. Within such homogeneous regions the error due to noise is likely to be smaller than the differences between the different climatologies.

Global cloud climatologies 'in preparation'

Several global-scale cloud climatologies are known to be currently in preparation. These include the following research projects.

1. London, Warren and Hahn at the University of Colorado are analysing conventional surface-observed cloud data sets. This climatology will include cloud type, amount and height at $5° \times 5°$ latitude–longitude resolution, diurnal (3-hourly) and inter-annual variations, and long-term trends. Data will include 1902 to the present day over land and 1854 to the present day over oceans. Global maps will be produced for total cloud amount and for each cloud type.

2. NASA's Goddard Institute for Space Studies is deriving a climatology of cloud fractional cover, visible optical thickness and cloud top height. The data used are the visible and infra-red scanning radiometer data from NOAA-5, and temperature and humidity profiles from the National Climate Center (NCC). This analysis reconstructs the viewing geometry of the satellite observations and then compares the observed radiances to theoretical radiances calculated assuming the field of view is either completely cloud-filled or cloud-free.

3. The ISCCP (e.g., Schiffer and Rossow, 1983) plans to produce a global satellite-derived cloud climatology. Provisional descriptions of these climatologies for comparison have been included in table 1.2. Most of the global cloud climatologies in preparation aim to produce a comprehensive three-dimensional analysis of cloudiness. When complete they will provide a useful source of cloud information for the remote sensing community.

1.3. Comparison of available cloud climatologies

The most widely used cloud climatologies are compared in figures 1.10 and 1.11. Most cloud climatologies presented in the literature are available only as maps. Quantitative comparison is difficult as numerical

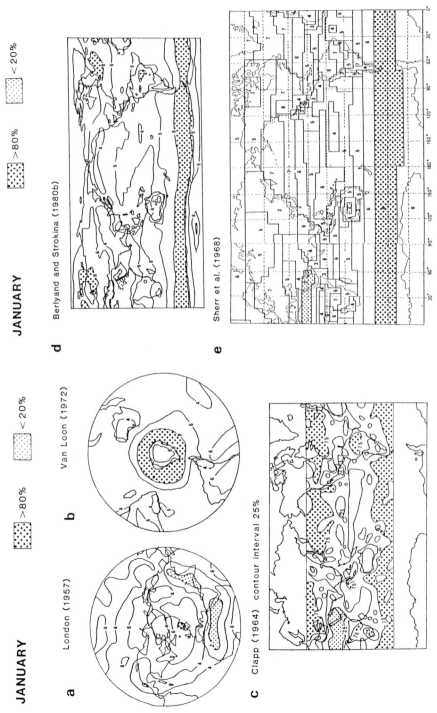

Figure 1.10. Global cloud climatologies for January, from the authors cited. Further details are given in table 1.3. The contour interval is 20% (2 = 20%, 4 = 40%, etc.), unless otherwise stated.

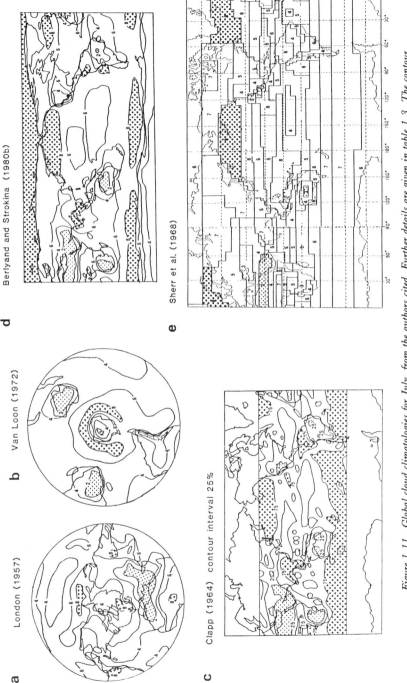

Figure 1.11. Global cloud climatologies for July, from the authors cited. Further details are given in table 1.3. The contour interval is 20% (2 = 20%, 4 = 40%, etc.), unless otherwise stated.

data are available for only two of the climatologies listed in table 1.2. The cloud amount data represented by the maps are for a variety of spatial and temporal averaging scales, different time spans and diurnal samples, and have also been derived from different types of cloud observations. The maps are presented on different cartographic projections, at different scales, using different contour intervals. The cloud climatologies are, therefore, difficult to compare, even by subjective, visual examination. The cloud climatologies available as maps of total cloud cover for January and July conditions are listed in table 1.3. The projection, type of observation, time of observation and time span are also given. Total cloud cover is the only parameter archived in most climatologies. It is also likely to be the most accurately derived parameter. The minimum number of cartographic projections have been used in this comparison. The main projections are mercator, polar-stereographic and the simple cylindrical projection. Use of different projections makes comparison of specific cloud climatologies very difficult. In figures 1.10 and 1.11 the global maps have been reduced to a similar size and redrawn, where necessary, with the Greenwich meridian on the left margin. The hemispheric polar stereographic maps have been reduced to a standard size.

The 'traditional established' climatologies (e.g., figures 1.10(a) and 1.11(a)) indicate very little variation of cloud amount over the oceans. Variations of cloud cover are restricted to continental and coastal

Table 1.3. Characteristics of cloud climatologies presented as maps in figures 1.10 and 1.11. Data compiled from the sources referenced in Hughes (1984).

Author[a]	Projection	Spatial average	Time of day[b]	Comments
(a) London, 1957	Polar stereo-graphic	5° lat & 10° long	1200 GMT	Northern Hemisphere only
(b) Van Loon, 1972	Polar stereo-graphic	?	?	Southern Hemisphere only
(c) Clapp, 1964	Mercator	5°	Sp	Only contour interval of 25 % available for this climatology
(d) Beryland & Strokina, 1980b	Simple cylindrical	5°	day-time	
(e) Sherr *et al.*, 1968	Mercator	?	DA	Not a contour map of cloud amount, but the mean cloud amount used to derive the cloud climatic regions (e.g., 2 = 20 %, etc)

[a]See Hughes (1984) for original references.
[b]DA = diurnal mean; Sp = satellite pass time.

locations. This is clearly a result of the availability of continental surface observations and paucity of oceanic surface observations. The cloud distributions over the Southern Hemisphere appear uniform and constant. The sparsity of surface observations for this hemisphere seems to be the cause of this rather than a lack of variability. The climatologies based on satellite nephanalyses by Clapp (1964) (figures 1.10(*c*) and 1.11(*c*)) show much more structure in the oceanic cloud amount.

The Berlyand and Strokina (figures 1.10(*d*) and 1.11(*d*)) cloud climatology seems likely to become one of the most widely used climatologies prior to the establishment of the satellite-derived climatologies such as the ISCCP. Berlyand and Strokina incorporated many data sources, including the aforementioned cloud climatologies, in the construction of their climatology. In addition, daytime observations from over 3600 surface stations were incorporated. The maps for January and July conditions in figures 1.10(*d*) and 1.11(*d*) probably represent the most comprehensive attempt at a global cloud climatology of total cloud amount currently available. The greater cloud amounts (than for instance figures 1.10(*a*) and 1.11(*a*)) at most geographical locations are the result of using daytime cloud observations only (assuming a night-time minimum in cloud cover). For example, the January estimates of cloud cover over the Sahara Desert and Arabian peninsula are higher than all other climatologies. The distribution of cloud cover in the Southern Hemisphere resembles the cloud climatology of Van Loon (1972) (figures 1.10(*b*) and 1.11(*b*)).

The 29 homogeneous cloud climatological regions defined by Sherr *et al.* (table 1.2), shown in figures 1.10(*e*) and 1.11(*e*), were designed specifically to satisfy the requirements of the remote sensing community. The modelled January and July conditions appear similar to the climatologies based solely on surface observations and those of Berlyand and Strokina.

The areas of most disagreement between the climatologies are the oceanic and polar regions (figures 1.10 and 1.11). This indicates the limited observational data for these areas and possible inter-annual variations. As previously discussed, retrieval, from satellite data, of cloud amount overlying snow/ice surfaces is particularly difficult. Detailed further study is required to provide accurate polar cloud amount statistics (see also Chapter 8).

Despite the considerable differences in the observational techniques and time-span of observations, the pattern of total cloud amount for all the climatologies is fairly consistent (figure 1.12). The magnitude of cloud amount and the finer spatial detail are very much more variable however. The differing observational systems, differing resolution (time and space) of the original data and differing summarizing procedures can result in apparently different cloud amount distributions. These comparisons

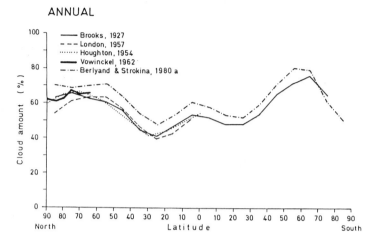

Figure 1.12. Meridional profiles of zonally and annually averaged cloud amount for selected cloud climatologies (see table 1.2).

suggest that it may not be possible to establish a unique global cloud climatology (Hughes, 1984).

Information relating to cloud amount, and cloud amount by type and at given heights is urgently required by many research groups. In particular, the success of many remote sensing programmes is dependent on the availability of such information. The lack of such an agreed global cloud climatology is a serious problem (e.g., Chapters 3, 5, 7 and 8).

1.4. Early Earth surveillance from space

As early as 1949 pictures of clouds over the Earth taken from vertical sounding rockets had been published. These very early atmospheric images demonstrated the feasibility of weather observation from satellites and underlined the considerable importance that the new technology could have for the rapidly developing science of meteorology. The major thrust into the new technology of remote sensing was made in the 1950s and 1960s by the meteorological community. By the early 1950s classified reports were being presented on the possible applications of satellite sensed information to the US and UK governments. In 1954 a US Navy 'aerobee' rocket returned images of the south-western USA revealing a storm passing over the coastal region from the Gulf of Mexico. By the late 1950s intensive development work was being undertaken by the space agencies in both the USA and USSR. In April 1959 responsibility for the

Table 1.4. American and Soviet meteorological satellites, 1960–1970.

Satellite	Launch date	Angle of orbit to equator (°)	Average height (km)
TIROS-I	1 Apr 1960	48	742
TIROS-II	23 Nov 1960	48	676
TIROS-III	12 Jul 1961	48	764
TIROS-IV	8 Feb 1962	48	777
TIROS-V	19 Jun 1962	58	782
TIROS-VI	18 Sep 1962	58	698
TIROS-VII	19 Jun 1963	58	649
TIROS-VIII	21 Dec 1963	58	753
Nimbus-I	28 Aug 1964	99	675
TIROS-IX	21 Jan 1965	96	1 640
TIROS-X	1 Jul 1965	99	797
ESSA-1	3 Feb 1966	98	769
ESSA-2	28 Feb 1966	101	1 384
Nimbus-II	15 May 1966	100	1 125
ESSA-3	2 Oct 1966	101	1 436
A.T.S.-I	7 Dec 1966	Geosync. orbit (151°W)	35 900
ESSA-4	26 Jan 1967	102	1 381
Cosmos-144	28 Feb 1967	81.2	625
ESSA-5	20 Apr 1967	102	1 387
Cosmos-156	27 Apr 1967	81.2	630
Cosmos-184	25 Oct 1967	81.2	635
A.T.S.-III	5 Nov 1967	Geosync. orbit (44–95°W)	35 900
ESSA-6	10 Nov 1967	102	1 445
Cosmos-206	14 Mar 1968	81.0	630
Cosmos-226	12 Jun 1968	81.2	626
ESSA-7	16 Aug 1968	102	1 448
ESSA-8	15 Dec 1968	102	1 436
ESSA-9	26 Feb 1969	102	1 465
Meteor-I	26 Mar 1969	81.2	678
Nimbus-III	14 Apr 1969	100	1 100
Itos-1	17 Jan 1970	102	1 141
Nimbus-IV	8 Apr 1970	100	1 090
NOAA-1	11 Dec 1970	102	1 140

first operational meteorological satellite system, which was to become known as TIROS, was transferred to NASA.†

The first TIROS satellite was successfully placed in orbit on 1 April 1960. TIROS-I was followed during the 1960s by a considerable array of American and Soviet meteorological satellites (see table 1.4). The earliest developments were focussed on the polar orbiting satellites but by the mid-1960s the Advanced Technology Satellite, A.T.S.-I, which was the first geosynchronous satellite, was successfully placed in Earth orbit.

Remote sensing of the Earth's surface, of oceans and of the cryosphere

†Names and acronyms related to remote sensing are listed in tables 1.1 and 1.5.

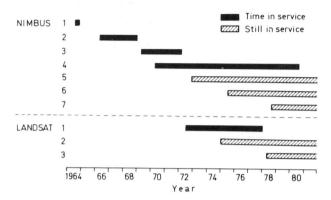

Figure 1.13. Decades of experimental performance of the Nimbus and Landsat satellites (copyright American Institute of Aeronautics & Astronautics, paper no. 82-0384).

followed after more than 15 years of development of meteorological satellites. The potential utility of satellite sensing was more clearly obvious and more immediately applicable to the research and operational requirements of the meteorologists. Scientists concerned with Earth resources management and monitoring had to wait until 1972 for the launch of the first operational satellite designed specifically for surface rather than atmospheric monitoring. This satellite system, called ERTS (Earth Resource Technology Satellite), was renamed Landsat two years after the first satellite was launched on 23 July 1972. It is interesting to note that the ERTS satellite is a direct adaptation of the Nimbus meteorological satellite system. The basic Nimbus configuration which had been developed throughout the 1960s (see figure 1.13) was modified by the addition of a wide-band recording and transmission system, the RBV (Return Beam Vidicon) and the MSS (Multi-Spectral Scanner). The orbit and attitude systems were also modified providing an approximately 9.30 a.m., local time, overpass rather than the standard Nimbus 12-noon pass. Since the mid-1970s there has been both independent growth and considerable interaction between the Earth science and meteorological applications of remotely sensed data retrieved from satellite systems.

1.5. Current and projected satellite missions

Satellites for Earth sensing can be considered as belonging to one of two groups: meteorological satellites or Earth resources satellites. Various national and international agencies and organizations are responsible for, or associated with, satellite systems. Most of these organizations are listed in table 1.5. Table 1.1 (p. 10) lists the names and acronyms of many of the currently available and planned satellite systems. Conventions develop about the method of writing satellite names which are usually related to their derivation. For example, ERTS remains upper case since it is an acronym standing for Earth Resource Technology Satellite whereas the operational satellites are renamed Landsat, which is a land resources satellite but its name is not an acronym. Similarly the term Nimbus is a name (actually of a rain cloud) rather than an acronym. These conventions do not always hold and also suffer from the mild abuse of scientists and public relations officers wishing to 'baptize' their system or experiment with a memorable name; for instance, in table 2.2 (pp. 70–71) the DUCKEX (Duck Island — 'eggs'?) experiment.

As described in Section 1.4 the Earth resources satellites (beginning with ERTS-I in 1972) were a direct development of the earlier meteorological satellite systems. As considerable advances are made in both research areas, the area of overlap between Earth resources and atmospheric science becomes greater. The sensors carried on development and operational satellites differ considerably from one another (see Chapter 2). The two fundamental wavelength regions centred at the peak of the Planck curves of the Sun and the Earth are generally retained. There is therefore a visible wave-band sensor usually extending over, at

Table 1.5. Organizations associated with satellite remote sensing.

Acronym	Full name
COSPAR	Committee for Space Research, ICSU
ERIM	Environmental Research Institute of Michigan, USA
ESA	European Space Agency
FNOC	Fleet Numerical Oceangraphic Center (US)
ICSU	International Council of Scientific Unions
IUCRM	Inter-Union Commission on Radio Meteorology, of URSI and IUGG
IUGG	International Union of Geodesy and Geophysics
JOC	Joint Organizing Committee of GARP (see table 2.2(b))
NASA	National Aeronautics and Space Administration (US)
NCC	National Climatic Center (US)
NESS	National Environmental Satellite Service, of NOAA (US)
NET	Nimbus Experiment Team responsible for CZCS Sensor
NMFS	National Marine Fisheries Service (US)
NOAA	National Oceanic and Atmospheric Administration (US)
NWS	National Weather Service (US)
PEG	Pacific Environmental Group (US)
SCOR	Scientific Committee for Oceanic Research, of ICSU
SURGE	SEASAT Users' Research Group of Europe
SWFC	South West Fisheries Center (US)
URSI	Union Radio Scientifique International, of ICSU
WMO	World Meteorological Organization

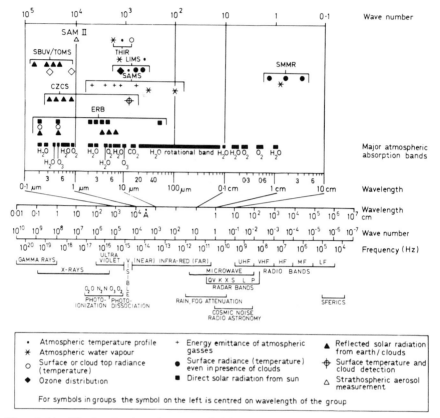

Figure 1.14. Wavelength regions of sensors on the Nimbus-7 satellite. Data products are listed in table 1.8 (copyright American Institute of Aeronautics & Astronautics).

least, the wavelength region $0 \cdot 5 - 0 \cdot 6 \mu m$,† and one centred on the window region of the Earth's atmospheric absorption spectrum in the thermal infra-red, extending from 8 to $13 \mu m$. A common choice of bands is $0 \cdot 55 - 0 \cdot 75 \mu m$ and $10 \cdot 5 - 12 \cdot 5 \mu m$. The considerable range and variety of sensors likely to be flown on satellites is illustrated in figure 1.14, which shows, as an example, the sensor response regions of the Nimbus-7 experiments. Considerable care must be exercised when comparisons are drawn between data sets taken from different satellite systems apparently operating in similar wavelength regions, as the response curve of the sensors may differ: see Chapter 2. Most current technology and almost

---

† The term 'visible' is applied rather broadly as the human eye is sensitive to radiation in the wavelength region $0 \cdot 45 - 0 \cdot 70 \ \mu m$ while 'visible' sensors can extend from $0 \cdot 45$ to $1 \cdot 5 \ \mu m$.

METEOSAT

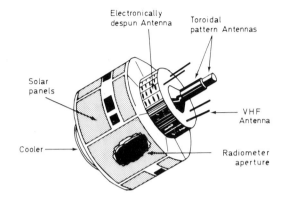

Electronically
despun Antenna

Toroidal
pattern Antennas

Solar
panels

VHF
Antenna

Cooler

Radiometer
aperture

NIMBUS

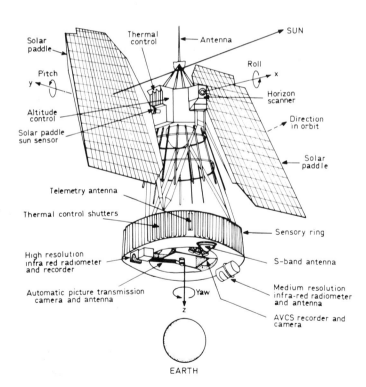

Solar
paddle

Thermal
control

Antenna

SUN

Pitch
y

Roll
x

Altitude
control

Horizon
scanner

Solar paddle
sun sensor

Direction
in orbit

Solar
paddle

Telemetry antenna

Thermal control shutters

Sensory ring

High resolution
infra red radiometer
and recorder

S-band antenna

Automatic picture transmission
camera and antenna

Yaw

z

Medium resolution
infra-red radiometer
and antenna

AVCS recorder and
camera

EARTH

Figure 1.15. Schematic diagrams of the METEOSAT and Nimbus satellites.

all the applications described in this book use data from passive sensors. Future satellites are likely to carry active sensing systems based, for example on lasers. Recent developments in remote sensing techniques designed to establish chemical species and the extent of pollution in the troposphere are already using active sensors: see Chapter 4.

The two types of satellite orbits are a polar orbiting configuration or a geostationary orbit. Polar orbiters have been used since the early 1960s (see table 1.4). A schematic diagram of a polar orbiter, the Nimbus spacecraft, is shown in figure 1.15. The polar orbiter is inserted into a low orbit approximately 500–1500 km above the surface of the Earth and the satellite path is approximately perpendicular to the equator. Characteristic orbital periods are of the order of 90–100 minutes, each orbit taking the satellite very close to both poles. Satellite orbits are usually designed so that the polar orbiters are sun synchronous; in other words, as the Earth rotates each orbit is over an area experiencing the same local time as the area viewed on the previous orbit, i.e., the Earth–Sun–satellite configuration remains constant. Around 15 orbits of the meteorological polar orbiters cover the illuminated globe. One-half of each orbit of these low-level satellites is over the dark or night side of the planet. In comparison with the meteorological polar orbiting satellites the higher resolution and hence much narrower image swathes of the ERTS/Landsat requires a much longer period to 'patchwork' together total coverage of the Earth's surface. In the case of the ERTS orbit a total of 18 days is required before the initial surface area is again under surveillance. The sub-satellite trace pattern for the ERTS satellite is, however, very similar to that of all polar orbiting satellites: see figure 1.16.

The geostationary satellites are placed in orbits very much higher than those of the polar orbiters. It is necessary that their position is approximately 35400 km above the Earth's surface and in the plane of the equator. Their orbit is constructed so that they move in the same sense as the Earth's rotation. In this way their position can remain geosynchronous. Thus these satellites are stationary with respect to a particular longitude and the equator. They therefore view the same area (the full disc of the Earth is observable since their orbital height is so great) of the Earth at all times. National and international space agencies aim to insert geostationary satellites into orbits so located that the area viewed is of immediate interest to the sponsoring countries (see figure 1.16). For example GOES East and West view the eastern and western halves of the United States, while the European Space Agency's METEOSAT-II is located at 0°, 0° (a schematic diagram of the METEOSAT spacecraft is shown in figure 1.15). The advantage of a geostationary orbit is that a continuous record can be made of the atmospheric or surface phenomena under investigation. The disadvan-

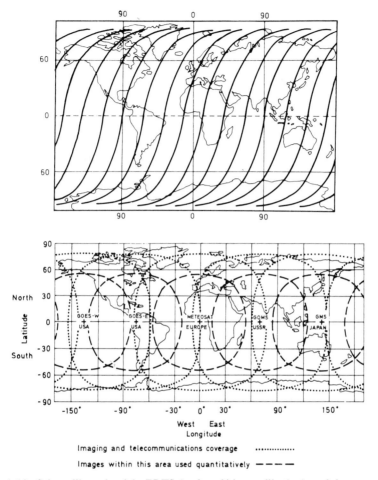

Figure 1.16. Sub-satellite paths of the ERTS-1 polar orbiting satellite (top); and the coverage of the globe provided by five geostationary satellites (bottom).

tages are that the much greater height of the orbit reduces the resolution which can be provided by the satellite and that curvature of the Earth makes sensing of latitudes higher than about 50° rather difficult (see figure 2.2, p. 50).

Most Earth resources applications currently require a resolution too high to be satisfied by data sensed from geostationary satellites. The meteorological community, however, already makes use of the complementary nature of geostationary and polar orbiting satellites. Table 1.6 lists the products and services currently available from the US geostationary satellites (GOES) together with products which are operationally possible but not yet implemented and products which could

Table 1.6. GOES geostationary satellite products.

Product and/or service current which is expected to continue during GOES-G and -H operations

(a) Meteorological products and/or services
1. GOES imagery (IR and visible)
2. Movie loops with annotation and/or narrative
3. United Press and Associated Press, print and captions
4. Local television support
5. Hurricane intensity classification
6. Satellite Interpretation Messages (SIM)
7. SELS and SIGMET charts
8. Special message to the Agency for International Development (AID)
9. Local thunderstorm, high wind, icing warning service
10. Meteorological consultation
11. Cloud motion winds: full disc
12. Cloud motion winds: United States Pacific Coast (200-mile limit)
13. Cloud motion winds: Gulf of Mexico
14. Cloud motion winds: Gulf of Alaska
15. Computer-derived cloud motion vectors (picture/pair winds)
16. Cloud top height data
17. Rainfall estimates (Scofield/Oliver techniques)
18. Surface frontal and pressure analysis
19. Moisture analysis
20. NMC support
21. Satellite cloud PROGS for NMC aviation weather branch
22. NMC support (satellite cloud PROGS for quantitative precipitation)
23. Support to NMC (surface and upper air analyses)

(b) Oceanographic and hydrologic products and/or services
1. Ocean current messages
2. Oceanographic consultation
3. Flash flood precipitation, amounts and estimates
4. Rainfall estimates
5. Snow melt enhanced IR images
6. Great Lakes ice analysis
7. Regional snow maps
8. River basin snow maps

(c) Agricultural products and/or services
1. Experimental: (i) freezeline analyses and tracking
               (ii) solar insolation estimates

(d) Other products and/or services
1. Numerical grid corrections
2. WEFAX program
3. EDIS archive
4. Data collection systems
5. VISSR data base

Category 1. Product and/or service that is not now done operationally which could be developed by the time of the GOES-G launch

(a) Meteorological products and/or services
1. GOES-VIS and multispectral imagery (MSI)
2. Rapid scan GOES imagery
3. Moisture

Table 1.6 continued

---

   4. *Thickness fields*
   5. *Thickness change*
   6. *Geostrophic vorticity*
   7. *Geostrophic winds*
   8. *Maximum convective activity*
   9. *Low level cloud top heights: Gulf of Mexico coastal zone*
  10. *Coastal zone winds*
  11. *Fog area and dissipation maps*

(b) Oceanographic and hydrologic products and/or services
   1. Flash flood guidance messages
   2. Flash flood monitoring messages
   3. Sea surface thermal composite: coastal zone of the United States
   4. Sea surface thermal composite: Great Lakes of the United States
   5. Sea surface thermal composite: movable display
   6. River basin snow maps, digital

(c) Agricultural products and/or services
   1. Precipitation estimates (AgRISTARS)
   2. Freeze warning
   3. Insolation (solar radiation incident at the surface AgRISTARS)

(d) Other products and/or services
   1. NWS–TDL archive
   2. Archive of VAS dwell-sounding submode and multispectral submode imagery
   3. Archive of VISSR digital data and imagery

Category 2. Products and/or services which are feasible but require development, testing and user acceptance

(a) Meteorological products and/or services
   1. GOES imagery with conventional meteorological data superposed
   2. Cloud cover maps
   3. Thickness fields near tropical storms
   4. Moisture convergence
   5. Oceanic temperature/moisture soundings from VAS 25°N – 50°N
   6. Surface air temperature and changes (3m level)
   7. Cloud cover analysis
   8. Clear air turbulence
   9. Hurricane wind field estimates
  10. Plot of wind estimates: high and low level

(b) Oceanographic and hydrologic products and/or services
   1. Three-hourly global rainfall estimates

(c) Agricultural products and/or services
   1. Land surface temperatures
   2. Minimum surface temperature: seven-day composite
   3. Maximum surface temperature: seven-day composite

(d) Other products and/or services
   1. VAS digital data base(s)
   2. Earth radiation budget maps
   3. Aerosol optical thickness maps

Table 1.7. Polar orbiting satellite products.

Sounding products
1. Layer-mean temperatures (K)
   (*a*) Layer precipitable water (mm)
   (*b*) Surface–700 mb; 700–500 mb; above 500 mb
   (*c*) Tropopause pressure (mb) and temperature (K)
   (*d*) Total ozone (Dobson Units)
   (*e*) Equivalent blackbody temperatures (K) for 20 HIRS2 stratospheric channels
   (*f*) Cloud cover
2. Thickness (m) and layer-mean temperatures (K) between selected standard pressure levels
   (*a*) Precipitable water (mm)
   (*b*) Tropopause pressure (mb) and temperature (K)
   (*c*) Cloud cover
3. Clear radiances
   (*a*) (mW m^{-2} sr^{-1} cm^{-1})
4. Earth-located, calibrated radiances from SSU plus selected HIRS2 and MSU channels
5. Earth-located sensor output, with calibration parameters appended, for HIRS2 and MSU (DPSS Level I-b data base).

Oceanographic and hydrologic products	*Accuracy goals*
1. Sea surface temperature observations	±1·5°C absolute
	±1·5°C relative
2. Sea surface temperature regional-scale analysis	As above
3. Sea surface temperature global-scale analysis	As above
4. Sea surface temperature climatic-scale analysis	As above
5. Sea surface temperature monthly observation: mean	As above
6. Weekly composite surface water temperature analysis	1·5°C absolute
	0.5°C relative
7. Ice concentration and coverage analysis[a]	±5 km
8. Surface water temperature analysis[a]	As for 1
9. Snow and ice melting conditions[a]	Contour location within ±5 km

Earth heat budget product list
1. Heat budget parameters
   (*a*) Daytime longwave flux
   (*b*) Night-time longwave flux
   (*c*) Reflected energy or equivalent (albedo, absorbed)
   (*d*) Available solar energy (calculated field to be included in output form)
   (*e*) Angular data

Mapped/gridded initial imagery products (non-quantitative)[b]
1. Hemispheric-mapped polar mosaics IR/VIS
2. Mercator-mapped mosaics IR/VIS
3. Polar-mapped 'chips' IR/VIS
4. Mercator-mapped 'chips' IR/VIS
5. Polar-mapped composites IR/VIS (minimum brightness/maximum temperature)
6. Pass-by-pass gridded imagery VIS/IR (1 satellite)
7. Imagery from Limited Area Coverage (LAC) data: both recorded and direct readout (ungridded)

---

[a]From hand analysis of full-resolution imagery.
[b]From one satellite.

become operational if both financial support and user requirements made them worthwhile. Table 1.7 lists data products currently available from polar orbiting satellites.

The current satellite programmes which are likely to be supported during the course of the next two decades are Earth resources, polar orbiting meteorological, geostationary meteorological and space platforms and laboratories.

Earth resources satellites

It is anticipated that Landsat-4 now in orbit will be followed by Landsat-E (which will become Landsat-5 on commissioning). The launch date will be chosen so that Landsat-5 can fulfil a clearly defined task in the mid- to late 1980s. The Nimbus polar orbiting satellites are being updated and developed specifically to augment the Earth resources capability of the US space agency. For example, the Electrically Scanning Microwave Radiometer (ESMR) on board Nimbus-5 provided polar images from which ocean ice cover and depth have been calculated. The success of this experiment led to the addition of a further ESMR experiment (ESMR F) on Nimbus-6. Table 1.8 lists the Nimbus-7 data products and their likely applications.

Polar orbiting meteorological satellites

The primary operational meteorological satellite flown by NASA is the TIROS-N series of polar orbiter satellites. These NOAA satellites are named by letters prior to launch and are re-designated with a number on becoming operational. For example NOAA-A became NOAA-6, NOAA-B was lost in a crash, ocean landing, C became 7 and D and E are now in orbit but have not yet been commissioned. Currently NOAA-6 and 7 are operational; NOAA-H, I and J are contracted to be built. It is hoped that two satellites will operate concurrently, one with a 0730 and one a 1430 LST overhead pass. The Defense Meteorological Satellite Program (DMSP) is the operational military satellite system, which usually comprises two or more polar orbiting satellites carrying meteorological sensors with a surface resolution of around 3 km. The DMSP satellites provide the major input source for the US Air Force Global Weather Central (AFGWC) which provides worldwide meteorological and space environmental support to all the US and some other national defence agencies. It is anticipated that these satellites will continue in service.

Table 1.8. Nimbus-7 data products and their uses[a].

Sensor	Film and tapes output products	Scientific parameters	Application
ERB	Daily, monthly and seasonal world grids; monthly and seasonal contour maps; zonal statistics	Earth fluxes; solar fluxes; zonal insolation	Climatology; ocean/atmosphere dynamics; weather modelling; terrestrial reflectance studies
SMMR	Orbital observations; bi-daily and monthly colour contour maps	Sea ice parameters; ocean surface conditions; atmospheric conditions; land parameters; glacial features	Ocean dynamics; ice dynamics; ocean /atmosphere interactions; cryospheric dynamics; climatology and weather modelling
SAM-II	Daily aerosol profiles; seasonal and annual contour maps and atmospheric cross-sections	Aerosol backscatter profiles; optical properties of stratospheric aerosols	Atmospheric sinks; Earth radiation budget studies; aerosol injection dynamics
LIMS	Daily atmospheric profiles; daily, monthly and seasonal contour maps and atmospheric cross-sections	Gas concentrations and temperature profiles in the stratosphere	Atmospheric pollution monitoring; photo-chemical studies; atmospheric gas dynamics; climatology
SAMS	Daily atmospheric profiles; daily, monthly and seasonal contour maps and atmospheric cross-sections	Gas concentrations and temperature profiles in the stratosphere and mesosphere	Atmospheric pollution monitoring; photo-chemical studies; atmospheric gas dynamics; climatology; wind dynamics
SBUV	Daily profiles of O_3; daily monthly and seasonal contour maps; solar spectra; zonal O_3 statistics	O_3 profiles; total atmospheric O_3; solar irradiances; terrestrial radiances	O_3 dynamics/modelling; climatology and meteorology; O_3 solar relationships
CZCS	2-minute images	Temperature; spectral radiances; chlorophyll; sediment	Geodynamics of coastal regions; chemical and thermal pollution studies; fishery resources; deep ocean monitoring; oil spill monitoring
THIR	Daily montages of temperature	Surface temperature; cloud top temperature	Effects of cloudiness on other Nimbus-G instrument data

[a]See table 2.2, p. 70 for an explanation of sensor acronyms.

Table 1.9. Current geostationary meteorological satellite systems (see also figure 1.16).

System operator	Satellite name	Location (longitude)	Launch/commissioning date
European Space Agency	METEOSAT-II	0°	Oct 1981
Japan	GMS-2	140°E	Aug 1981
USA	GOES-West[a]	135°W	Oct 1981
	GOES-East	75°W	Jun 1981
	GOES-Central (WEFAX only)	107°W	Jun 1977
USSR	GOMS	76°E	Circa 1983
India	INSAT-1	74°E	1982 (replaced 1983)

[a]Temporary, possible total, failure reported in December 1982.

Geostationary meteorological satellites

Table 1.9 lists the current and proposed international geostationary meteorological satellite systems. It is anticipated that some of the national and international agencies which have launched these geostationary satellites will provide replacement satellites as required. In view of the fact that the ESA replacement METEOSAT-II had to be launched ahead of the anticipated date, it is not certain whether it will continue in service until the new generation of ESA satellites, planned for the period 1987–1997, becomes available. The USA have one remaining geostationary satellite (GOES-F), which is likely to be placed in orbit in mid-1983. Further GOES satellites (G, H and and I) are currently being planned. Table 1.6 lists both operationally available and proposed user products from the GOES satellite system.

Space platforms and space laboratories

The successful advent of the Space Shuttle in 1981 points to the increasing use of manned space platforms for meteorological and environmental monitoring. A schematic diagram of the operational process of the US Space Shuttle is shown in figure 1.17. Payloads are currently under development in research and development projects associated both with Earth resources management and atmospheric physics. National and international space agencies are likely to fly payloads on board the Space Shuttle. It is anticipated that if funding continues to be made available space platforms could be extended, leading to the ability to carry much greater payloads with a wider range of applicability (figure 1.18).

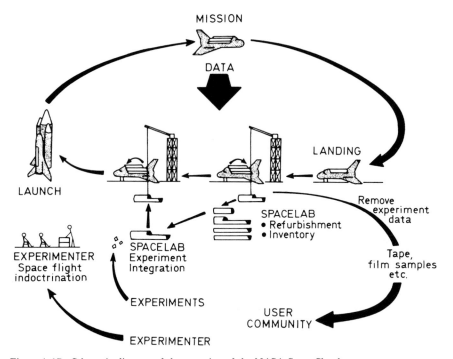

Figure 1.17. Schematic diagram of the operation of the NASA Space Shuttle.

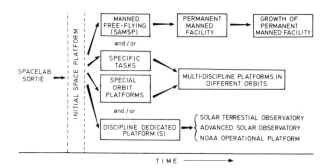

Figure 1.18. Possible evolution of space platforms in the future.

1.6. Overview of the book

The Earth's environment can now be monitored from space. All remote sensing activities have to consider the effect of the atmosphere and often the additional effects of clouds on the retrieved data. In Chapter 2 we discuss the theory of radiative transfer in some detail. This chapter, however, provides a more complete development than will be required for any one of the applications described in the rest of the book. Some basic definitions and generalized formulae are introduced here.

Our climate is fundamentally dependent upon the atmospheric state, the oceanic circulation and the extent and condition of the cryosphere. Arguably, the single most important parameter to be established from satellite surveillance of the Earth is the cloud–radiation budget interaction. It is possible that the Earth's surface temperature may alter following a change in any of the boundary conditions controlling the Earth's climate (e.g., the solar constant or the amount of CO_2 in the atmosphere). This may, in turn, cause a change in the amount of cloud cover. Clouds have two important effects on the radiation budget of the Earth–atmosphere system: (i) clouds increase the albedo of the system because of their scattering properties at solar radiation wavelengths, thus reducing the amount of absorbed radiation; and (ii) clouds decrease the longwave radiation loss to space because of their absorption properties at infra-red wavelengths. Cloud amount changes are thus a potential mechanism able to amplify or dampen any externally caused climatic change. It is, therefore, important to analyse the effect of clouds on the Earth's radiative regime.

In Chapter 3 a review is presented of the data pertaining to the radiation budget of the Earth as observed from space. The interactions of clouds with the radiation streams are considered in detail.

Chapter 4 provides a complete review of the important components of the chemistry and photochemistry of the Earth's troposphere. The vital role played by water vapour and liquid water in the processes described is emphasized. Remote sensing techniques are now being developed which permit the retrieval of chemical data. These new methodologies are also reviewed.

Vertical temperature sounding of the atmosphere and the detection of variations in surface properties are now being conducted using satellite-borne instruments. Chapters 5 and 7 contain details of the methods of retrieving information from this new source. In both chapters the need to identify and remove cloud-contaminated data from the data base is underlined. In the case of the retrieval of information about land and ocean surfaces (Chapter 7), early sensors and retrieval techniques are shown to be inadequate. Considerable caution must be exercised when deriving surface details from satellite data.

Chapter 6 builds directly on the historical survey of cloud information described in this introduction. Cloudiness data are of intrinsic interest to many scientists. The methods designed to construct global-scale cloud climatologies are therefore of interest *per se* as well as permitting removal of cloud 'contamination' for many remote sensers.

The final chapter describes some of the most recent techniques to be used in association with satellite data and returns to the main theme of the book: clouds and their interaction with the environment. Understanding the interaction between cloud and the cryosphere edge, a dynamic area of the Earth's climate, may be crucial to a complete appreciation of the atmospheric environment of Earth.

Throughout this text we seek to investigate from space some of the attributes of our cloudy planet that makes it both an hospitable and fascinating environment in which to live.

References and further reading

The Biosphere, 1970, Scientific American (San Francisco: W. H. Freeman), 134 pp.

Barrett, E. C. and Curtis, L. F., 1976, *Introduction to Environmental Remote Sensing* (London: Chapman and Hall), 336 pp.

Cess, R. D., Briegleb, B. P. and Lian, M. S., 1982, Low-latitude cloudiness and climate feedback: comparative estimates from satellite data. *Journal of the Atmospheric Sciences* **39**, 53–59.

Clapp, P. K., 1964, Global cloud cover for seasons using TIROS nephanalysis. *Monthly Weather Review* **92**, 495–507.

Cracknell, A. P., 1981, *Remote Sensing in Meteorology, Oceanography and Hydrology* (Chichester: Ellis Horwood), 542 pp.

Eyre, J. R. and Jerrett, D., 1982, Local-area atmospheric sounding from satellites. *Weather* **37**, 314–23.

Hughes, N. A., 1984, Global cloud climatologies: an historical review. *Journal of Climate and Applied Meteorology* **23**, 720–747.

London, J., 1957, "A study of atmospheric heat balance". Report contract AF19(122)–165, College of Engineering, New York University (NTIS No. 117227).

NASA, 1976, *Mission to Earth: Landsat views of the world* (Washington D.C.: NASA), 456 pp.

Rossow, W. B., Henderson-Sellers, A. and Weinreich, S. K., 1982, Cloud feedback — a stabilizing effect for the early Earth? *Science* **217**, 1245–47.

Schiffer, R. A. and Rossow, W. B., 1983, The International Satellite Cloud Climatology Project (ISCCP): the first project of the World Climate Research Programme. *Bulletin of the American Meteorological Society* **64**, 779–85.

Van Loon, H., 1972, Cloudiness and precipitation in the Southern Hemisphere. In: "Meteorology of the Southern Hemisphere", edited by Newton, C. W. *Meteorological Monographs No. 13.*

2

Radiation, the atmosphere and satellite sensors

John R. Schott
Rochester Institute of Technology
Rochester, NY, USA
and
Ann Henderson-Sellers
University of Liverpool
Liverpool, UK

2.1. Introduction

Remote sensing from space for Earth observation involves the observation of electromagnetic (EM) radiation which has undergone considerable interaction with the atmosphere. This may take the form of subtle interactions with the elements that comprise the clear-sky regions of the troposphere, or it may be very strong direct interactions with clouds. Whether the final goal is observation of Earth surface features or the atmosphere itself, it is important to understand the properties of electromagnetic radiation as EMR is affected by what is being viewed and by the atmosphere through which the viewing takes place. This chapter will present the basic principles of propagation of electromagnetic energy, including the effect of the atmosphere and clouds on satellite viewing systems. Also included is a consideration of satellite sensing systems, incorporating the essential elements of calibrating sensors to the absolute energy levels observed and the much more difficult step of isolating what effect the atmosphere has had on the observed signal. Finally, projected future satellite sensor systems are discussed.

Transfer of radiant energy through the Earth's atmosphere is an exceedingly complex process. Section 2.2 deals with the essential basic concepts and definitions and then proceeds to describe clear-sky radiative transfer in detail. Most of the theory introduced here can be obviated when a specific task, such as the sensing of a selected land surface type in a known climatological region, is identified. Thus Section 2.3 and the rest of this book refer only briefly to the theory presented here. The case of cloudy situations is still more complex. Remote sensers requiring surface information must eliminate all cloud contaminated data (Chapter 7) while other scientists may wish to draw on the theory of radiative transfer to assist in identification of cloud type, height and amount (Chapter 6). In general, then, the fairly detailed treatment of the theory of radiative transfer presented in Section 2.2 has been included for completeness but could, if desired, be omitted on first reading.

2.2. Theoretical background — the nature of radiation

All objects with temperatures above absolute zero emit EM energy (figure 2.1). The magnitude and spectral distribution of this energy can be defined by Planck's distribution equation for a perfect radiator (see table 2.1)

$$M_\lambda = \pi B_\lambda(T) = \frac{2\pi hc^2}{\lambda^5} \left[\exp\left(\frac{hc}{\lambda kT}\right) - 1 \right]^{-1}$$
(2.1)

where B_λ is Planck's function; M_λ is the spectral radiant exitance or radiant emittance, in $W\,m^{-2}\,\mu m^{-1}$; h is Planck's constant, $6 \cdot 626 \times 10^{-34}\,J\,s$; c is the speed of light, $2 \cdot 99792 \times 10^8\,ms^{-1}$; λ is the wavelength, in m; k is the Boltzmann constant, $1 \cdot 38054 \times 10^{-23}\,J\,K^{-1}$; and T is the temperature in K.

By fixing the temperature in Planck's equation, it is possible to plot the spectral radiant exitance from a surface as a function of wavelength, as shown in figure 2.2. By comparing magnitudes of spectral irradiance ($W\,m^{-2}\,\mu m^{-1}$) emitted by blackbodies at temperatures of 6000 K and 300 K (figure 2.2) it can be seen that at the top of the 'atmospheres' of the Sun and the Earth, the solar irradiance (at all wavelengths) is greater than the terrestrial irradiance. The lower 6000 K curve in figure 2.2(b) shows the solar flux at the top of the Earth's atmosphere (i.e., the irradiance values have been reduced by the square of the ratio of the Sun's radius to the mean Earth–Sun distance). The Planck curves for the Sun and the Earth (see figure 1.6, p. 7) are almost separate, and thus solar and terrestrial radiation can be considered to be almost independent of one another, overlapping slightly in the region around $2-4\,\mu m$. The effect of absorption and scatter by particulates, gases and water vapour on these fluxes is shown by the curves beneath the solar and terrestrial blackbody curves in figure 2.2. The primary absorbers at short wavelengths are O_2, O_3 and water vapour (cf. figure 1.6), while in the infra-red spectral region CO_2 also becomes important. A second curve (labelled diffuse solar beam) is also drawn under the solar curve to illustrate the effects of clouds

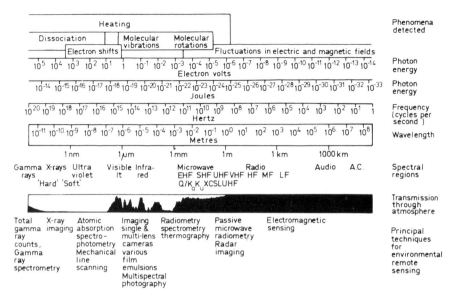

Figure 2.1. The electromagnetic spectrum (from Barrett and Curtis, 1976; see Chapter 1).

on received solar radiation. It should also be noted that figure 2.2(*b*) seems to suggest that very much more radiation is absorbed by the Earth than is emitted. However, there are two further effects which reduce the received solar radiation. The solar beam is intercepted by an area of πR^2 (where R is the Earth's radius) at any time whereas the Earth is emitting over the whole of a surface area of $4\pi R^2$. Additionally, approximately 30 % of the solar radiation incident at the top of the atmosphere is reflected away rather than being absorbed into the Earth–atmosphere system. Figure 1.5 (p. 6) is a schematic illustration of the balance between absorbed solar radiance and emitted infra-red radiation. The Earth's radiation budget is discussed more fully in Chapter 3.

Since the photospheric temperature of the Sun is approximately 6000 K and the Sun functions like a near perfect emitter, the Sun's spectral radiant emittance can be approximated by the spectral distribution of a Planckian radiator at 6000 K. The total radiant exitance from a surface at a given temperature can be found by computing the area under the appropriate curve in figure 2.2(*a*). This can be expressed mathematically as the integral of M_λ with respect to wavelength for all wavelengths, i.e.

$$M = \int_0^\infty M_\lambda \,\mathrm{d}\lambda = \int_0^\infty \frac{2\pi hc^2}{\lambda^5} \frac{1}{\left[\exp\left(\dfrac{hc}{\lambda kT}\right) - 1\right]} \,\mathrm{d}\lambda \qquad (2.2)$$

$$= \sigma T^4$$

Table 2.1. Definition of useful radiometric and remote sensing terms.

Symbol	Definition	Units
t	Time	s
h	Planck's constant: $6 \cdot 626 \times 10^{-34}$	J s
c	Speed of light: 3×10^8	m s^{-1}
λ	Wavelength	m or μm
k	Boltzmann constant: $1 \cdot 38054 \times 10^{-23}$	J K^{-1}
T	Temperature	K
ν	Frequency	Hz
σ	Stefan–Boltzmann constant: $5 \cdot 6697 \times 10^{-8}$	W m^{-2}K^{-4}
ξ_α	Absorption efficiency	
C_α	Absorption cross-section	m^2
r	Molecular radius	m
β or γ	Absorption or scattering coefficient	m^{-1}
τ	Atmospheric transmission	
z or ΔL	Path length	m
m	Molecular density	m^{-3}
W	Vertical water vapour amount	kg m^{-2}
τ'	Optical depth	
α	Absorptivity (also used as fractional cloud cover in Chapters 5 and 7)	
r	Reflectivity	
ϵ	Emittance or emissivity	
Z	Solar zenith angle (i.e., the angle between the Sun and the normal to the Earth)	rad
θ	Angle between sensor and the normal to the Earth	rad
θ'	Elevation of the source relative to the normal to the surface	rad
θ''	Angle between the normal to the target surface and the sensor	rad
ϕ	Azimuthal angle between the source and the sensor projected into the plane of the object	rad
Q	Radiant energy	J
Φ	$\dfrac{dQ}{dt}$ = radiant flux	J s^{-1} (or W)
M	$\dfrac{d\Phi}{dA}$ = radiant exitance or radiant emittance = flux radiated per unit area from a surface	W m^{-2}
E	$\dfrac{d\Phi}{dA}$ = irradiance = flux per unit area onto a surface	W m^{-2}
Ω	Area intercepted by the solid angle divided by the square of the radial distance to the area	sr
I	$\dfrac{d\Phi}{d\Omega}$ = radiant intensity = flux per unit solid angle from a source. Typically a point source or an elemental area of an extended source	W sr^{-1}
L	$\dfrac{d^2\Phi}{d\Omega dA \cos\theta}$ = radiance = radiant flux per unit area per unit solid angle at any point in space	W m^{-2} sr^{-1}
$B_\lambda(T)$	$B_\lambda(T) = \dfrac{2hc^2}{\lambda^5} \dfrac{1}{\exp(hc/\lambda kT) - 1}$	W m^{-2} sr^{-1} μm^{-1}
$B_\nu(T)$	$B_\nu(T) = \dfrac{2h\nu^3}{c^2} \dfrac{1}{\exp(h\nu/kT) - 1}$ or $2hc^2\nu^3 \dfrac{1}{\exp(hc\nu/kT) - 1}$	W m^{-2} sr^{-1} Hz^{-1} or W m^{-2} sr^{-1} (cm^{-1})$^{-1}$

Table 2.1 continued

Symbol	Definition	Units
δ	Cloud climate sensitivity parameter defined by equation 3.4: $$\delta = \frac{\partial Q}{\partial A_c} - \frac{\partial F}{\partial A_c}$$	
ϵ	Cloud climate sensitivity parameter defined by equation 3.5: $$\epsilon = \frac{\partial F}{\partial Q} \simeq \frac{\Delta F}{\Delta Q}$$	
ϵ	T_B/T = emissivity of sea ice (isothermal): T_B = brightness temperature; T = surface physical temperature (equation 8.1).	
T_B	$(C_w \epsilon_w T_w + C_I \epsilon_I T_I) \exp(-\tau') + T_A$ = microwave radiative transfer formula for water and ice mixture at (over) $1 \cdot 55 \mathrm{GHz}$.	
C_I	$\dfrac{T_B - T_0}{\epsilon_I T_{\mathrm{eff}} - T_0}$: Ice concentration at $1 \cdot 55 \mathrm{GHz}$ (equation 8.3).	
$d(t)$	Earth–Sun distance correction factor $= \bar{d}_0{}^2/d_0^2(t)$, where $d_0(t)$ is the Earth–Sun distance at time t; and $\bar{d}_0$ the mean value	
$\varrho(Z,\nu,\theta,\phi)$	An isotropic scattering function ($\varrho > 1$ when scattering is more than isotropic and < 1 when less than isotropic)	
a	Air mass through which target is observed	

in $\mathrm{W\,m^{-2}}$, where M is the radiant emittance; and σ is the Stefan–Boltzmann constant, $5 \cdot 6697 \times 10^{-8}\,\mathrm{W\,m^{-2}\,K^{-4}}$.

This fundamental fourth-power relationship between temperature and radiant exitance is known as the Stefan–Boltzmann equation. Since the Planckian distribution is a well behaved function having a single maximum, the value of the maximum point can be found by setting the derivative equal to zero and solving for wavelength. This yields Wien's displacement law through

$$\mathrm{d}M_\lambda/\mathrm{d}\lambda = 0 \text{ for } \lambda_{\max} = A/T \tag{2.3}$$

where A is a constant, $2898\,\mu\mathrm{m\,K}$; and T is the temperature in K.

Wien's displacement law has particular application to optimization of sensing systems designed to observe EM radiation from a given source. For example, it shows that sunlight with a distribution temperature near $6000\,\mathrm{K}$ will peak in the visible portion of the spectrum (figures 2.1 and 2.2), whereas objects near Earth ambient temperature of $300\,\mathrm{K}$ will have peak radiant exitance values in the thermal infra-red near $10\,\mu\mathrm{m}$ (figure 2.1).

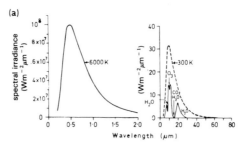

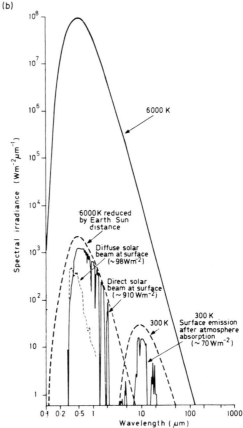

Figure 2.2. (a) *Spectral irradiances emitted from blackbodies at temperatures of 6000K and 300K. The fluxes correspond roughly to those at the top of the photosphere of the Sun and the atmosphere of the Earth respectively. The effect of absorption of the Earth's atmosphere (cf. figure 1.6, p. 7) is also shown.*

(b) *Spectral irradiance curves for (i) the Sun (6000K); (ii) the solar flux at the mean Earth–Sun distance; (iii) the Earth (300K). The effects upon both solar and terrestrial fluxes of absorption by the atmosphere are shown. The diffuse curve illustrates the additional absorption and scatter due to clouds.*

Figure 2.3. Full disc METEOSAT visible (0·4–1·1 μm) image for 5 November 1978 at 1145 GMT (courtesy of the European Space Operations Centre).

The Sun is the principal source of EM radiation, so it is natural to think first of viewing objects irradiated by the Sun. Figure 2.3 is a visible (0·4–1·1 μm) waveband full-disc image of the Earth from the geostationary satellite METEOSAT-I, i.e., a 'picture' of the Earth resulting from the reflection of short-wave radiation which is similar to that which we would ourselves see from space. The areas of the globe that appear bright are those with a high albedo that reflect most of the solar radiation incident upon them. These are the polar regions, the areas of persistent low stratus cloud off the western edges of most continents, mid-latitude depression belts and the region close to the equator of the Inter-Tropical Convergence Zone (ITCZ). (For a detailed discussion of the global-scale energy balance see Chapter 3 and for an analysis of ocean vs. land surface cloud regimes see Chapter 7). However, it is clear that objects not in direct sunlight (e.g., in shadow) can be observed through reflected sky light. Finally, each object observed must also be radiating energy according to the Planck radiation equation. Thus the EM radiation observed will be a combination of reflected energy from direct sunlight (specular irradiation), reflected energy from scattered sunlight (sky light or diffuse irradiation) and a self-emitted component.

It is necessary to note that alternative forms of equation 2.1 can be given in terms of the frequency ν of the radiation rather than the wavelength λ. If the relationship between frequency and wavelength is given by $\lambda\nu = c$, equation 2.1 can be rewritten as

$$M_\nu = \pi B_\nu(T) = \frac{2\pi h\nu^3}{c^2}\,\frac{1}{\left[\exp\left(\dfrac{h\nu}{kT}\right) - 1\right]} \tag{2.1a}$$

Alternatively, if the frequency, ν, is expressed in wavenumbers, equation 2.1 becomes

$$M_\nu = \pi B_\nu(T) = 2\pi hc^2\nu^3\,\frac{1}{\left[\exp\left(\dfrac{hc\nu}{kT}\right) - 1\right]} \tag{2.1b}$$

All three methods of writing the Planck function and the radiance are used in this text (compare this chapter with Chapter 7), in order to acquaint the reader with the different usages.

Up to this point it has been assumed that EM radiation propagates unobstructed through the atmosphere. In fact, the atmosphere plays a significant role in determining how much energy reaches the Earth from the Sun as well as determining where in the EM spectrum the observed energy will be found.

Absorption and scattering

The propagation of radiation through the atmosphere is affected by two processes: absorption and scattering. In absorption, a fraction of the energy passing through a volume element of the atmosphere is absorbed by atmospheric constituents and re-emitted at different wavelengths. Most of the absorption is spectrally selective absorption by the permanent gases CO_2 and O_2 and water vapour with lesser amounts due to atmospheric contaminants.

Atmospheric absorption can be thought of as absorption of photons of radiation striking atmospheric constituents. If the atmospheric constituents are considered as spheres of radius r then they will present an area πr^2 to the incident radiation for absorption. However, different constituents will have different absorption efficiencies (ξ_α). These can be used to modify the geometric cross-section of the particles to obtain an effective cross-section known as the absorption cross-section, C_α

$$C_\alpha = \pi r^2 \, \xi_\alpha \qquad (2.4)$$

Summing each constituent per unit volume weighted by the absorption cross-sections results in the fractional area absorbing per unit volume. This is equal to the fraction of the energy that will be absorbed per unit length according to

$$\beta_\alpha = \sum_i m_i \pi r_i^2 \xi_{\alpha i} = \sum_i m_i C_{\alpha i} \qquad (2.5)$$

where β_α is the absorption coefficient, in m^{-1}; m_i is the number of elements of the ith constituent per unit volume, in m^{-3}; and r_i is the mean radius of the elements in the ith constituent, in m. The transmission through an absorbing medium can be expressed according to Lambert as

$$\tau_\alpha = \exp(-\beta_\alpha z) \qquad (2.6)$$

where τ_α is the transmission through the total path considered; and z is the path length considered, in m.

It can be seen that the radiant flux incident on an absorbing atmosphere will be reduced in passing through the atmosphere by a factor τ_α or

$$\Phi_\alpha = \Phi_0 \tau_\alpha = \Phi_0 \exp(-\beta_\alpha z) \qquad (2.7)$$

where Φ_α is the flux through the atmosphere; and Φ_0 is the flux incident on the atmosphere.

In addition to absorption, a fraction of the radiant energy passing through the atmosphere is scattered. The scattering function takes on different forms depending largely on the size of the scattering element compared to the wavelength of the radiation being scattered. For example, for gas molecules typically having radii much smaller than the wavelength of the incident radiation, Rayleigh scattering occurs. Here the fraction of the energy scattered at an angle θ relative to the direction of propagation is expressed as

$$\beta_{sca}(\theta,\lambda) = \frac{2\pi^2}{m\lambda^4}(n(\lambda) - 1)^2(1 + \cos^2\theta) \qquad (2.8)$$

where $\beta_{sca}(\theta,\lambda)$ is the scattering coefficient in the direction θ, in m^{-1}; m is the molecular density, i.e., number of molecules per unit volume, in m^{-3}; λ is the wavelength, in m; and $n(\lambda)$ is the spectral index of refraction of the molecules.

It is important to note here that under the conditions where Rayleigh scattering holds (small, non-absorbing, randomly spaced, low density molecules) scattering varies inversely as the fourth power of the wavelength of the incident flux. This accounts for the increased scattering of blue light by the atmosphere causing the sky to appear blue. Regrettably, the larger particulates in the atmosphere shift the scattering function from the well defined form expressed by Rayleigh. As the size of the particles approaches the wavelength of the incident light (i.e., aerosols, haze, water vapour, etc.), more and more light is forward-scattered and the inverse fourth-power dependence on wavelength becomes less pronounced. Scattering in this region is called Mie scattering and is not simply defined (cf. Van de Hulst, 1981). As the particle size continues to increase in size compared to the wavelength (i.e., smoke, dust, etc.), a point is reached where the scattering is reasonably independent of wavelength or direction. This is called non-selective or isotropic scattering. The general shapes of each of the scattering functions are shown in

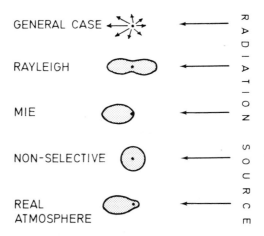

Figure 2.4. In general, for radiation incident from the right the intensity scattered in each direction is indicated by the length of the arrow from the scattering particle. The loci of scattering intensity are shown for four scattering particle types.

figure 2.4 along with an approximate shape for the combined scattering function of a real atmosphere. The wavelength dependence of scattering for real atmospheres has been shown by Curcio (1961) not to be an inverse fourth power as suggested by Rayleigh scatter, but to vary roughly from values of -2 to $+0\cdot07$ for the exponent.

If the radiation scattered is assumed to be lost from the propagating beam, then the integral of the combined scattering function over all angles yields the fractional loss due to scattering per unit path, i.e.

$$\beta_{sca} = \int_0^\pi \beta_{sca}(\theta)\,d\theta = \sum_i m_i C_{\beta i} \qquad (2.9)$$

where β_{sca} is the scattering coefficient representing the fractional loss in signal due to scattering per unit length, in m^{-1}; $\beta_{sca}(\theta)$ is the scattering function per unit length per angle θ from the propagation direction, in m^{-1}; m_i is the number of scattering elements per unit volume having scattering cross-section $C_{\beta i}$, in m^{-3}; and $C_{\beta i}$ is the scattering cross-section analogous to the absorption cross-section of equation 2.4, in m^2.

The transmitted flux Φ through a non-absorbing medium can then be expressed as

$$\Phi = \Phi_0 \tau_\beta = \Phi_0 \exp(-\beta z) \qquad (2.10)$$

where τ_β is the transmission through the path length z.

Thus, the extinction coefficient describing the total fractional loss per unit length in an absorbing and scattering atmosphere can be expressed as

$$\beta_{ext} = \beta = \beta_\alpha + \beta_{sca} = \sum_i m_i C_{iext} \qquad (2.11)$$

where β_{ext} is the extinction coefficient per unit length, in m^{-1}; and C_{iext} is the extinction cross-section of the ith element constituting the atmosphere, in m^2.

It follows that

$$\Phi = \tau_\alpha \tau_\beta \Phi_0 = \Phi_0 \exp[-(\beta_\alpha + \beta_{sca})z] = \tau \Phi_0 \qquad (2.12)$$

where Φ is the flux transmitted through the path length z; and τ is the transmission over the path length z for an absorbing and scattering medium.

In many cases, it is desirable to combine $(\beta_\alpha + \beta_{sca})z$ into one term τ', called the optical thickness or optical depth. The contribution to optical depth due to absorption and scattering from each element in the atmosphere can then be defined and summed to yield the total optical depth from which the transmission through the path can be found by

$$\tau = \exp(-\tau') \qquad (2.13)$$

It should be noted here that the definitions of τ and τ' are not consistent throughout atmospheric science. Members of the remote sensing community are more often concerned with the total transmission τ than the optical thickness, τ', and it is in order to simplify equations that the notation defined here has become common in remote sensing literature.

Energy – matter interactions

Energy reaching a surface is either absorbed, transmitted or reflected. The absorptivity, transmissivity and reflectivity are defined as the fraction of the total incident flux which is absorbed, transmitted or reflected. Thus

$$\Phi_0 = \Phi_r + \Phi_\tau + \Phi_\alpha$$

or

$$1 = \frac{\Phi_r}{\Phi_0} + \frac{\Phi_\tau}{\Phi_0} + \frac{\Phi_\alpha}{\Phi_0} = r + \tau + \alpha \qquad (2.14)$$

where Φ_r, Φ_τ and Φ_α are the fluxes reflected, transmitted and absorbed, respectively; and r, τ and α are the reflectivity, transmissivity and absorptivity. The emissivity is a measure of how effectively an object radiates compared to the perfect Planckian radiator of equation 2.1. That is

$$\epsilon = M_\epsilon(T)/M_{bb}(T) \qquad (2.15)$$

where ϵ is the emissivity; $M_\epsilon(T)$ is the actual radiant exitance from the object at temperature T; and $M_{bb}(T)$ is the theoretical exitance from a perfect Planckian radiator at temperature T.

From equation 2.15 it is clear that the Planckian radiator has an emissivity of 1. Kirchoff's law states that for objects in thermodynamic equilibrium the absorptivity equals the emissivity, which from equation 2.14 indicates that the Planckian radiator must have zero reflectivity, i.e.

$$1 - \alpha - r - \tau = 0 = 1 - 1 - r - \tau = 0$$

and neither r nor τ can be negative, so

$$r = \tau = 0$$

Such an object with zero reflectivity and unit emissivity is called a blackbody and is synonymous with the ideal Planckian radiator.

For opaque objects, the transmission will equal 0 and equation 2.14 reduces to $\alpha + r = 1$ for objects in the visible, where the primary concern is with reflected light. Equation 2.14 reduces to $\epsilon + r = 1$ in the infra-red region, where self-emission is important. In the visible region, where self-emission radiation levels are very small compared to incident solar radiation levels, the primary interest is with reflected energy. The directional reflection function of a surface can take on many forms. At one extreme is specular reflection, where the reflected energy is reflected directly from the surface of the reflecting object. This is the case for very smooth surfaces (mirrors are ideal specular reflectors). At the other extreme is a perfectly diffuse reflector which has the property that the reflected radiance is the same in all directions. Most real world reflectors are nearly diffuse or nearly specular. These terms are illustrated in figure 2.5. The complete reflectivity function of a surface can only be defined if its reflection as a function of the relevant Sun–object–sensor angles are defined (cf. figure 2.6), i.e.

$$r = r(\theta, \theta', \phi) \tag{2.16}$$

where θ is the angle between the sensor and the normal to the surface; θ' is the elevation of the source relative to the normal to the surface; and ϕ is the azimuthal angle between the source and the sensor projected into the plane of the object observed. In addition to direct solar reflection, the radiation scattered by the atmosphere diffusely irradiates the surface and

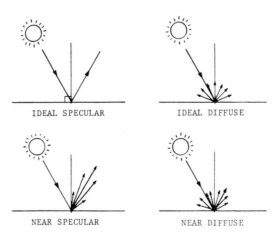

Figure 2.5. Directional reflection functions.

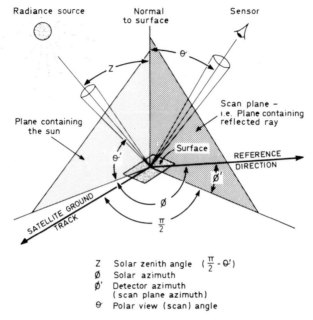

Z   Solar zenith angle $(\frac{\pi}{2} - \Theta')$
Ø   Solar azimuth
Ø'   Detector azimuth
     (scan plane azimuth)
Θ'   Polar view (scan) angle

Figure 2.6. Relative positions of source, sensor and target surface, for the simplified case of the target surface being level with the Earth's surface.

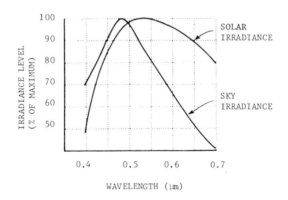

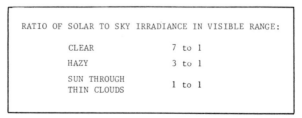

Figure 2.7. Approximate solar and sky irradiance levels under various atmospheric conditions.

is also reflected. This diffuse irradiation can be very significant in the visible region of the EM spectrum, as illustrated in figure 2.7. This figure illustrates that, particularly in the blue portion of the spectrum, the irradiance from the sky can be significant, even exceeding solar irradiance, under sufficiently hazy conditions.

Reflected radiance

Drawing on the concepts developed above, the radiance from a reflecting surface can be expressed as

$$L_r(\theta,\theta',\phi) = E_s r \pi^{-1} + E_{sky} r \pi^{-1} \tag{2.17}$$

in $W\,m^{-2}sr^{-1}$, where $L_r(\theta,\theta',\phi)$ is the radiance reflected from the surface; E_s is the irradiance from the Sun onto the surface, in $W\,m^{-2}$; r is the directional reflectance; and E_{sky} is the irradiance from the sky, in $W\,m^{-2}$. The solar irradiance on a surface can be expressed as a function of the irradiance from the Sun outside the atmosphere and the atmospheric transmission through the slant path to the target. This can be written mathematically as

$$E_s = E_s' \cos\left(\frac{\pi}{2} - \theta'\right) \exp(-\tau'/\cos Z) \tag{2.18}$$

where E_s' is the solar irradiance outside the atmosphere on a plane perpendicular to the axis of propagation, in $W\,m^{-2}$ (note: this is assuming a collimated beam with no loss due to spreading through the atmosphere); $\pi/2 - \theta'$ is the angle between the normal to the surface and the Sun (cf. figure 2.6); τ' is the optical depth through the atmosphere normal to the Earth at the target; Z is the angle between the Sun and the normal to the Earth; and $\cos Z$ accounts for the increased optical depth for non-vertical solar irradiation (note: if the target is level with the Earth's surface, then $\pi/2 - \theta'$ and Z are the same angle).

The sky irradiance is expressed as a function of the downwelling radiance resulting from scattering of incident solar irradiance. Since $dE = L_d \, d\Omega \cos\theta$, the sky irradiance can be written as

$$E_{sky} = \int_\Omega L_d(\theta,\phi) \cos\theta \, d\Omega$$

and substituting $d\Omega = \sin\theta \, d\theta \, d\phi$ gives

$$E_{sky} = \int_0^{2\pi} \int_0^{\pi/2} L_d(\theta,\phi) \cos\theta \sin\theta \, d\theta \, d\phi \tag{2.19}$$

where L_d is the downwelling radiance from the sky coming from the direction angles θ and ϕ, in $Wm^{-2}sr^{-1}$; and the integral incorporates the downwelling radiance from all directions in the hemisphere above the reflecting surface. L_d is a complex function of the scattering function $\beta(\theta)$ of equation 2.9. Substituting equations 2.19 and 2.18 into 2.17, a more complete representation of the reflected radiance from the surface is obtained, i.e.

$$L_r = L_r(\theta,\theta',\phi) = [E_s'\cos\left(\frac{\pi}{2}-\theta'\right)\exp(-\tau'\sec Z)r(\theta,\theta',\phi)$$

$$+ \int_0^{2\pi}\int_0^{\pi/2} L_d r(\theta,\theta',\phi)\cos\theta\sin\theta\,d\theta\,d\phi]/\pi \quad (2.20)$$

Expressing the reflected radiance in this fashion very clearly points out the importance of atmospheric extinction and scattering in governing the radiance from a surface. For example, equation 2.20 indicates that if the optical depth increases (as a result of more haze or a longer slant path for the solar radiation), then the reflected radiance will be more and more a function of the sky light irradiance, since the term $\exp(-\tau'\sec Z)$ will decrease and L_d will increase. This is verified by the empirical results illustrated in figure 2.7. In many cases the diffuse downwelling radiance incident on a reflecting surface is not all from the sky. In the case of a highly varying terrain, the element being observed may also be irradiated by reflected radiance from adjacent terrain elements, as illustrated in figure 2.8. In this case, the irradiance incident on the reflecting element

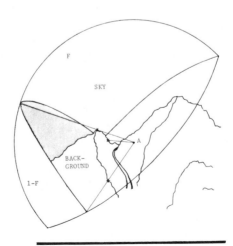

Figure 2.8. Fractions of the hemisphere above point 'A' which are sky (F) and background (1 − F).

can still be expressed as in equation 2.19 if it is noted that the down-welling radiance L_d will have components due to radiance reflected from neighbouring surfaces. In many cases the downwelling radiance from the sky (L_d) and non-sky background (L_B) can be approximated by constants, and a simple shape factor approach used to estimate downwelling radiance. In this case, equation 2.20 can be expressed as

$$L_r = \left\{ \left[E_s' \cos\left(\frac{\pi}{2} - \theta'\right) \exp(-\tau' \sec Z) \right] \middle/ \pi + FL_d + (1 - F)L_B \right\} r \quad (2.21)$$

where F is the fraction of the hemisphere above the reflecting surface which is sky (cf. figure 2.8).

Propagation of reflected radiance to the satellite

The sensor is not at the Earth's surface, positioned to observe the reflected radiance described by equation 2.20 directly, but more typically, the observed radiance is that which reaches a satellite. Therefore, the propagation of the EM signal from the reflecting surface to the sensor together with the attendant atmospheric effects must be considered.

Molecules can possess energy in four forms: translational, rotational, vibrational and electronic, of which all but translational energy are quantized. Electronic energy levels are widely spaced and transitions are generally in the ultraviolet and visible regions (see figure 2.1). Changes in the vibrational state of a molecule generally involve the absorption/emission of radiation of wavelengths typical of the near and middle infra-red. Changes of only rotational states are associated with absorption/emission in the far infra-red and microwave spectral regions. Water vapour and O_3 possess a permanent dipole moment resulting from a charge displacement, in the former between the oxygen and hydrogen atoms (figure 1.2, p.3). Interactions are therefore complex, with many vibration and rotation features (see figures 2.1, p. 47 and 1.6, p. 7). Purely symmetric gaseous molecules such as O_2 and N_2 have no permanent dipole moment and interact with terrestrial radiation only weakly through higher order effects. Certain symmetrical molecules (e.g., CO_2) can possess a transient dipole moment, resulting in vibrational features in the middle and near infra-red part of the spectrum, but no rotation features. Atmospheric sounding depends on the way in which individual gases absorb and re-emit terrestrial radiation. Cloud cover severely inhibits retrieval of atmospheric temperature information. It is only in wavelength regions longer than approximately $100 \mu m - 1 mm$, i.e., the microwave region, where absorption of radiation by clouds becomes relatively unimportant and retrieval of surface and atmospheric information can be made without loss in the presence of non-precipitating cloud cover. The ubiquitous nature of clouds in the Earth's troposphere means that these wavelength regions are the subject of intense investigation and many of the specific programmes described in the following chapters draw upon information from microwave and radar sensors which are used to complement the fundamental wavelength regions of the visible and terrestrial infra-red. Atmospheric temperature sounding from space (Chapter 5) depends on the fact that most of the radiation measured in selected spectral intervals in the infra-red and microwave regions has been emitted by the atmosphere rather than directly from the surface. A multispectral radiometer can be used to retrieve vertical temperature profiles by careful selection of the wavelength regions in which radiation is to be measured. The radiation emitted to space at a given wavelength λ is determined primarily by three factors.

1. The temperature profile of the atmosphere.
2. The mixing ratio of molecules absorbing and emitting radiation at a given wavelength λ (generally, λ for temperature sounding experiments are chosen so that the absorbing species are oxygen and carbon dioxide for which the mixing ratios can be considered roughly constant with height).
3. The transmittance function of the atmosphere above the level under observation (i.e., the amount of absorption of the detected radiation which has been caused by its passage through the atmosphere from pressure level, to the satellite).

For a level in the atmosphere high enough so that the transmittance is close to 1, little radiation is also emitted because the source function of the atmosphere is proportional to its absorption coefficient. The height at which appreciable emission occurs depends on the absorption coefficient, i.e., on the wavelength. If on the other hand it is desired to 'view' a level of the atmosphere close to the surface, it is necessary to choose a range of frequencies where the atmospheric absorption is minimal. In general, the radiation observed at wavelength λ achieves a maximum contribution from an intermediate level in the atmosphere for which the atmospheric transmittance is about e^{-1}. Careful selection of the wavelength region centred on λ can mean that temperature information can be retrieved primarily about a particular (preselected) level in the atmosphere. As is described in detail in Chapter 5, the radiation stream measured with a multispectral radiometer designed for temperature profile retrieval is such that detailed vertical structure is not easily determined. Satellite temperature soundings can provide smoothed vertical profiles over areas previously completely omitted from the global weather data set and represent a considerable advance for meteorological forecasting.

Operational atmospheric sounding is currently being performed on the TIROS-N series of polar orbiter satellites. These instruments are referred to as TOVS (the TIROS-N Operational Vertical Sounder). The instrument scan and the methodology used for atmospheric thickness and thermal wind retrievals are described with reference to Eyre and Jerrett (1982) in Chapter 1. In Chapter 5 current and future applications of vertical temperature soundings are described in detail. The atmospheric sounding instruments, like many of the applications packages on board current satellite systems, were designed primarily for use in operational weather forecasting.

Thus the propagation of EM radiation to satellite altitudes results in loss of signal strength due to extinction through the atmosphere and also to an increase in the observed radiance due to scattering of radiation upwards to the sensor.

The upwelling radiance scattered to the sensor is a function of the atmospheric constituents as well as a direct function of the Earth surface features being observed. This scattered radiance or atmospheric 'flare' term is a significant factor in reducing image quality for Earth observation. In propagating from the reflecting surface to the sensor, the reflected radiance is modified according to

$$L = L(\theta,\theta',\phi) = L_r \exp(-\tau' \sec \theta'') + L_u(\theta'',Z,\phi) \qquad (2.22)$$

where L is the radiance observed at the sensor; θ'' is the angle between the normal to the Earth surface and the sensor (note: if the reflecting surface is parallel to the Earth's surface, then $\theta'' \equiv \theta$); and L_u is the upwelling radiance due to atmospheric scatter and is a function of the scattering function $\beta(\theta)$ as described in equation 2.9. It is also important to note that some of the flux scattered will not be direct solar irradiance but flux reflected from the observed or other surfaces and then scattered upwards to the sensor. Thus, the scattered radiance over a region with low surface albedo will be somewhat less than the scattered radiance over a region with high surface albedo independent of the reflectivity of a specific target. Combining equations 2.21 and 2.22 yields a complete expression from the reflected radiance reaching a satellite sensor in the form

$$L = \left[E'_s \pi^{-1} \cos\left(\frac{\pi}{2} - \theta'\right) \exp(-\tau' \sec Z) + FL_d \right.$$

$$\left. + (1-F)L_B \right] r \exp(-\tau' \sec \theta'') + L_u \qquad (2.23)$$

The energy paths described by equation 2.23 are illustrated in figure 2.9. This figure emphasizes the role the atmosphere plays in governing the radiance observed by a satellite sensor.

To this point, the reflected radiance has been treated as if it were

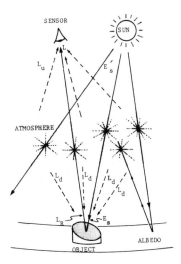

Figure 2.9. Energy paths between source, object and sensor.

$$E_s = E'_s \cos(\pi/2 - \theta') \exp(-\tau' \sec Z)$$

$$L_r = [E'_s \pi^{-1} \cos(\pi/2 - \theta') \exp(-\tau' \sec Z) + \int_\Omega L_d d\Omega] r$$

$$L = [E'_s \cos(\pi/2 - \theta') \pi^{-1} \exp(-\tau' \sec Z) + \int_\Omega L_d d\Omega] r \exp(-\tau' \sec \theta'') + L_u$$

monochromatic. In fact, the region of the EM spectrum selected for observations is largely dependent on the spectral properties of the source, the target and the atmosphere. Figure 2.10 shows the spectral irradiance from the Sun outside the Earth's atmosphere on the same graph as the approximate transmission vertically through the atmosphere. This figure indicates that there are certain spectral regions where surface observation is impossible due to the nearly complete extinction of radiation, primarily by molecular absorption. It also illustrates that the maximum amount of

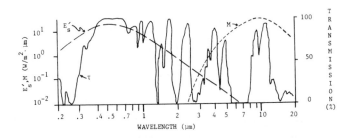

Figure 2.10. Spectral irradiance from the Sun outside the Earth's atmosphere (E'_s), radiant exitance of the Earth (M), and atmospheric transmission (τ) relative to wavelength (figure 2.2, p. 50).

energy from the Sun available for reflection conveniently coincides with the atmospheric 'window' between $0 \cdot 3 \mu m$ and $1 \cdot 3 \mu m$. This spectral coincidence is what motivates the use of this region for most remote sensing of reflected radiance. Within this large spectral window many narrower windows can be chosen for observing variations in the spectral reflectance functions of the Earth's features. Numerically, this means that all the variables in equation 2.23, except the orientation angles, must be replaced with the equivalent spectral values integrated over the wavelength interval observed. For example

$$L_u = L_{u \Delta \lambda} = \int_{\lambda_1}^{\lambda_2} L_{u \lambda} d\lambda \qquad (2.24)$$

where $L_{u \Delta \lambda}$ is the upwelling radiance over the wavelength interval $\Delta \lambda$ between λ_1 and λ_2 (note: for convenience the wavelength subscripts will be included only where they are necessary for clarity); and $L_{u \lambda}$ is the upwelling spectral radiance at wavelength λ, in $W m^{-2} sr^{-1} \mu m^{-1}$.

At longer wavelengths, as can be seen from figure 2.10, the solar energy available to be observed by reflection decreases while the self-emission from objects near Earth ambient temperature begins to increase. In the atmospheric window between $3 \mu m$ and $5 \mu m$ both solar reflection and self-emission can be important. Close to the next major atmospheric window, the solar energy available for reflection becomes quite small while the self-emitted component becomes very significant. To make the expression for the radiance reaching a sensor valid for all wavelengths, it is necessary to add a self-emitted component to equation 2.23 to obtain

$$L = \left\{ \left[E_s' \pi^{-1} \cos\left(\frac{\pi}{2} - \theta' \right) \exp(-\tau' \sec Z) + F L_d + (1 - F) L_B \right] r \right.$$
$$\left. + \epsilon L_T \right\} \exp(-\tau' \sec \theta'') + L_u \qquad (2.25)$$

where L is the radiance reaching the satellite sensor in the bandpass of interest, in $W m^{-2} sr^{-1}$; ϵ is the emissivity of the surface in the bandpass of interest; and L_T is the blackbody radiance from a surface at temperature T over the wavelength interval observed, in $W m^{-2} sr^{-1}$. More properly, the emitted radiance from the surface, ϵL_T, is

$$\int_{\lambda_1}^{\lambda_2} \epsilon_\lambda L_{T \lambda} d\lambda$$

if ϵ_λ is not a constant with respect to wavelength. It must be recognized that in the infra-red portion of the spectrum the solar irradiance term in equation 2.25 becomes quite small and the terms L_d and L_u are less a function of scattered solar irradiation and due more to self-emission by the atmospheric constituents, most notably water vapour. In a similar fashion, the radiance from the background L_b comes more from self-emitted radiation from background objects than from reflected solar radiation. In the region of the $8-14\mu$m atmospheric spectral window, it is the temperature of objects that primarily governs the radiance reaching the satellite. In the thermal infra-red, therefore, it is no longer the reflectivity variations of the Earth's surface that are being observed, but rather, primarily, the temperature variations (cf. figure, 2.1 p. 47). In fact, in general it is the temperature of surface features that are observed if sensing is undertaken in the thermal and far infra-red regions.

Sensing systems

So far in this chapter the major concern has been identification of the radiance reaching a satellite sensor. However, these sensors typically do not record radiance directly but, rather, they record some function which can, with varying degrees of effort, be related to radiance. To illustrate this it is useful to consider a real 'image', i.e., a positive photographic transparency taken from space (e.g., an image taken by NASA astronauts on the Skylab programme or by scientists aboard the Space Shuttle). The observable or measurable quantity associated with this transparency is its optical density, D

$$D = -\log_{10}\tau_T \qquad (2.26)$$

where τ_T is the fraction of the flux transmitted through the transparency compared to the flux incident on the transparency. The density at any point on the image is in turn a function of the log of the exposure reaching that point. This relationship between density and exposure H is a function of the type of film and the chemical processing that transforms the latent image to the photographic transparency. The form of this relationship can be monitored by exposing a portion of the film to known exposure increments before processing by means of a sensitometric control wedge. After processing, the density of these steps can be plotted against the log of the exposure used to generate them to obtain the H and D curve for the particular piece of film. The exposure at any point on an image can then be determined by measuring its density and interpolating between steps on the H and D curve to obtain a $\log H$ value and, therefore, an exposure value. The film's exposure can be expressed as

$$H = Et \qquad (2.27)$$

where E is the irradiance on the film, in $\text{W}\,\text{m}^{-2}$; and t is the exposure time, in s. The irradiance on the film can, in turn, be related to the radiance reaching the camera through the G number of the camera system, where

$$L = EG \qquad (2.28)$$

where $G = 4F\#^2/\tau_L\pi$, in sr^{-1}; $F\#$ is the F number of the lens and equals the focal length divided by the aperture diameter; and τ_L is the transmission of the lens system on the camera. Thus, it can be seen that through internal calibration of the camera and film processing system it is possible to compute the radiance reaching a camera on a satellite platform from the density value on the image. It must be noted that the exposure off the optical axis of a camera system is reduced somewhat due to geometrical considerations, and corrections for this 'fall-off' effect must be made for targets imaged off axis. Selection of spectral bands can be accomplished by placing appropriate filters in front of the camera system or by using colour or colour infra-red films. In the latter case, the band selection is governed by the spectral response of photographic dye layers and considerable spectral 'cross-talk' between bands can occur. In order to obtain precisely controlled multispectral data using photographic systems, multiple images must be collected using multiple film filter combinations. A major weakness of this approach is that precise spatial registration of the separate images can be extremely difficult. This difficulty, combined with the spectral limitations of photographic emulsions (film is only sensitive up to about $1 \cdot 1\,\mu m$), has prompted the use of electro-optical imaging systems. These systems have an additional advantage in that it is easier to retrieve an electronic signal from space than it is to recover a roll of exposed film. While a detailed discussion of electro-optical sensing systems is beyond the scope of this text, one simplified version of a multispectral scanner and the procedures for its internal calibration are presented here. Figure 2.11 is a simplified diagram of one type of multispectral electro-optical imaging system. The data are collected by a rotating mirror that sweeps line after line of data into the focussing optics. The lines are advanced in the along-track direction by the forward motion of the satellite. The focussing optics are all first surface mirrors to avoid attenuation of UV or infra-red radiation. A beam splitter is used to direct a fraction of the signal through a filter to isolate a thermal infra-red spectral band. This filtered flux comes to focus on a cooled thermal infra-red detector. These detectors must be cooled to temperatures near absolute zero to minimize noise. It should be recognized that they are designed to respond to very low levels of self-emitted energy so that any object whose temperature is not close to absolute zero acts as an undesirable source of radiant flux on to the detector. The flux that passes through the beam splitter is spectrally separated by a prism or a diffraction grating and observed using a variety of detectors. In this case, six spectral bands are

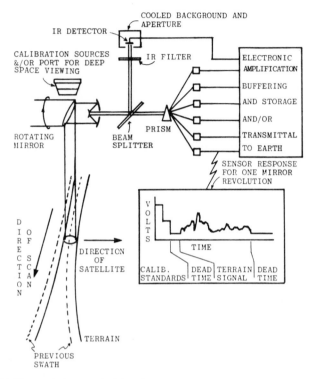

Figure 2.11. Multispectral electro-optical imaging system.

used to observe the reflected solar energy. In addition to scanning the Earth, the rotating mirror also looks back inside the sensor package where sources of known radiance are located. These are lamps of varying brightness for the reflected radiance standards and blackbodies of known temperature for the thermal radiance standards. In some cases during the back scan, the detectors are permitted to view deep space or a stellar object of known radiance for thermal or reflected radiance calibration. The electronic signals from the detectors are amplified and recorded for eventual transmission to Earth. The calibration of these sensors typically involves detailed pre-launch tests to ensure that the output of the sensors is a first-order linear function of the incident radiance. Once this is established, if the sensors 'see' two objects of known radiance in the back scan of each scan line, then the straight-line relationship between incident radiance and output signal can be defined uniquely.

   It is also important to note that the spectral response of the sensor is wavelength dependent. Thus the wavelength range quoted for a sensor may be neither the full range over which the sensor is responsive nor the

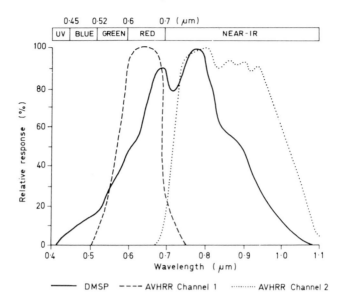

Figure 2.12. Response curves of visible sensors on the NOAA AVHRR system and the DMSP visible sensor.

range over which the sensor is 100 % responsive. For example, figure 2.12 shows the response function of two of the NOAA AVHRR channels and the DMSP visible channel. The latter sensor differs considerably from the generally accepted view of a sensor being either 'on' (i.e., responsive) or 'off' (i.e., having zero response). The fall-off in sensor response can compound a variable surface reflectivity or the effects of the atmosphere to produce wave-band responses that are easily misinterpreted. For example, the response curve of AVHRR-2 decreases longwave of $0 \cdot 8 \mu m$, a point at which vegetation shows a sudden increase in reflectivity. This upturn in the vegetation reflectivity curve is likely to be underestimated by measurements made in this channel.

Most current multispectral scanner systems use a slight variation of the system illustrated in figure 2.11. These systems have a wider field of view on the collection optics and project an image of the ground on to a focal plane. Filtered detectors at the focal plane define the area sampled. As the image is swept across the focal plane, multiple scan lines are sampled. The same area is sampled as its image is sequentially swept across each spectral filter through the use of timing offsets (see figure 2.13). A list of many of the commonly referenced sensor systems is given in table 2.2.

Satellite sensing also often employs sensors that are similar to conventional television systems. The video output from these systems can be radioed conveniently to Earth. Since radiometric calibration of video sen-

Table 2.2. Sensor and experiment names and acronyms.

Acronym	Name
Sensors	
ALT	Radar altimeter on SEASAT for ocean topography measurements.
AMTS	Advanced Meteorological Temperature Sounder. Proposed multichannel infrared instruments for atmosphere and ocean surface temperatures.
ARGOS	Platform positioning to 1 km and data relay system on NASA and NOAA satellites operated by Service Argos, France
ARIES	Two-channel infra-red scanner operated by Laboratoire de Meteorologie Dynamique, France.
AVHRR	Advanced Very High Resolution Radiometer. Visible and infra-red scanner replacing VHRR, giving more channels and higher radiometric precision, on TIROS-N, NOAA-6 and later NOAA satellites.
AXBT	Airborne Expendable Bathy-Thermograph. Radio or wire connected droppable probe for ocean temperatures to 300 m depth.
CZCS	Coastal Zone Colour Scanner. Precise, narrowband optical scanner for ocean colour measurements on Nimbus-7.
ERB	Earth Radiation Budget measurements.
ESMR	Electronically Scanning Microwave Radiometer on Nimbus-5 and -6 for ice and rain-rate mapping.
HIRS	High Resolution Infra-Red Sounder on Nimbus-6, TIROS-N and later NOAA satellites.
IRIS	Infra-Red Interferometer Spectrometer Sounder on Nimbus-3 and 4.
LAMMR	Large Antenna Multichannel Microwave Radiometer. Similar to the SEASAT SMMR but with a 4 m dish antenna.
MSS	Multispectral Scanner on Landsat. High resolution, broad band, optical and near infra-red scanner designed for land applications.
MSU	Microwave Sounder Unit on TIROS-N and later NOAA satellites.
OCS	Ocean Color Scanner. An airborne prototype of the CZCS flown on a U2 aircraft.
OLS	Operational Linescan System. Visible and infra-red scanner on DMSP satellites.
RAMS	Random Access Measurement System on Nimbus-6. Platform positioning to 2 km, and data relay.
SAMS and SAM-II	Mechanically scanning radiometer on Nimbus satellites.
SAR	Synthetic Aperture Radar. Side-looking mapping radar with along-track resolution greatly enhanced by coherent signal processing. Used in space on SEASAT.
SARSAT	Search and Rescue Satellite Aided Tracking, for positioning of aircraft in emergencies, flying on NOAA-E.
SAR-580	The ERIM X-L band SAR transferred to a Canadian Convair 580 aircraft and operated for SURSAT and other programmes.
SASS	SEASAT-A Scatterometer System. Measured directional scattering properties of the surface for wind velocity measurements over the ocean.
SBUV	Solar Backscatter Ultraviolet instrument, on NOAA-E.
SLAR	Side-looking Airborne Radar. The term describes both real and synthetic aperture systems, but usually implies a real aperture non-coherent mapping radar as opposed to SAR.
SMMR	Scanning Multichannel Microwave Radiometer on SEASAT and Nimbus-7. Uses a 1·25 m dish antenna to map ocean, ice and atmosphere parameters.
SR	Scanning Radiometer. Relatively low resolution visible and infra-red scanner used on early Nimbus satellites.
SSMI	Special Sensor Microwave Imager.
SSU	Stratospheric Sounder Unit. As for MSU.
THIR	Temperature, Humidity Infra-red Radiometer on Nimbus satellites.
VHRR	Very High Resolution Radiometer. Visible and infra-red scanner giving about 1 km resolution on NOAA satellites. Now replaced by AVHRR.

Table 2.2 continued

Acronym	Name
VIRR	Visible and Infra-Red Radiometer. Gave coarse visible and infra-red feature location on SEASAT. Similar to SR.
VISSR	Visible and Infra-red Spin Scan Radiometer. Forms full disc images from SMS and GOES satellites.
VTPR	Vertical Temperature Profile Radiometer. Used on NOAA satellites for atmospheric profiling.

Experiments

CFOX	Canada France Ocean Optics Experiment. Canadian west coast and Arctic; Mediterranean.
DUCKEX	Duck Island Experiment on SAR imaging of the ocean, US east coast.
FGGE	First GARP Global Experiment. Intensive weather observations world wide, emphasizing southern oceans, international.
FLEX	International Fladenground Experiment in the North Sea.
GARP	Global Atmospheric Research Program, international.
GATE	GARP Atlantic Tropical Experiment. Air-sea interactions off West Africa, international.
GOASEX	Gulf of Alaska Experiment. SEASAT-related ocean measurements, US, Canada.
ICEX	Ice and Climate Experiment.
JASIN	Joint Air-Sea Interaction Experiment. International study centred in Atlantic north-west of Scotland.
LAMPEX	Large Area Marine Productivity Pollution Experiment. US East Coast.
MIZEX	Marginal Ice Zone Experiment.
MARSEN	Maritime Remote Sensing Experiment, North Sea 1979, international.
PROLIFIC	Programme Ligurean France Italy Canada 1978. Ligurean Sea, Mediterranean.
SURSAT	Surveillance Satellite experiments. Canada 1978, SEASAT and airborne SAR.
TOPEX	Ocean Topography Experiment. Precision radar altimeter and geoid measurements for ocean circulation mapping.

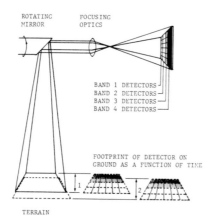

Figure 2.13. Multispectral focal plane imaging system.

sing systems can be extremely difficult, these systems are primarily utilized where only relative brightness values are required. Mention of video
output prompts a discussion of the types of output from most satellite sensor systems. These can be roughly divided into two groups: the 'imagery'
data, and the sounder data. Chapter 5 deals explicitly with information
derived from the vertical sounding instruments. The term 'image' has
been used since sounders were first put on spacecraft to help to distinguish
the two types of sensors. The term can be misleading, since it suggests
photography from manned spacecraft or television as on the early weather
satellites. Most imagery data, like sounder data, are now taken by scanning radiometers on satellites. The imagery data are usually digitized so
that they can be processed numerically in the same way as the narrow-
waveband sounder data. The data do not have to be displayed as an
image, although this is generally done. On the other hand, individual
bands from IR sounders can have data displayed in an image format so
that sounders too can provide images.

Resolution

One of the most important parameters relating to a satellite sensor is its
resolution. Three types of resolution are important: spectral, radiometric
and spatial. Spectral resolution is defined by how many spectral channels
are sampled and how narrow the bandpass is for each individual channel.
For example, if it were required to isolate a phenomenon that is
characterized by high absorption at $0\cdot556\,\mu m \pm 0\cdot0001\,\mu m$, a sensor with
four channels between $0\cdot4$ and $0\cdot8\,\mu m$, each $0\cdot1\,\mu m$ wide, would have
insufficient spectral resolution. Most imaging satellites have a limited
number of spectral channels (7 on NASA's Landsat-4 thematic mapper)
and utilize relatively large bandpasses to maintain high signal levels.

Radiometric resolution is a measure of how adequately the radiance
levels reaching a sensor can be measured. It is often expressed in terms
of Noise Equivalent Power, NEP, in W, or in the thermal infra-red in
terms of Noise Equivalent Temperature Difference NEΔT, in K. The
NEP is the minimum resolvable change in the radiant flux reaching the
sensor that can be recorded by the sensing system. For example, if a
sensor had an NEP of $1\cdot4\times10^{-6}$W and a reflectance change on Earth
resulted in a change of $0\cdot5\times10^{-6}$W at the sensor, the change would be
lost in the noise of the imaging, detection and amplification system. The
NEΔT works in exactly the same fashion, except that instead of changes
in incident flux it is expressed in terms of the temperature change that
would result in an equivalent change in the incident flux for the
wavelength interval being sensed.

Resolution is generally assumed to denote spatial resolution, which is the ability to differentiate small objects or to locate precisely changes in a feature. It can be expressed in terms of the numbers of line pairs per unit length that can be resolved into a distinct line pattern. The line pairs are alternating dark and light bars of equal width with a fixed contrast between the dark and light elements. The resolution of a satellite system is best expressed in terms of the Ground Resolution Distance (GRD). This is the reciprocal of the resolution expressed in units of distance at the Earth's surface. For example, a system that could resolve 20 line pairs per 100 m on the ground would have a GRD of 5 m. The GRD of a sensor system is a function of the resolution of the sensor as well as the atmospheric effects on the day the data are collected. Atmospheric turbulence will introduce geometric distortions reducing the spatial resolution below a simple geometric scaling to GRD.

The projected Instantaneous Field Of View (IFOV) is often mistaken for the resolution of a system. The IFOV is an angular measure of the field of view over which the satellite sensor accepts data for processing into each data point or pixel in an image (the term 'pixel' stands for *pic*ture *el*ement).

The projection of the angular IFOV to a ground diameter, side length or area is often called the projected IFOV and used instead of the resolution of a system. While IFOV is a very useful measure of system performance, it must be recognized that it is only indirectly a function of the spatial resolution of the sensor system.

Remotely sensed radiation fields in the presence of clouds

Clouds profoundly affect the radiation fields in the two primary wavelength regions used by remote sensers: the visible, and the thermal infra-red. The presence of clouds in the Earth's atmosphere can increase the optical thickness from a value of less than $0 \cdot 5$ to anything from $\tau' = 2$ for thin cirrus clouds to $\tau' > 50$ for clouds with considerable vertical extent. In the thermal infra-red, clouds absorb and re-emit radiation originally emitted from the surface and hence exert considerable control over the total emitted infra-red radiation to space. Scattering of radiation by air molecules (Rayleigh scattering) is of considerable importance in the ultraviolet and visible parts of the spectrum but it is scatter and absorption by cloud particles and droplets that form the predominant part of the system or planetary albedo.

Clouds are usually wet atmospheric aerosols composed of tiny spheres of liquid water ranging in radius from 2 to 200 μm. These water droplets fall with terminal velocities of the order of $2 \, \mathrm{cm \, s^{-1}}$. Concentrations of droplets within clouds are of the order of a few hundred million per cubic metre. Thus a small cubical cloud of sides 1 km contains of the order of

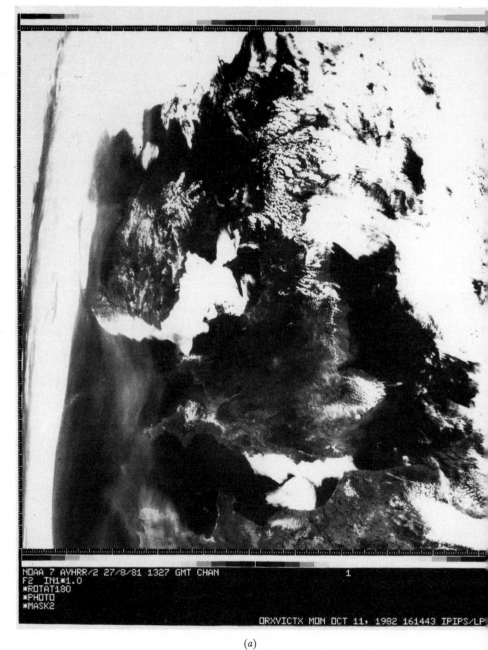

(a)

Figure 2.14. (a) Visible; and (b) thermal infra-red images from the NOAA-7 AVHRR showing the British Isles and N.W. France on 27 August 1980 (the grey scale on the infra-red image grades from white (cold) to black (hot); courtesy of P. Baylis, University of Dundee).

(b)

10^{17} droplets. Clouds are not simply random features covering around
50 % of the Earth's surface but have forms, heights and reflectivities
which are a function of both location and the prevailing weather (see,
e.g., figure 2.3, p. 51). For example, figure 2.14 shows a pair of NOAA-7
AVHRR images of the British Isles and north-western France. On this
summer's day (27 August 1980) cloud is seen to have formed
predominantly over the ocean areas, and particularly in the English
Channel. The infra-red image (figure 2.14(*b*)) illustrates the temperature
contrast between land and ocean areas that has led to this differential
cloud formation. The ocean surface temperature fields are discussed in
detail in Chapter 7. In contrast to this pair of images, the full-disc image
in figure 2.3 (p. 51) illustrates the climatological cloud regimes (see
Chapter 3 for a full discussion) prevailing in equatorial, low-latitude and
mid-latitude regions.

Cirriform clouds are composed of ice crystals, altiform are supercooled
water droplets existing at temperatures below 273 K, while stratiform
clouds are generally layered; the latter structure being indicative of stable
conditions and vertical velocities of the order of $0 \cdot 1 - 1 \cdot 0 \, \mathrm{m \, s^{-1}}$. Cumuliform clouds occur in unstable conditions and are associated with much
greater vertical velocities, of the order of $5 - 50 \, \mathrm{m \, s^{-1}}$. Nimbus clouds are
rain- or snow-producing. The WMO cloud classification follows a scheme
originally suggested by an English chemist, Luke Howard, in 1803. Four
basic types of cloud are identified, as follows.

1. Cirrus: filament (fibrous ice clouds).
2. Stratus: layer (usually air-mass uplift clouds).
3. Cumulus: heap or pile (convective clouds).
4. Nimbus: rain clouds.

Names are compounded (e.g., cumulonimbus) and prefixes (e.g., alto-)
used to indicate structure and height. Typical cloud heights vary with
latitude (see table 2.3).

The relationship between the particulate size (in this case a typical
cloud droplet) and the wavelength of the radiation determines the type of
interaction. For water clouds, droplet sizes are of the order of $10 \, \mu m$,
while for cirrus clouds the ice crystals have radii of the order of $30 \, \mu m$.

Table 2.3. Cloud type and approximate height as a function of latitude.

Height	Latitude			Cloud type
	Polar	Temperate	Tropical	
High	3–8 km	5–18 km	6–18 km	cirro
Middle	2–4 km	2–7 km	2–8 km	alto
Low	0–2 km	0–2 km	0–2 km	strato

Thus the radius is very much greater than the wavelength of the solar radiation and Mie scattering results (see figure 2.4, p. 54). Since clouds tend to be optically thick ($\tau' \approx 5$ for each $100 \, \text{m}$ thickness of cloud), multiple scattering ensues with an average of between 20 and 30 interactions between a solar photon and cloud particulates for each $\tau' = 5$. Thus the higher the value of optical thickness τ' the greater the multiple scattering and hence the higher the cloud albedo (see table 2.4).

Stratus clouds exhibit very different albedos depending on their vertical extent. For instance, stratus cloud of depth $100 \, \text{m}$ would have an albedo of $0 \cdot 2$; for $200 \, \text{m}$ the albedo increases to $\sim 0 \cdot 4$ and for greater than $300 \, \text{m}$ the albedo rises to between $0 \cdot 6$ and $0 \cdot 7$. Cirriform clouds, being much thinner, have very small optical thicknesses and hence much lower albedos. In the thermal infra-red part of the spectrum, i.e., at wavelengths longer than $4 \cdot 0 \, \mu\text{m}$ (figure 2.1, p. 46), absorption by liquid water and water vapour dominates any scattering. For instance, at $10 \, \mu\text{m}$ water clouds reflect less than $3 \, \%$ of the incident thermal radiation and ice clouds less than $1 \, \%$. Cirriform clouds (because of their lower water content) are both poorer absorbers and poorer scatterers, the latter being caused by their greater particle size. Absorption in the terrestrial wavelength regions depends on thickness. For instance, a cloud with vertical extent as small as $0 \cdot 05 \, \text{km}$ absorbs around $40 \, \%$ of the incident energy and $0 \cdot 5 \, \text{km}$ of cloud is probably absorbing close to $100 \, \%$.

The cloud feature which dominates the impact of cloudiness on the thermal infra-red radiation stream is the cloud-top height, which is a function of the vertical extent and type of the cloud itself. Low clouds radiate at temperatures not greatly dissimilar to the ambient surface temperature, and hence their impact on the emitted terrestrial radiation stream is comparatively small. Middle level clouds, on the other hand, can have temperatures considerably lower than the mean surface temperature and cumuliform clouds, while being small in horizontal extent, often have tops close to the tropopause.

Cirrus clouds interact with the radiation streams in such a way that their net effect is believed to be in a sense opposite to all other cloud types.

Table 2.4. Typical short-wavelength cloud albedos (approximate).

Cloud type	Albedo
Low	$0 \cdot 6$
Middle	$0 \cdot 5$
High	$0 \cdot 2$
Cumuliform	$0 \cdot 7$

It is generally accepted (see Chapter 3) that for low- and middle-level clouds the 'albedo' effect dominates the 'greenhouse' effect. This situation is reversed for cirrus clouds. These optically thin ice-crystal clouds interact with the solar and terrestrial radiation in such a way that their infra-red emissivity increases much faster as a function of τ' than does the albedo. Thus for a given value of τ' the impact on the solar radiation stream is less than the enhancement of the surface temperature due to the 'greenhouse' effect. Overall then cirrus clouds tend to warm the climate system. The interaction between clouds and the Earth's radiation is discussed in detail in Chapter 3.

2.3. *The effect of the atmosphere on the observed image*

In this section the effect that propagation through the atmosphere has on the image observed will be considered. In particular, an effort will be made to evaluate whether atmospheric effects are significant.

It is useful to simplify equation 2.23 into a more manageable form and assume, for convenience, that the concern is with level Lambertian (perfectly diffusing; see figure 2.5, p. 57) reflectors in relatively flat terrain. Under these conditions, equation 2.23 reduces to

$$L = Kr + L_u \qquad (2.29)$$

where

$$K = \left\{ E'_s \pi^{-1} \cos\left(\frac{\pi}{2} - \theta'\right) \exp\left[-\tau' \sec\left(\frac{\pi}{2} - \theta'\right) \right] + L_d \right\} \exp\left(-\tau' \sec\theta \right)$$

Thus, for a fixed Sun – object – camera orientation, the reflected radiance reaching a sensor is a simple function of the surface reflectivity, the scattered upwelling radiance L_u and a constant governed largely by the source strength modified by the extinction and scattering coefficients of the atmosphere. In trying to view the Earth, the observer is really trying to see the reflectance variations on the Earth's surface. Equation 2.29 indicates that the effect of the atmosphere in any given spectral region is to add a 'flare' term into the system which reduces the contrast or tends to reduce the apparent variations in reflectance. In addition, two competing processes are taking place. One is that the direct specular source strength E_s from the Sun is reduced by the atmospheric extinction; the other is the diffuse source strength L_d, which is increased by atmospheric scatter. The net effect of these competing processes depends on the conditions under which the Earth is viewed and the wavelength interval considered. The effect of viewing through an increasingly thick atmosphere is a decrease in observed contrast. For example, on a clear day, landscape features are observable on foreground hillsides but not on more distant, though similar, slopes. The dramatic decrease in contrast is the result of viewing through a longer path length. The Earth's atmosphere, because of its dynamic nature, introduces equivalent changes in images viewed from space from day to day, or in some cases, from point to point, within an image. Since these effects are the result of dynamic processes, they must be corrected on a target area or image specific basis if detailed quantitative analyses are to be performed (see also Chapter 7). In addition to the changes that occur in one spectral band, it must be noted that these changes do not occur equally in all spectral bands. The excess scatter of short-wavelength radiation, for example, will tend to add larger amounts of scattered path radiance in blue sensing channels as well as increasing the diffuse downwelled radiance more in the blue than in longer wavelength channels. One result of this phenomenon is that photographs from space appear bluer than might be expected and this bluish cast is accentuated on hazy days.

The magnitude of the flare term L_u can be considered in order to estimate its importance. If L_u were zero, the reflectivity of a surface could be expressed from equation 2.29 as $r = L/K$ or, alternatively, a 100% reflector would have K radiance units. If flare is now added into the system to the amount L_u then it is possible to express this as the equivalent of L_u/K reflectance units of flare. It has been suggested from empirical studies that the value of L_u/K could be expected to vary from about 10% in the red to as much as 30% in the blue portion of the spectrum. It is useful to consider what effect an added path radiance equivalent to a 10% reflector would have on a satellite image. If two objects have reflectance values of 5% and 10%, the contrast ratio for these objects is $10/5 = 2 \cdot 0$. If we imaged the objects in the absence of the atmosphere, for example, from a few metres above, a contrast ratio of $2 \cdot 0$ would be observed. If, however, they are observed through an atmosphere that has added scattered radiance equivalent to a 10% reflector to all objects imaged, the contrast ratio will be reduced to $(10 + 10)/(10 + 5) = 1 \cdot 33$. The effect of this upwelling radiance will be much more significant for low reflectance objects and for sensors employing shorter wavelengths. One important way to observe the effect the atmosphere has on the radiance reaching a satellite sensor is to consider what effect the normal variation in L_u/K will have on reflectance. The value of L_u/K can easily change by 20% even on those days which appear to be clear. This is the equivalent of a 2% reflectance change even if L_u/K is only the equivalent of a 10% reflector. In passing from a clear to a hazy sky condition, even larger changes can easily occur. In particular, in cases where low reflectance surfaces such as water or forests are being studied, the magnitude of the variation in perceived reflectance is nearly as large as the entire signal. This precludes quantitative analysis of the reflectance characteristics of these surfaces unless the magnitude of the atmospheric effects is well defined. Thus it can be seen that quantitative analysis of reflected radiance data from space requires precise assessment of the magnitude of the atmospheric effects. The effect of changes in the upwelling radiance have been emphasized here for simplicity. The values of the other atmospheric terms (e.g., K in equation 2.29) are also extremely important. It is also important to recognize that these phenomena are always present and can easily lead to misinterpretation of images analysed by subjective methods. Whether a quantitative analysis of digital satellite data or traditional image interpretation is attempted, it is extremely important to consider atmospheric effects if the data are to be interpreted properly. The question arises as to how important these atmospheric phenomena are in the long-wave infra-red region. It is in this EM region that the major efforts to measure the temperature of Earth surface features are appropriate. To simplify the considerations of the magnitude of potential errors, it will be assumed that a perfect blackbody is available as a target. In this case $r = 0$, so equation 2.25 reduces to

$$L = \tau L_T + L_u \qquad (2.30)$$

It has been shown that if the effect of the atmosphere is totally neglected (i.e., $\tau = 1$, $L_u = 0$) and an attempt is made to convert directly observed radiance to apparent temperature, it is easy to introduce errors as large as 13 K even on days that would normally be considered clear (i.e., giving good infra-red images). It must also be noted that if the atmospheric corrections appropriate for one day are applied on a second day, the errors may well exceed several K. Temperature is one of the fundamental factors governing many life processes as well as being a principal driving factor in most meteorological processes. In general, knowledge of true surface temperature to within approximately 1 K is necessary if meaningful analyses using these data are to be performed. Once again it can be seen that careful removal of atmospheric effects must take place if useful temperature data are to be extracted from satellite sensors. Additionally, it must be recognized that the errors estimated are minimal errors in most cases for ideal radiators. A real world surface will have errors compounded with the associated problems of the limited knowledge of surface emissivities and reflected radiance.

2.4. Sensing the Earth from space

So far, the nature of the atmospheric phenomena that degrade satellite imaging have been considered in terms of the effects that these phenomena have on the images. In this section some techniques that can be applied to calibrate satellite images to permit observation of true Earth surface characteristics such as reflectivity and temperature are considered. In addition, some of the satellites and sensor systems available for Earth observation are discussed.

Computation of atmospheric effects on the reflected radiance reaching a satellite

In order to simplify the discussion, it will be assumed that a full internal calibration of a sensor has been accomplished, and that conversion of an image brightness value to the radiance observed at the satellite (cf. Section 2.2, p. 46) is possible. This reduces the problem to finding the atmospheric and illumination coefficients that relate reflectivity to observed radiance. For many conditions of satellite observation, the conditions that apply to equation 2.23 hold and it is possible to express the relationship between radiance and reflectivity by a simple first-order linear model. This model is further simplified for satellite systems with a narrow total field of view such as the Landsat 1–3 sensors. Under this condition the values of K and L_u in equation 2.29 are approximately constant throughout the scene. This small total field is one of the major advantages of very high altitude satellites. This advantage is usually acquired at the cost of spatial resolution, since the IFOV must be extremely small to achieve small pixel values on the ground, and these small IFOV values mean decreased signal strength, and, therefore, a higher signal to noise ratio. However, when these conditions do hold, it is possible to regress the radiance observed for selected targets in each spectral band against the known reflectivity for those targets. The slope and intercept of this regression would yield the values of K and L_u respectively for the image. These coefficients could then be used to compute the reflectivity values of any unknown targets imaged. This procedure can be repeated for each spectral channel imaged by the sensing system. Here it is important to point out that the coefficients computed are only valid for that particular image because the atmospheric conditions will change with location and time. In fact, for a Landsat scene that covers over 100km on a side, it is possible for significant atmospheric changes to occur within a given scene. The major weakness in this approach is that known ground reflectance values are required. One way to obtain these values is to measure the reflectivity values *in situ* at the time of the satellite overpass. Conceptually this is an attractive approach requiring only a minimum of two points of significantly different reflectance. In practice it is a somewhat more cumbersome process requiring the measurement of several points to reduce errors. Furthermore, it is important to consider what a 'point' is for the satellite. The MSS sensor aboard the Landsat satellite is imaging a spot approximately 76m on a side. Attempting to approximate the *in situ* reflectivity of a natural surface large enough to isolate a specific spot this size requires considerable effort. This method is further complicated in many instances by inaccessible terrain which precludes sending in field teams with bulky spectroreflectometers.

Piech and Schott (1975) have suggested a method potentially appropriate for circumventing some of these limitations. They suggest an approach which has been demonstrated successfully using aerial images. This technique relies on imaging areas inside and just beyond shadow elements to compute L_u. It then assumes that the average value of many samples of certain man-made surfaces (e.g., concrete pavement) is a constant. Thus, since L_u is known

$$K = (L_c - L_u)/r_c \qquad (2.31)$$

where L_c is the mean radiance for many samples of the feature c; and r_c is a known mean reflectivity of some man-made feature c. Once again, this process can be repeated in each spectral channel sensed by the satellite. This technique, however, cannot be applied directly to many satellite systems because their resolution is inadequate for resolving specific shadow elements or specific man-made targets. It can, however, be used from aircraft to measure the reflectance of large natural targets also imaged by the satellite. This aerial approach thus becomes, in effect, an *in situ* reflectometer. In order to be effective, the spectral channels used in the aerial system must match those in the satellite. Also, the aircraft underflight must closely coincide with the satellite overpass because of relatively short-term changes in the reflectivity of natural surfaces. The need for a concurrent underflight can make this approach unattractive in many cases from an operational and/or cost standpoint.

The limitations imposed by ground truth or underflight sampling can be overcome through the use of atmospheric models. For example, it is possible to compute surface reflectance values from observed radiance data and a knowledge of the atmosphere. The numerical model could utilize standard atmospheres with additional information relating to the horizontal visual range values. Considerable error can be introduced in the tuning of the modelled atmosphere. The detailed observational programme required to reduce this error is likely to require underflights or large-scale ground-truth analysis.

In conclusion, it is possible to convert reflected radiances observed by satellites to the reflectivity of Earth objects if the limitations described above are recognized. This is a major accomplishment because reflectivity is a fundamental material property and spectral reflectivity in particular carries a great deal of information about the object and its condition. For example, spectral reflectivity can be used to isolate a corn field from a hay field and healthy corn from diseased corn. Such a process requires removal of atmospheric phenomena, especially sub-pixel clouds (see the discussion of sensing of land features in Chapter 7).

Computation of temperature from emitted radiance measurements

The computation of temperature from the emitted radiance reaching a satellite system is somewhat analogous to the problem of measuring reflectivity from space. Again, some simplifying assumptions will be made to expedite the discussion. Firstly, it will be assumed that the surfaces are Lambertian, have unit emissivity, and are located within a small view angle (i.e., τ and L_u are constants for the targets considered). In this case, equation 2.25 reduces to the same form as equation 2.30

$$L = \tau L_T + L_u \qquad (2.32)$$

Equation 2.32 indicates that the observed radiance is a first-order linear function of the surface radiance. If the parameters τ and L_u can be defined, then the temperature of the surface can be found from the observed radiance at the spacecraft. Equation 2.32 can be solved by simply measuring the temperature of at least two objects, computing the surface radiance in the spectral band observed and fitting a straight line through the surface radiance values and the corresponding radiance values observed at the sensor. Scarpace *et al.* (1975) applied this approach to aerial thermal infra-red imagery with considerable success (observed errors of $0 \cdot 1 \, \text{K}$). Their approach required uniform temperature targets covering areas significantly larger than the IFOV of the imaging system. For thermal infra-red satellites with projected IFOVs of $0 \cdot 5 - 1 \, \text{km}$ on a side, this poses a serious logistical problem.

   To overcome the difficulties of obtaining adequate ground truth, many users have turned to atmospheric propagation models. One of the most well known of these is incorporated into the US Air Force Geophysics Laboratory's LOWTRAN computer code (for example, interpretation of results from NASA's Heat Capacity Mapping Mission, HCMM, has successfully used the LOWTRAN package). This code incorporates several user-selectable standard atmospheres. Once an atmosphere is selected, the model incorporates molecular absorption, molecular scattering and aerosol extinction to arrive at atmospheric transmittance τ and path radiance L_u in a specified bandpass for a given geometric orientation. It will also permit the user to input a specific set of atmospheric conditions. The atmospheric conditions are obtained by radiosonde (balloon) data which include for each altitude sampled the atmospheric pressure, the temperature and the dew point. Radiosonde data can be taken in conjunction with a satellite overpass at a specific location. This, however, can be a very expensive and cumbersome task. Alternatively, radiosonde data can be obtained on a daily or twice daily basis for many major weather stations. Where possible, careful selection of the study site location can help to ensure that the available radiosonde data are sufficiently close to supply a reasonable estimate of the prevailing atmosphere at the time and location where the satellite data are obtained.

In one experiment, surface radiance values and associated surface temperature values were computed according to equation 2.32. When these data were compared to the known surface temperature values, residual errors of 9K were found. In some cases, this type of accuracy may be acceptable; however, as suggested earlier, in many cases more accurate temperature values are required. An underflight calibration technique has also been utilized to calibrate the HCMM thermal infra-red sensor. This underflight approach utilized a thermal infra-red imaging system radiometrically calibrated using the atmospheric profile approach described by Schott (1979). This technique permits calculation of atmospheric effects based on sampling the atmosphere by flying the thermal infra-red imaging system at several altitudes over the target. Using this approach, the integrated surface temperature and, therefore, the integrated surface radiance for very large features can be calculated precisely. These known radiance values in the spectral bandpass of the satellite then become the ground truth for a linear solution of equation 2.32. This technique can be used to compute surface temperatures from satellite data with expected errors of $1 \cdot 1\,K$. While this result indicates that the desired temperature measurement accuracies can be obtained using satellite data, it still has two limitations. Firstly, an underflight at the time of the satellite overpass is required, and, secondly, the results have only been demonstrated for surfaces with very high emissivities (i.e., near-blackbody targets). The errors associated with higher reflectivity objects which reflect significant amounts of downwelling radiance L_d still need to be evaluated.

Most of the discussion in this chapter has focussed on the analysis of clear-sky radiances. The presence of clouds in the atmosphere significantly increases both absorption and scattering at visible wavelengths. As the number of interactions between cloud particles and solar photons increases in cloudy conditions, most of the simple scattering models which assume single scattering break down. At this point only a complete solution of the radiative transfer equation can trace the path of a particular photon. Even very sophisticated computer routines which can perform these calculations rely upon assumptions about the cloud structure, droplet size distribution and, often, the vertical distributions of particulates and humidity within the cloud. The optical thickness τ' increases rapidly and emitted thermal infra-red radiation can be assumed to come from close to the cloud top for all but cirrus clouds.

In general, remote sensing applications can deal with the effect of clouds by removing these 'contaminated' data from the archive either completely or in order to treat it in some different fashion. For example, surface sensing is clearly impossible (except at micro-wavelengths) in the presence of cloud; cloud-clearing techniques used in conjunction with vertical sounder data have already been mentioned in this chapter, and Chapters 5 and 7 deal with the effects of cloud contamination on other data sets.

Information about cloud condition is, of course, of considerable value for many climatological and environmental applications. Chapters 3, 6 and 8 deal specifically with cloudiness data retrieved from satellite data. In Chapter 6 cloud data are used to derive information about atmospheric wind fields. Generally, however, an image and sometimes even an individual pixel is neither completely cloud-filled nor completely cloud-free. Sub-grid scale clouds are exceptionally difficult to detect (Chapter 7). They are probably ubiquitous and their existence mars the derivation of remotely sensed information relating to both the surface and to the cloud fields themselves.

2.5. *Future satellite and sensor systems*

In this chapter, the phenomena that cause the emission and reflection of radiation and atmospheric modification of EM signals have been described. Some of the methods available for computing the magnitude of these effects on the signal reaching satellite sensors have also been discussed. In this section mention will be made of some satellite sensors likely to be made available for Earth observation. The world-wide space community has launched over 100 satellites a year since the early 1970s, many of which have Earth observation sensors of some type (see Chapter 1). Even a brief description of the variety of satellites and sensor packages is beyond the scope of this text. However, an attempt will be made to describe some of the principal sensor systems currently in use or planned for use in the near future.

Probably the most widely accessed satellite data come from NOAA's meteorological satellites. The ITOS series satellites are one of the principal sources of satellite imagery. The ITOS satellites are in Sun-synchronous orbits carrying three principal payloads. These include the scanning radiometer with a $7 \cdot 5$ km projected IFOV for both the visible $(0 \cdot 5 - 0 \cdot 7 \mu m)$ and thermal infra-red $(10 \cdot 5 - 12 \cdot 5 \mu m)$ channels. The scanning radiometer provides twice daily (day/night) global coverage on an operational basis. The advanced very high-resolution radiometer AVHRR aboard the NOAA satellites is essentially a high-resolution counterpart of the scanning radiometer providing approximately 1 km resolution of selected areas imaged by the scanning radiometer. Because of the higher data density and limited recording space on the satellite AVHRR data are somewhat limited in many areas, particularly where line-of-site communication with NOAA receiving stations is unobtainable. The final ITOS Vertical Temperature Profile Radiometer (VTPR), which is designed to measure the temperature of the atmosphere from the Earth's surface to $30 \cdot 5$ km, senses the radiance in several CO_2 absorption channels as well as the radiance in an atmospheric window to determine the apparent Earth surface temperature. Drawing on the differential absorption of CO_2 in the selected channels, it is possible to compute a vertical temperature profile of the atmosphere. More recently, HIRS and MSU sounder units have been carried on the NOAA satellites. The other major meteorological satellites are the NOAA Geostationary Operational Environmental Satellites (GOES) and their Japanese, Soviet and European Space Agency counterparts (see table 1.9, p. 41). These satellites are geosynchronous, with sensors continuously scanning the Earth. The GOES satellites have visible and thermal infra-red scanning radiometers with projected vertical IFOV values of $0 \cdot 9$ km and $7 \cdot 2$ km respectively. These satellites can provide repetitive coverage of the hemisphere viewed approximately every 30 minutes to monitor dynamic processes such as cloud movements, storms, floods, etc.

In addition to the imaging satellites, many atmospheric sampling satellites exist which are engaged either in experimental or operational efforts to monitor atmospheric conditions. By careful selection of atmospheric absorption or emission bands, these experiments are attempting to develop operational methods of computing such things as air temperature and humidity profiles, and the concentration of various gases or aerosols as a function of altitude. Some of these sensors view the Earth directly, while others view the Sun through the Earth's atmosphere, functioning much as a laboratory absorption spectrophotometer does.

In addition to the meteorological satellites, probably the most widely used Earth observation satellites are the Landsat series satellites. These satellites are in Sun-synchronous orbits with 18-day repeat coverage and have been in nearly continuous operation since 1972. The most widely used sensor aboard Landsat is the Multispectral Scanner (MSS). It is a four-band system with nominal spectral bands at $0 \cdot 5 - 0 \cdot 6 \mu m$,

$0·6-0·7\mu$m, $0·7-0·8\mu$m and $0·8-0·9\mu$m. The projected IFOV of the satellites is 76 m with a swathwidth of approximately 185 km. The Landsat data are available either as hard-copy image products or computer-compatible digital tapes. The long-term operation, high spatial resolution, world-wide coverage and relative ease of data accession have made Landsat MSS data the most widely utilized data for Earth resource observation.

In addition to the MSS sensor which is now considered operational, the most recent Landsat (Landsat-4) also carries a new sensor known as the Thematic Mapper (TEM). The TEM is an experimental sensor designed to improve the spatial resolution and spectral range of the Landsat satellites in order to expand the application of the data. The TEM has seven spectral bands at $0·45-0·52\mu$m, $0·52-0·60\mu$m, $0·63-0·69\mu$m, $0·76-0·90\mu$m, $1·55-1·75\mu$m, $1·08-2·35\mu$m and $10·4-12·5\mu$m. The first six bands have projected IFOVs of 30 m, while the thermal infra-red has a projected IFOV of 120 m. The ground coverage for the TEM is identical to the MSS coverage. This means that TEM digital images contain almost an order of magnitude more data than a comparable MSS scene. These extremely large data volumes prohibit onboard storage of image data. To overcome this difficulty, NASA plans to place the Tracking Data Relay Satellite System (TDRSS) satellites in very high orbits. These satellites will receive data from imaging satellites even when they are out of sight of receiving stations and transmit the data to a receiving station. Thus, in effect, the imaging sensor is never out of communication with a receiving station. These TDRSS satellites will probably become the major down-links for all future imaging satellites, making the present world-wide receiving stations obsolete. All the high-density data can be down-linked to one or two special purpose receiving stations. The Landsat-E satellite, which will become Landsat-5 on launch, will be configured very much like Landsat-4, but it will probably be the first Landsat satellite placed in space by the Space Shuttle.

The Shuttle can be expected to add a new dimension to satellite imaging systems, while revitalizing some other systems (see Figure 1.17, p. 42). The relative ease of placing satellites in space should make it easier to test new ideas for Earth observation sensors. In addition, with the potential for data retrieval, more photographic sensing can be expected. The camera systems carried by NASA's manned spaceflights provided a fascinating, but quite limited, high spatial resolution look at the potentials of photographic sensing from space. The Shuttle, with its large payload capacity and retrieval capabilities, should make it feasible to generate significant amounts of high-resolution satellite photography.

In addition to the widely available data from the US satellites, the next decade will see the launching of the French SPOT satellite (approximate launch 1984), Japan's ERS (approximate launch 1986), and the European Space Agency's ERS (approximate launch 1987), as well as satellites of India and several other nations. The details of the individual sensor configurations which will be carried on these satellites is beyond the scope of this text. However, some general characteristics appear to be developing for these sensors. Many of them will utilize the new linear array technology instead of the more traditional system with a limited number of detectors and scanning mirrors (i.e., the across-track data will come from a bank of hundreds to thousands of detectors rather than a scanning mirror). Typically, the systems will be multispectral with projected IFOVs of tens of metres. In addition, many of the sensors (starting with France's SPOT satellite) will be pointable either along- or across-track, permitting collection of stereo imagery or increased coverage of a special target area.

Thus, by the late 1980s, there could well be five to ten, or even more, high-resolution, multispectral Earth observation satellites orbiting the Earth. Each of these satellites may well be collecting 16×10^6 pixels per frame in each of five to ten spectral channels with as many as 100 frames acquired per day. That represents upwards of 10^{11} pixels of data a day without even considering meteorological satellites, special purpose sensors, duplicate sensors on the same craft, etc. In view of the massive amount of data potentially available, a major aim of research science is to identify and optimize the methods of information retrieval. As has been emphasized throughout this chapter, the atmosphere plays a crucial role in controlling what information can be extracted from satellite information. In addition, satellite data can play a critical role in defining atmospheric conditions which control, directly or indirectly, the environmental state of the Earth and most of our life processes.

References and further reading

Curcio, J. A., 1961, Evaluation of atmospheric aerosol particle size distribution from scattering measurement in the visible and infrared. *Journal of the Optical Society of America* **51**, 548–51.

Goody, R. M., 1964, *Atmospheric Radiation, Volume I: Theory* (Oxford: Clarendon Press).

Grum, F. and Becherer, R., 1979, *Optical Radiation Measurements, Vol 1, Radiometry* (New York: Academic Press).

Lillesand, T. M. and Kieffer, R. W., 1979, *Remote Sensing and Image Interpretation* (New York: John Wiley and Sons).

Piech, K. R. and Schott, J. R. 1975, Atmospheric corrections for satellite water quality studies. *Proceedings of the SPIE Scanners and Imagery Systems for Earth Observation* **51**, 84–89.

Piech, K. R. and Walker, J. E., 1971, Aerial color analyses of water quality. *Proceedings of the American Society of Civil Engineers, Journal of Surveying and Mapping Division* **97**, 185–197.

Reeves, R. G. (ed.), 1975, *Manual of Remote Sensing* (Falls Church, Va.: The American Society of Photogrammetry).

Scarpace, F. L., Madding, R. P. and Green, T., III, 1975, Scanning thermal plumes. *Photogrammetric Engineering and Remote Sensing* **41**, 1223–31.

Schott, J. R., 1979, Temperature measurement of cooling water discharged from power plants. *Photogrammetric Engineering and Remote Sensing* **45**, 753–61.

Slater, P. N., 1980, *Remote Sensing: Optics and Optical Systems.* (Reading Mass.: Addison-Wesley).

Stimson, A., 1974, *Photometry and Radiometry for Engineers* (New York: John Wiley and Sons).

Van de Hulst, H. C., 1981, *Light Scattering by Small Particles* (New York: Dover Publications, Inc.).

3
The Earth's radiation budget and clouds

Ann Henderson-Sellers
University of Liverpool
Liverpool, UK

3.1. Satellite observations of the Earth's radiation budget

The network of Earth polar-orbiting and geostationary satellites provides a vast data source for the meteorologist, climatologist and environmental scientist. The large quantity of available data has necessitated the design of automated systems to interpret and analyse the raw radiation data. Radiance measurements are generally made in at least two wavelength regions: in the 'visible' part of the short wavelength region (around $0 \cdot 5 - 0 \cdot 6 \mu m$) and in the 'window' region (approximately $10 - 12 \mu m$) of the long-wave thermal infra-red part of the spectrum. Two important meteorological and climatological variables are the albedo (or reflectivity) of a particular area and the net radiation budget. Both of these parameters are potentially derivable from satellite radiance data. The number of channels, spectral regions of the sensors, spatial resolution and temporal sampling vary from satellite to satellite. For instance, the geostationary satellite METEOSAT-II possesses a spectral band in the region of water vapour absorption, centred at a wavelength of $6 \cdot 3 \mu m$, in addition to channels in the visible and thermal infra-red; and radiance measurements of the full disc are made every 30 minutes, compared with polar orbiting satellites which generally only sample a specific location twice a day.

Automated processing of the large volume of satellite radiance data now available can be undertaken in two modes. Operational processing must include assumptions and approximations so that the assigned task can be completed by the time the next data set arrives. Research analysis can be conducted at a more leisurely pace but significant problems may still be encountered. A number of automated methods of deriving cloud and radiation budget statistics directly from satellite radiance data have been proposed. Automated processing is crucial if the full volume of data received is to be included in the cloud and radiation budget archives. Automated cloud retrieval algorithms can be grouped into three types: (i) threshold methods; (ii) statistical analyses; and (iii) radiative inversion calculations (see Chapter 1). Cirrus cloud remains extremely difficult to identify from satellite-based information. The presence of cirrus cloud in the region of interest introduces a large range of variability into the satellite signals, making analysis by a fully automated system extremely difficult. Additionally, cirrus cloud is often obscured to the ground-based observer by other cloud layers and therefore ground-truth calibration of satellite-based cloud determination is difficult to achieve. Failure to note the presence of cirrus cloud could lead to mis-identification of other elements (particularly other cloud types but also land surface features; see also Chapter 7). Additionally, as described earlier, it is now generally agreed that an increase in the amount of cirrus cloud leads to surface and tropospheric warming while an increase in the percentage of cover at other levels tends to cool both surface and system.

Until a successful method of determining this important atmospheric configuration is achieved, a fully automated, accurate, nephanalysis is impossible. This conclusion has important repercussions for current and planned programmes designed to construct global cloud climatologies.

Compilations of estimates of the Earth's planetary radiation budget obtained from operational meteorological satellites have been performed by several authors (e.g., Winston et al., 1979; Stephens et al., 1981). In these investigations the spectral intervals of measurement, continuity and length of record, time of observation and spatial resolution differ significantly. For example, the Stephens et al. (1981) data set extends an earlier analysis with the addition of the Nimbus data for a total of 48 months from the years 1964–1977. Ohring and Gruber (1983) review many of these data sets. In the first two sections of this chapter the Winston et al. (1979) data are used as an example of such compilations.

In the composition of the Winston et al. (1979) data set, measurements by the two-channel scanning radiometer on board the NOAA polar-orbiting satellites were used from the period June 1974 to March 1978. The satellites have a period of approximately 117 minutes with equator crossing times at 2100 local standard time (LST) northbound and 0900 LST southbound. The radiometer senses energy in the visible ($0 \cdot 5 - 0 \cdot 7 \mu m$) portion of the spectrum and in the infra-red window region ($10 \cdot 5 - 12 \cdot 5 \mu m$). The spatial resolutions of the sensors are approximately 4 km for the visible and approximately 8 km for the infra-red at the sub-satellite point. The components of the Earth–atmosphere radiation budget were computed by the USA's National Earth Satellite Service (NESS). The NESS operational processing includes correction of the infra-red radiances to a nadir view at the top of the atmosphere and the visible observations are divided by $\cos Z$, where Z is the solar zenith angle.

The components are the outgoing long-wave radiation, the albedo and the related absorbed solar radiation, and the net radiation, which is defined as the absorbed solar radiation minus the outgoing long-wave radiation. An atlas summarizing mean monthly, seasonal and annual average conditions has been produced by Winston *et al.* (1979). Details of the derivation of the flux values from the operational data, the processing of the data to form suitable outputs such as maps and averages, and the creation of an archive are given in Ohring and Gruber (1983).

Assumptions and approximations are clearly necessary in the production of any operational archive. The NESS procedures include spatial averaging and:

1. *For albedo* the assumption that reflectance is isotropic and independent of solar zenith angle. NESS also assume that the narrow wavelength region sensed by the radiometer ($0 \cdot 5 - 0 \cdot 7 \, \mu m$) is a reasonable proxy for the full solar spectrum. This assumption is currently being debated in the literature (see Briegleb and Ramanathan, 1982; and the discussion in Section 3.4). Additionally no diurnal information is available, since short-wave radiation is retrieved from an overhead pass only once a day.

2. *For outgoing long-wave radiation* the total flux is estimated from the window ($10 \cdot 5 - 12 \cdot 5 \, \mu m$) measurement using a regression model derived from calculations using 99 different model atmospheres. Long-wave radiance measurements are available at each of the two overhead passes. Ohring and Gruber (1983) argue that at fairly coarse time and space scales these errors will be reduced in importance.

3.2. Annual radiation budget measurements and the seasonal cycle

Ohring and Gruber (1983) have undertaken a detailed analysis of the satellite radiation measurements described in Winston *et al.* (1979). They find that the global average value of albedo is $0 \cdot 314$, absorbed solar energy is $232 \, W \, m^{-2}$, and the net radiation is $-13 \, W \, m^{-2}$. (They suggest that if an adjustment of these values to a solar constant of $1376 \, W \, m^{-2}$, as measured by the Earth Radiation Budget (ERB) experiment on Nimbus-7, were to be made, the adjusted values would give an albedo of $0 \cdot 306$, absorbed solar energy of $239 \, W \, m^{-2}$ and net radiation of $-5 \, W \, m^{-2}$. This adjustment would reduce the bias which they recognize in the original data.)

Figure 3.1 shows the annual cycle in the radiation budget components —albedo, outgoing long-wave radiation, absorbed solar radiation and the net radiation—for the sample period of 45 months. The albedo exhibits a pronounced annual cycle with a maximum in the Northern Hemisphere during the winter and a minimum during the summer. The outgoing

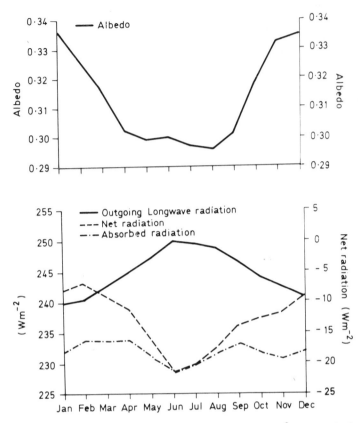

Figure 3.1. Annual cycle of albedo, outgoing long-wave radiation (Wm⁻²), absorbed solar radiation (Wm⁻²) and the net radiation (Wm⁻²), based on a 45-month SR data set (after Ohring and Gruber, 1983).

long-wave radiation also exhibits an annual variation; however, its maximum in the Northern Hemisphere is during the summer. The out-of-phase nature of the relationship between albedo and outgoing long-wave radiation is a result of the large annual cycle of temperature experienced over the land as well as the effects of clouds and snow cover. The amplitude of the annual cycle of absorbed solar radiation is about $5\,W\,m^{-2}$ and is small compared with the outgoing long-wave radiation, and much smaller than that implied by the variation in albedo. The phase of the variation is also inconsistent with the phase of the albedo, both exhibiting minimum values during the summer months. This is probably caused by the occurrence of the minimum albedo at the time when the Sun is furthest from the Earth. This relative difference of incoming solar energy ($\pm 3 \cdot 4\%$) is enough to dampen the amplitude and reverse the phase of the annual cycle in absorbed solar energy from that implied by the albedo.

Maps of the mean winter (December, January and February) albedo, outgoing long-wave radiation and net radiation are presented in figure 3.2. Each field is shown in a mercator projection from 60°N to 60°S and then in a polar stereographic projection from 50° latitude poleward for the Northern Hemisphere (left side) and Southern Hemisphere (right side).

The ITCZ is clearly evident in the figure, as are the high albedo regions in the eastern Pacific and eastern Atlantic associated with the extensive stratocumulus clouds usually present there. The desert areas of North Africa also exhibit high albedo, some areas slightly in excess of 0.4.

Outgoing long-wave radiation exhibits similar distributions but there is an out-of-phase relationship with the albedo. The exception to this occurs over the high albedo stratocumulus clouds and the Sahara desert. In the case of the cloudy area the closeness of the clouds to the ocean surface gives them a high emitting temperature, thus resulting in large outgoing long-wave radiation. The desert areas are also high temperature surfaces, thus emitting relatively large amounts of long-wave radiation. This accounts for the minimum or slightly negative net radiation in these areas.

Net radiation achieves its greatest surplus, in excess of $100\,\mathrm{W\,m^{-2}}$, over the sub-tropical high pressure zones of the Southern Hemisphere.

The albedo, outgoing long-wave radiation and net radiation for the summer months (June, July and August) are displayed in figure 3.3.

The low-level stratocumulus clouds on the eastern boundaries of the oceans, prominent during the winter season, are also present during these months. They are now more evident in the eastern North Pacific. However, the net radiation (Figure 3.3(c)) is almost the reverse of that found during the winter: north of 10°S there is a net radiation surplus (except for the stratocumulus cloud and desert regions) and between 10°S and 30°S there is a deficit of net radiation. There is generally a surplus of net radiation in the 30°N–60°N zone except for two small areas located at about 180°, 50°N and 45°W, 50°N. In the Southern Hemisphere there is a deficit of net radiation everywhere.

With no illumination over the south polar region the net radiation follows the pattern of the outgoing long-wave radiation. Over the north polar region there is a deficit in net radiation that more or less follows the pattern of the albedo map. The high albedos result in little absorbed energy and hence there are large deficits in net radiation in these regions. It is also probable that the narrow wavelength region used $(0.5-0.7\,\mu\mathrm{m})$ to construct the short-wave fluxes is responsible for a significant underestimation of the broad-band flux (Section 3.4). This effect would further increase the deficit in net radiation. Temporal changes can be

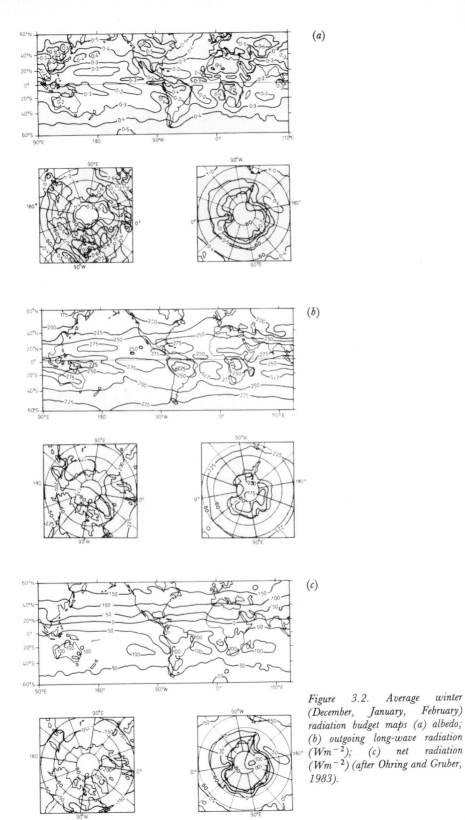

Figure 3.2. Average winter (December, January, February) radiation budget maps (a) albedo; (b) outgoing long-wave radiation (Wm^{-2}); (c) net radiation (Wm^{-2}) (after Ohring and Gruber, 1983).

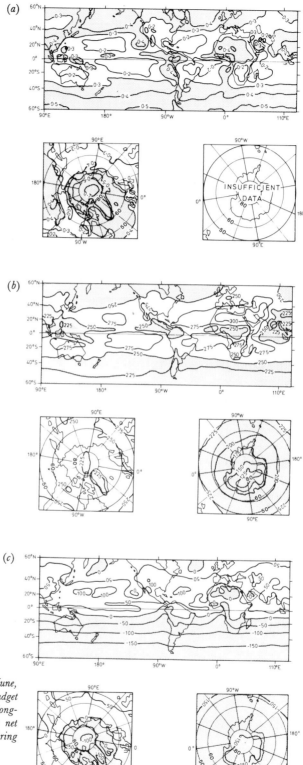

Figure 3.3. Average summer (June, July, August) radiation budget maps: (a) albedo; (b) outgoing long-wave radiation (Wm⁻²); (c) net radiation (Wm⁻²) (after Ohring and Gruber, 1983).

followed both by comparison of these stationary seasonal plots or by a time variation analysis.

The seasonal cycle of the climatic variables at the surface of the Earth is controlled, to a large extent, by the seasonal cycle in the surface heat and water budgets. In turn, these are related to the seasonal cycle of the components of the radiation budget of the Earth–atmosphere system. The annual cycle of radiation budget components as observed from satellites for several different time and space regions are discussed by Ohring and Gruber (1983). The temporal variations of the net radiation balance components are shown in figures 3.4–3.7, which show the time–latitude distribution of zonally averaged monthly mean albedo, outgoing long-wave radiation flux, absorbed solar radiation and net radiation for the period June 1974 to February 1978.

The outgoing long-wave radiation (figure 3.5) exhibits many of the same characteristics as the albedo (figure 3.4) but in an inverse sense. This is because cloud cover generally extends into the middle and upper troposphere, giving rise to the condition of high albedo and low outgoing long-wave radiation. Snow and ice give rise to the same relationship.

The annual variation of absorbed solar energy (figure 3.6) is pronounced outside the zone $5°-10°N$; the maximum of absorbed energy occurring during the summer months of each hemisphere, regardless of latitude. The phase of the variation clearly follows the course of the Sun during the year, as seen by the solar declination plotted in figure 3.6.

The net energy flux (figure 3.7) also exhibits a pronounced annual variation with phase relationships similar to the absorbed solar energy, i.e., maximum of net radiation during the summer months of both hemispheres. The maximum and surplus of energy also follow the solar declination. Ohring and Gruber (1983) examined the reason for this and concluded that the annual variation of net radiation is dominated by the variation in absorbed solar energy, which has a much greater amplitude than the variation of outgoing long-wave radiation.

Table 3.1 (from Ohring and Gruber, 1983) compares characteristics of the Earth's radiation budget components from the Winston *et al.* (1979) data with those of various other authors. The NOAA/NESS high-resolution data have been averaged over $10°$ latitude belts and all albedo values have been scaled to $1353 \mathrm{W m}^{-2}$, the value used in the Winston *et al.* (1979) computations, in order to facilitate comparisons. Figure 3.8 compares annual average meridional profiles of albedo. All three estimates shown are similar between $65°N$ and $65°S$. Poleward of those latitudes the much larger differences may be the result of the different wavelength regions used in the original data retrieval (cf. Section 3.4, p. 110).

Figure 3.9 shows profiles for the outgoing long-wave radiation. The NOAA/NESS values are systematically higher, with a difference in the

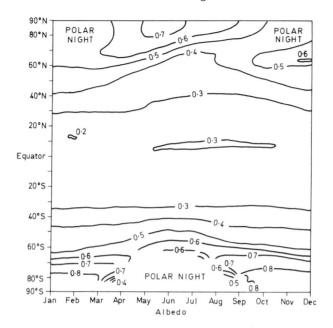

Figure 3.4. Time–latitude section of the mean monthly albedo (after Ohring and Gruber, 1983).

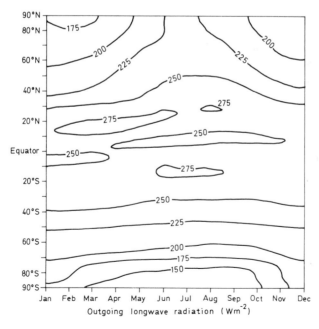

Figure 3.5. Time–latitude section of the mean monthly outgoing long-wave radiation (Wm^{-2}) (after Ohring and Gruber, 1983).

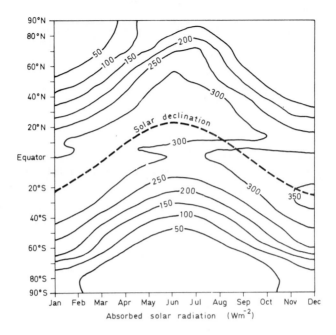

Figure 3.6. Time−latitude section of the mean monthly absorbed solar radiation (Wm^{-2}) (after Ohring and Gruber, 1983).

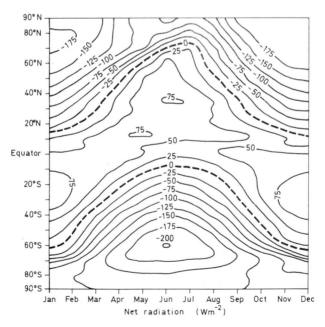

Figure 3.7. Time−latitude section of the mean monthly net radiation (Wm^{-2}) (after Ohring and Gruber, 1983).

global average of $14 \mathrm{Wm}^{-2}$. Ohring and Gruber (1983) suggest that the systematic difference may be related to the procedures used in deriving the regression coefficients which are used to calculate the full infra-red flux from the $10 \cdot 5 - 12 \cdot 5 \mu \mathrm{m}$ window observations. They draw attention to preliminary calculations which suggest that incorporation of partly cloudy conditions and spectrally varying cirrus emissivity can account for about 35 % of the difference.

In summary, various independent compilations of the Earth's radiation budget as viewed from space lead to the conclusions that the annually and globally averaged albedo is approximately $0 \cdot 31$ and the outgoing long-wave radiation is about $250 \mathrm{W m}^{-2}$.

Table 3.1. Comparisons of characteristics of the scanning radiometer (SR), Ellis and Vonder Haar (EV) and Campbell and Vonder Haar (CV) data sets (after Ohring and Gruber, 1983, which lists the original references).

	SR	EV	CV
Radiometer-type	High-resolution scanner	Flat plate medium-resolution scanner	Wide-angle
Spectral range (μm):			
short-wave	$0 \cdot 5 - 0 \cdot 7$	$0 \cdot 2 - 4 \cdot 0$	$0 \cdot 2 - 3 \cdot 8$
long-wave	$10 \cdot 5 - 12 \cdot 5$	$4 \cdot 0 - 50$	$3 \cdot 8 - > 50$
Years covered	1974–1978	1964–1971	1975–1977
Total months of data	45	29	24
Satellites	NOAA-2, 3, 4, 5	Experimental Nimbus-2, 3 ESSA-7 ITOS-1 NOAA-1	Nimbus-6
Local equator crossing time	0900	0830–1500 (depending on satellite)	1200
Resolution:			
original	4 km, short-wave 8 km, long-wave	Full disc ($\sim 10°$–$20°$, half power); medium resolution ~ 100 km	Full disc ($\sim 10°$ half power)
data set	$2 \cdot 5°$ lat/long	Zonal average; 10–$20°$ lat	Zonal average; $10°$ lat

3.3. Regional and global climatologies from satellites

Satellites provide an opportunity to obtain new climatological information of at least two types: firstly, for inhospitable regions of the globe

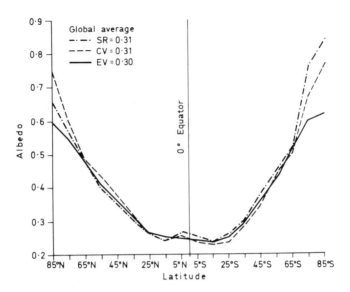

Figure 3.8. Comparison of NOAA/NESS (SR), Ellis and Vonder Haar (EV) (1976) and Campbell and Vonder Haar (CV) (1980) estimates of the annual average zonal albedo (from Ohring and Gruber, 1983; q.v. for original references).

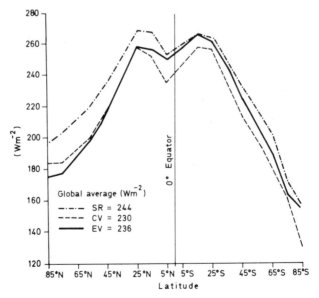

Figure 3.9. Comparison of NOAA/NESS (SR), Ellis and Vonder Haar (EV) (1976) and Campbell and Vonder Haar (CV) (1980) estimates of the outgoing long-wave radiation (Wm^{-2}) (from Ohring and Gruber, 1983; q.v. for original references).

where meteorological stations are scarce, and satellites provide primary atmospheric information; and secondly, the traditional surface-based climatological statistics can now be supplemented and compared with information derived directly from radiative fluxes measured by satellites at the top of the atmosphere. In this section atmospheric information for the Arctic is discussed as an example of the new regional information becoming available. A new oceanic global cloud climatology derived from satellite radiance data is described and finally traditional regional climatologies are compared with satellite-derived radiation measurements.

The Arctic atmosphere sensed from space

The presence of aerosols in the Arctic atmosphere has been known for some time. The extent of these aerosols has been shown for widely separated stations along the North American coast of the Arctic Ocean and in Svalbard. Local pollutants are believed to be a negligible source. There have been few measurements of the aerosol content of the air over the Arctic Ocean itself and information is almost invariably restricted to measurements of the ground-level concentration and properties of the aerosols. Only one set of measurements of the vertical distribution of aerosols has been found. Single-station observations clearly cannot provide information about the extent in time and space of this phenomenon. It would clearly be preferable to utilize satellite data in a regional survey of Arctic aerosol loading. A preliminary investigation has recently been initiated using the Defense Meteorological Satellite Program (DMSP) imagery.

During spring-time in the Arctic, the pack ice begins to break up and areas of open water (as leads and polynyas) form, so giving a large contrast in surface albedo with the surrounding snow-covered ice. Figure 3.10 illustrates one of the images (15 May 1979) investigated, which shows the large albedo contrast between the two surface types. A recent analysis has explored this albedo contrast.

An image processor was used to evaluate the ratio between brightness of the snow-covered surface and open water for a large number of cases of clear and obviously cloudy skies. The derived ratios are plotted against solar zenith angle in figure 3.11. All the cases studied are for the period May/early June and lie between $70°$ and $78°N$. The ratios can be seen to lie in the range $3 \cdot 5 - 4 \cdot 8$. There are insufficient data to draw any conclusions regarding the zenith angle dependence of the brightness ratios.

Clear-sky snow:sea albedo ratios at the top of the atmosphere were calculated using a 24 spectral band delta Eddington radiative transfer

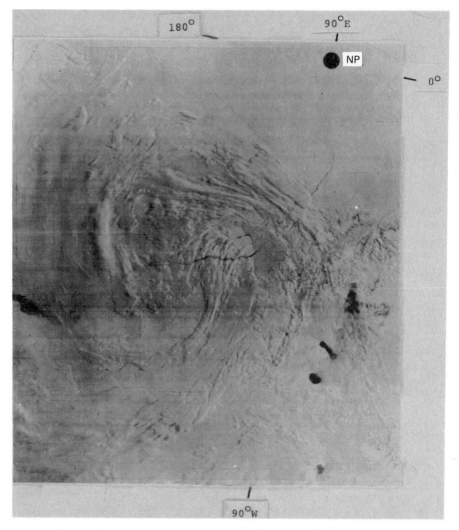

Figure 3.10. DMSP satellite visible image for 15 May 1979 in the region of the North Pole, showing the contrast between sea and snow surfaces. A lead (dark linear feature) in the ice can be traced under a cloud formation. Compare this visible image with the DMSP image illustrated in Figure 1.9 (p. 20) (courtesy of D. Robinson, Lamont Doherty Geological Observatory).

package. The DMSP sensor response over the region $0 \cdot 4 - 1 \cdot 1 \, \mu m$ was incorporated into the radiation calculations and theoretical ratios calculated to compare with the observations shown in figure 3.11. Calculations were performed using data from an established spectral snow albedo model interpolated on to the 24 spectral band grid. Two

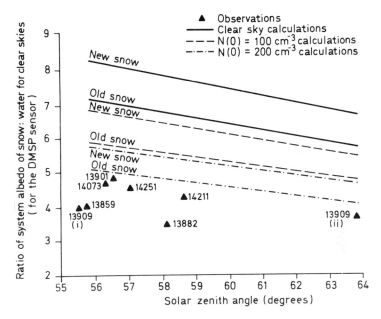

Figure 3.11. Variation in ratio of the system albedo over snow to system albedo over water. Observations are marked by triangles and numbers refer to the identification number given in the original reference. Radiative transfer calculations for clear skies and two degrees of aerosol loading (see Shine et al., 1984 for details) are given by the lines. Calculated values account for DMSP sensor response (e.g., figure 2.12, p. 69).

types of snow were considered with grain radii of 100 and 1000 μm (the latter with a small soot content of $0 \cdot 05$ ppm), respectively, and sea surface albedos were derived using the expression given by Briegleb and Ramanathan (1982).

The discrepancy between the observed and calculated ratios in figure 3.11 could be the result of additional scattering/absorbing particles in the atmosphere that were not included in the radiation calculation. The presence of scattering/absorbing aerosols will lower the calculated albedo ratios, since over a low-albedo sea surface the additional scatterers will increase the system albedo while over a high-albedo snow surface the additional absorbers will prevent radiation reaching the ground, so decreasing the system albedo. However, quantitative modelling of the effect of aerosols is hampered by the paucity of data relating to the aerosols and their height distribution. Shine *et al.* (1984) have recently reported on preliminary calculations in which a continental-type aerosol is incorporated into the radiative calculations. The effect of the addition of these aerosols to the atmosphere on the snow : sea albedo ratios is also shown in figure 3.11. For a ground concentration of $200\,\mathrm{cm}^{-3}$ the albedo (as seen by the DMSP sensor) over a sea surface is increased by about 0·05 while the albedo over snow is decreased by 0·05.

The implication of these calculations is that aerosols are an Arctic-wide phenomenon, present in sufficient quantities to alter the radiation budget of the region significantly. The preliminary analysis described here suggests that ratioing of selected brightness data could provide a means of remotely sensing the aerosol loading of the Arctic atmosphere on a space scale impossible from surface stations and with the added advantage of yielding the bulk properties of a vertical column of the atmosphere.

A new preliminary global oceanic cloud climatology

The most widely accepted cloud climatology is that of London (1957; see Chapter 1). This gives cloud type, height and total cloud cover as a function of latitude for the Northern Hemisphere and was derived prior to satellite measurements from a compilation of conventional surface-based observations. The oceanic cloud data are particularly uncertain. More recent global-scale climatologies based on satellite data include oceanic areas, e.g., Clapp (1964; see Chapter 1). However, these climatologies either contain little information on inter-annual variations or are limited to brief periods of time and therefore possibly unrepresentative. Hughes and Henderson-Sellers (1983) have examined the relationship between the system albedo over oceanic areas (derived from NOAA scanning radiometers) and the US Air Force 3-D nephanalysis total cloud amount data. They calculated a predictive relationship between these variables from which a new oceanic global cloud climatology was produced.

Hughes and Henderson-Sellers (1983) analysed concurrent 3-D nephanalysis total cloud amount and system albedo data over selected regions for March, June, September and December, 1977. The immense volume of data processing required to analyse the 3-D nephanalysis cloud amount data makes a larger-scale study of the cloud amount itself prohibitive. The analysis permits both calculation of a latitudinal cross-section of cloud regimes and an examination of seasonal variations over the UK and north-west Europe. Since the scheme described relates satellite-observed albedo to cloud amount, the authors considered it important to include cloud amount data that were not themselves solely determined from satellite radiance measurements. The relationship between mean monthly system albedo and cloud amount data for oceanic areas only was examined. Hughes and Henderson-Sellers (1983) found that mean diurnal monthly cloud amount and 0900 LST mean monthly cloud amount have similar functional dependencies on the system albedo and that there is no marked seasonality in the relationship. A predictive relationship between cloud amount and system albedo was determined

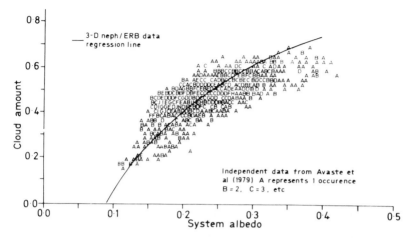

Figure 3.12. Scattergraph of scanning radiometer system albedo data (x axis) and Sadler's (1969) cloud amount data (y axis) (from Avaste et al., 1979) with the regression formula (after Hughes and Henderson-Sellers, 1983; which includes the original references).

which gives a cloudy sky albedo of 0·62 and a clear sky albedo of 0·09 (coefficient of determination = 0·85). The cloud and surface albedo values compare favourably with others in the published literature. Hughes and Henderson-Sellers (1983) compared their results with earlier studies (figure 3.12). The satisfactory level of agreement between the data sets seems to confirm the validity of using archived system albedo data to derive cloud amount.

The predictive relationship of Hughes and Henderson-Sellers (1983) has been used to establish a new oceanic global cloud climatology averaged over three years and for four months representative of the four seasons (January, April, June and October), presented in figure 3.13.

Substantial seasonal variations in effective cloud amount do occur. The near-equatorial cloud band attains its southern-most position in April. Cloudiness minima persist throughout the year in the oceanic areas dominated by the sub-tropical highs, especially in the North and South Pacific Oceans where lowest cloud amount occurs in July. Other areas undergoing considerable seasonal variation occur in the mid-Atlantic Ocean and Indian Ocean. Despite the changes in cloud amount occurring seasonally, the location of the areas of cloudiness minima and maxima remain relatively constant.

Regional climatologies from space

Ohring and Gruber (1983) presented a fascinating comparison between the classical climatological analyses of Budyko (1974) and satellite-derived

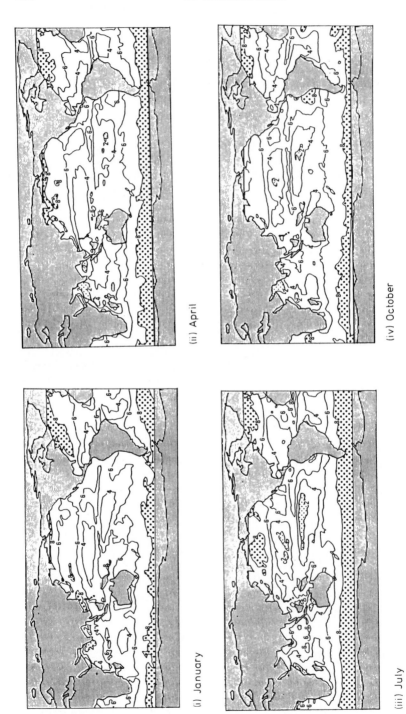

(i) January

(ii) April

(iii) July

(iv) October

Figure 3.13. Calculated global distribution of oceanic monthly mean cloud amounts for (i) January; (ii) April; (iii) July; (iv) October using system albedo data averaged over the years 1975, 1976, 1977 (after Hughes and Henderson-Sellers, 1983).

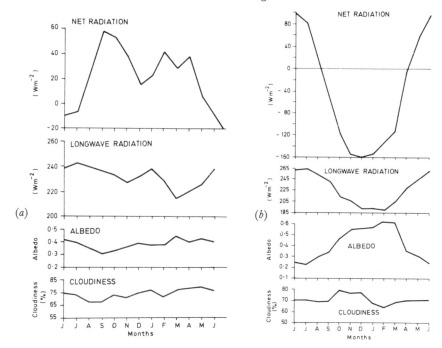

Figure 3.14. (a) *Seasonal variation of the climatological parameters of the Earth–atmosphere system for an equatorial continental location (0°, 67·5°W; São Gabriel, Brazil). (b) As for (a) but for a mid-latitude continental location (52·5°N, 82·5°E; Barnaul, USSR) (both after Ohring and Gruber, 1983).*

climatological information. They considered the annual cycle of the energy budget at the top of the atmosphere for a series of the areas considered in Budyko's (1974) traditional climatology. The satellite data which Ohring and Gruber (1983) analysed are taken from NOAA scanning radiometer observations for 2·5° by 2·5° areas centred as closely as possible on the climatic locations chosen by Budyko.

Ohring and Gruber (1983) considered five different climates.
1. Equatorial continental.
2. Equatorial monsoon.
3. Tropical continental.
4. Subtropical continental.
5. Mid-latitude continental.

In each case they relate the curves of albedo, emitted long-wave radiation and net radiation to the annual cycle of insolation, surface temperature and surface-observed cloudiness. Two examples of this new type of satellite-based climatology are discussed here in detail. The climatological curves described are shown in figure 3.14.

Equatorial continental climate (São Gabriel, Brazil)

In figure 3.14(*a*), the net radiation curve exhibits features associated with the cycle of solar radiation typically encountered in an equatorial climate. There are two maxima, in September and February. Ohring and Gruber (1983) suggest that the displacement of the second maximum from the expected month of March, when the Sun is over the equator, is probably the result of the occurrence of a peak in the albedo curve in March; the albedo maximum being due, in turn, to the increase in cloudiness at this time. This interpretation is consistent with the long-wave radiation curve, which shows a minimum in March and must be controlled primarily by the changes in cloudiness, since surface temperatures in equatorial regions exhibit very little seasonal variability. The annual variation of net radiation is small. However, Ohring and Gruber (1983) drew attention to the difference of about $25\,\mathrm{W\,m^{-2}}$ in net radiation between June and December—two months which have the same solar declination angle. The albedos are seen to be very similar for the two months. The long-wave radiation difference can only account for approximately $7\,\mathrm{W\,m^{-2}}$ of the observed $25\,\mathrm{W\,m^{-2}}$. Thus the majority of the difference must be due to the variation in Earth–Sun distance between June and December.

Mid-latitude continental (Barnaul, USSR)

Ohring and Gruber (1983) suggest that for this region the winter maximum in albedo ($0\cdot60$) is the result of snow cover rather than cloudiness, which is at a minimum during this season (see figure 3.14(*b*)). There is a very large annual range of long-wave emission ($>70\,\mathrm{W\,m^{-2}}$) which is a direct consequence of the large annual cycle in surface temperature.

The type of climatological information derived by these techniques from satellite radiance measurements could be used both to supplement traditional climatological analysis and to test climate models directly.

3.4. Clouds and the sensitivity of the climate system

The Earth is believed to be in radiative equilibrium over a period of a few years. This radiative balance can be expressed by the equation

$$F = (S/4)(1 - a) \tag{3.1}$$

where F is the emitted long-wave radiation; S is the solar constant; and a is the planetary albedo. Satellite observations clearly represent the most appropriate method of validation of equation 3.1, but currently available observations are not of high enough accuracy. Equation 3.1 also represents a convenient starting point for examining the sensitivity of the Earth's climate to alterations in external and internal climatic forcing factors. It is useful to consider a global sensitivity parameter

$$\beta = S_0 \frac{dT_s}{dS} \tag{3.2}$$

where S_0 is the current value of the solar constant; and T_s is the global mean surface temperature. $\beta/100$ may be interpreted as the change in the global mean surface temperature that would be produced by a 1 % change in the solar constant.

The Earth's radiation budget is inextricably linked to the total cloud amount and the cloud type and character. It is quite possible that because of changes in surface temperature resulting from a change in the solar constant (or in any of the other parameters controlling the Earth's climate), there will be a change in the amount of cloud cover, A_c. Clouds have two important effects on the radiation budget of the Earth–atmosphere system: firstly, clouds increase the albedo a of the system; and secondly, clouds also decrease the long-wave radiation loss to space (see Chapter 2).

The net radiation passing through each horizontal unit area at the top of the Earth–atmosphere system may be written as

$$\text{Net} = Q - F \tag{3.3}$$

where Q is the solar radiation absorbed by the system and F is the emitted infra-red radiation. A cloud sensitivity parameter δ may be defined for determining the effect of a change in cloud amount on net radiation

$$\delta = \frac{\partial \text{Net}}{\partial A_c} = \frac{\partial Q}{\partial A_c} - \frac{\partial F}{\partial A_c} \tag{3.4}$$

where $\partial Q/\partial A_c$ represents the albedo or short-wave effect of the clouds; and $\partial F/\partial A_c$ represents the greenhouse or long-wave effect of the clouds; see figure 3.15. Therefore if, for example, δ is less than zero, the effect of the changing the albedo is greater than of changing the outgoing long-wave radiation.

There have been several attempts to evaluate δ using satellite data and the results have been conflicting. Cess (e.g., Cess *et al.*, 1982) found δ to be close to zero, indicating that cloudiness may not be a significant climate feedback mechanism. Other workers (e.g., Ohring and Clapp, 1980) found that the albedo effect was more important, giving δ less than zero. Cess *et al.* (1982) analysed satellite data from two different satellite systems and found conflicting results. Data derived from the NOAA/ NESS satellites indicated δ was less than zero while those derived from

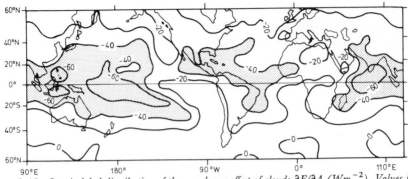

Figure 3.15. Quasi-global distribution of the greenhouse effect of clouds $\partial F/\partial A_c (Wm^{-2})$. Values less than $-40\,Wm^{-2}$ are stippled (after Ohring and Gruber, 1983).

Nimbus series satellites indicated δ greater than zero. The authors attributed much of the difference between the results to the effect of the different spectral resolutions of the instruments on board the satellites. The albedo of clouds and natural surfaces is a strong function of wavelength, so that results from narrow-band sensors are likely to give erroneous estimates of broad-band flux and the parameter δ. In assessing the radiative effect of cloud cover changes, δ is the important parameter. For satellite-based analysis it is not straightforward to separate changes in fluxes due to cloud cover variations and those due to other effects (such as cloud height distribution changes, surface albedo changes and variations in the vertical temperature and moisture fields). It may therefore be more helpful to follow the analysis of Cess et al. (1982) which is based on a quantity ϵ where

$$\epsilon = \frac{\partial F}{\partial Q} \simeq \frac{\Delta F}{\Delta Q} \qquad (3.5)$$

i.e., $\epsilon = 1$ corresponds to compensation in changes in F and Q (from whatever cause) and $\epsilon < 1$ signifies the albedo change is dominant. An illustration of the importance of the waveband response of different sensors is outlined here. The analysis assesses the dependence of the denominator in the approximation to equation 3.5, ΔQ, on the satellite sensor response.

Consider a radiometer with, for example, the narrow band of the NOAA-6 AVHRR channel one, which extends from $0 \cdot 5$ to $0 \cdot 7\,\mu m$, compared with the Nimbus broad band (full solar spectrum) radiometer. Calculations made with a 24-band spectral model and using a standard mid-latitude summer atmosphere with sea surface albedos and vegetation albedos taken from Briegleb and Ramanathan (1982) are used to simulate the radiances observed by these two instruments. A cloud optical thickness was selected so that with 50 % cloud cover a planetary albedo of about $0 \cdot 3$ would result, to correspond with the observed globally

Table 3.2. Planetary albedo and short-wave energy (Wm^{-2}) absorbed by the Earth–atmosphere system: comparison of broad band with values calculated assuming a narrow-band ($0 \cdot 5$–$0 \cdot 7 \mu$m) sensor is representative of broad-band values. Mean annual insolation (cosine of solar zenith angle = $0 \cdot 5$; daylength = 12h) over sea and vegetated surfaces with a standard, mid-latitude summer atmospheric profile. Cloud optical thickness = 20.

Surface	Albedo		Absorbed solar radiation (Wm^{-2})	
	Broad band	Narrow band	Broad band	Narrow band
Sea surface:				
clear	$0 \cdot 131$	$0 \cdot 128$	297	298
total overcast	$0 \cdot 596$	$0 \cdot 685$	138	108
Vegetated surface:				
clear	$0 \cdot 187$	$0 \cdot 136$	278	296
total overcast	$0 \cdot 599$	$0 \cdot 686$	137	107

averaged planetary albedo (e.g., Winston *et al.*, 1979). Over a sea surface, the required optical thickness was found to be 20 for a cloud between 900 and 950mb, corresponding to a liquid water content of about $0 \cdot 23$gm^{-3}. Table 3.2 shows the planetary albedos calculated for the entire solar spectrum ($0 \cdot 25$–$4 \cdot 0 \mu$m) and for the limited band sensor, for a cosine of solar zenith angle of $0 \cdot 5$ and daylength of 12 hours (i.e., corresponding to mean annual insolation). The solar constant was taken as 1368Wm^{-2}. Table 3.2 also shows the radiation absorbed by the Earth–atmosphere system assuming that the calculated albedo values are representative of the whole solar spectrum.

For the sea surface, the narrow-band sensor underestimates the clear-sky albedo by $0 \cdot 003$. This is due to the strong increase in planetary albedo at wavelengths less than $0 \cdot 5 \mu$m due to Rayleigh scatter. The error is negligible compared with other uncertainties and satellite random errors. For cloudy skies the narrow-band sensor overestimates the albedo by $0 \cdot 089$. This is because the cloud albedo falls off sharply in the near infra-red. The overestimate increases with cloud optical thickness because greater near infra-red absorption accentuates the albedo contrast across the spectrum; for a cloud with an optical thickness of 50 the overestimate is $0 \cdot 1$. From the absorbed radiation given in table 3.2, the error in ΔQ resulting from the use of a narrow-band sensor can be estimated, and it is found that $\Delta Q_{narrow} = 1 \cdot 20 \Delta Q_{broad}$, where ΔQ_{narrow} is the change in absorbed shortwave energy calculated using the narrow spectral band and ΔQ_{broad} the same quantity for the entire solar spectrum.

It is interesting to consider high latitude regions, noting that of all natural surfaces, snow and ice exhibit the strongest variation in albedo with wavelength, with the albedo decreasing markedly in the near infra-red. Snow albedo is also dependent to a large degree on the state of the snow surface. Calculations have also been performed for a cosine of solar

A. Henderson-Sellers

Table 3.3. As table 3.2 for typical high-latitude summer conditions (cosine of solar zenith angle = 0·3; daylength = 24h) for new and old snow surfaces using a standard sub-arctic summer atmospheric profile.

Surface	Albedo		Absorbed solar radiation (Wm^{-2})	
	Broad band	Narrow band	Broad band	Narrow band
New snow:				
clear	0·662	0·855	139	60
overcast	0·733	0·863	110	56
Old snow:				
clear	0·552	0·774	184	93
overcast	0·697	0·819	124	74

zenith angle of 0·3 and a day length of 24h (typical of summertime conditions at high latitudes). Two different snow conditions were used, one representing new snow (with a clear-sky surface albedo of 0·83) and one for old snow (with a clear-sky surface albedo of 0·68).

The results presented in table 3.3 show very large discrepancies between the albedo for the whole spectrum and the albedo for the narrow-band sensor. Under clear skies these amount to around 0·2 and under cloudy skies to 0·1. These differences are sufficient to cause results from a narrow-band sensor to underestimate the radiation absorbed by the Earth–atmosphere system by a factor of 2 (cf. earlier discussions of figures 3.2, 3.3 and 3.8). Further, the change in albedo with cloud amount is considerably different for the two wavelength bands, giving

$$\Delta Q_{narrow} = 0·11 \Delta Q_{broad} \quad \text{(new snow)}$$

and

$$\Delta Q_{narrow} = 0·31 \Delta Q_{broad} \quad \text{(old snow)}$$

It is possible, therefore, to conclude that the narrow-band radiometer will underestimate the changes in albedo and absorbed short-wave energy on changing cloud amount. An estimate of ϵ using a narrow-band radiometer such as considered here at high latitudes would be far too high. Earlier results imply that ϵ increases with latitude (Hartmann and Short, 1980). The results reported here indicate that some of this effect may be due to the sensor response if the radiometer senses only in the 0·5–0·7 μm region although the narrow-band case considered here gives rise to more extreme results than, for instance, the broader NOAA SR channel. Thus comparisons between cloud–climate feedback evaluations (e.g., table 3.4) must be viewed with some caution.

If climatic change is accompanied by significant variations in cloud amount then it is of great importance to know the relationship between

Table 3.4. Summary of estimates of $\partial F/\partial A_c$ (Wm^{-2}) and $\partial F/\partial Q$ (after Ohring and Gruber 1983).

Method	Investigator[a]	$\partial F/\partial A_c$	$\partial F/\partial Q$ ($\equiv \epsilon$)
Simulation			
1-D, global average model:			
single cloud	Schneider (1972)	-75	
	Ramanathan (1976)	-71	
	Coakley (1977)	-51	
multiple clouds	Cess (1974)	-68	
	Wang & Domoto (1974)	-61	
Global average of zonal energy budget,			
multiple clouds	Hoyt (1976)	-34	
2-D, zonal average model, single cloud	Ohring & Adler (1978)	-33	
3-D, GCM, from $F(T_s, A_c)$ zonal regression	Coakley & Wielicki (1979)	-71	
260 representative points on Earth monthly means $F(T_s, A_c)$ regression	Budyko (1969)	-73	
Observation			
$F(T_s, A_c)$ regression:			
annual, zonal means	Cess (1976)	-91	
seasonal, zonal means	Ohring & Clapp (1980)	-60	
monthly, zonal mean	Warren & Schneider (1979)	-58	
Comparison of clear and cloudy sky values of F	Ellis (1978)	-40	
$F[a(A_c)]$ regressions, monthly means, regional	Ohring *et al.* (1981)	-35	$0 \cdot 33$
$F(a)$ regressions, daily values, regional	Hartmann & Short (1980)		$0 \cdot 4$
$F(Q)$ regressions, monthly means, low latitudes			
Ellis & Vonder Haar (1976), satellite data			$1 \cdot 1$
Campbell & Vonder Haar (1980*b*), satellite data	Cess *et al.* (1981)		$1 \cdot 0$
Gruber & Winston, satellite data			$0 \cdot 5$
$F(Q)$ regressions, daily values, regional Nimbus-7, scanning radiometer (preliminary)	Ohring & Ganot (1982)		$0 \cdot 51$

[a]For original references see Ohring and Gruber (1983).

the changing opacity of the atmosphere in the terrestrial infra-red and the change in albedo at solar wavelengths. Any attempt to assess the strength and sign of the cloud climate feedback (on both a global and a regional scale) from satellites must take into account the satellite sensor response. While satellite radiance measurements provide the best opportunity of establishing the direction and magnitude of the cloud–climate feedback sensitivity, considerable caution must, clearly, be exercised when results are interpreted.

3.5. The importance of cloud statistics for climate modelling

It is essential that the performance of climate models be tested. One of the critical aspects of all model types is the way in which the atmospheric radiative transfer processes are parameterized. Satellite observations such as those described earlier in this chapter offer an important means of validating model results. Clouds are the most transient phenomenon incorporated into climate models. The feedback effects which they may control could be very great (see Section 3.4). Their presence is now included in all types of climate models.

Cloud climatologies are required for climate models either as input data for models that require prescribed cloudiness, e.g., Meleshko and Wetherald (1981), or for validation of models which predict cloud configurations, e.g., Hansen *et al.* (1983). It is important that climate modellers know how their choice of cloud climatology affects interpretation of their model results. Additionally, it is important to know how model-generated clouds compare with either surface- or satellite-observed clouds. The comparison and validation of cloud climatologies becomes vital if models are sensitive to the cloud specification.

Following the discussion of cloud information in Chapter 1 a number of conclusions can be drawn regarding the currently available global cloud climatologies.

1. There is, at present, no unique and/or agreed global cloud climatology.

2. The magnitude and distribution of cloud amount in the climatologies is strongly dependent on the spatial and temporal sampling and averaging scales inherent in the observations and over which averages are made.

3. A cloud climatology based on satellite observations is not comparable with a climatology based on conventional surface observations. Climatologies including both types of observations have to be interpreted with considerable care and may not be comparable with a climatology derived from only one type of observation.

4. The major limitations of the cloud climatologies are non-comparability and lack of accuracy. The cloud climatologies are especially poor in oceanic and polar regions. All climatologies show differing magnitudes and distributions of cloud amount. Climatologies incorporating satellite data tend to display greater spatial heterogeneity, especially in oceanic regions. The location of continental minima of cloud is the most consistent feature of all climatologies. The most variable estimates of cloud amount tend to occur for cloud amount between 40 and 60%.

5. Zonal averages fail to reveal the complexities of the geographical distribution of cloud amount. Compare, for example, figure 1.12 with figures 1.10 and 1.11, pp. 28, 24, and 25. The zonal averages are also affected by the spatial and temporal averaging strategies.

These conclusions emphasize the limitations of the current cloud climatologies. These cloud data sets provide a wide range of cloud amount estimates for the climate modelling community.

Cloud prediction in climate models

There are various schemes of cloud prediction in climate models, each with a different interpretation of cloudiness. The prediction of cloud parameters in numerical models is hindered by the following problems.

1. The cloud formation and dissipation processes are poorly understood.

2. Most clouds are sub-grid scale, horizontally, vertically and temporally. A given value of cloud cover for a grid area may refer to very different patterns of cloud varying from patchy, fair-weather cumulus, to a partly overcast, party clear situation of an approaching frontal system.

3. The 'true' cloud climatology is not known, nor the extent of the geographical and seasonal cloudiness variations.

The climate models parameterize the processes that cannot be explicitly resolved by the limited computational capability of the model. Thus the small-scale processes, such as cloud formation and dissipation, are related to measurable or explicitly computable parameters on larger scales, or their effects are specified more or less arbitrarily. Thus, cloudiness is either internally generated by the model or externally prescribed from climatological statistics.

Generally, two types of clouds are modelled: large-scale clouds and convective clouds. Most existing cloud prediction schemes are based on the reasoning that clouds will form either: (*a*) if the large scale relative humidity exceeds a critical value; or (*b*) if the level of vertical convective activity, as measured by the vertical velocity, exceeds a certain index.

It is important to note that these parameters themselves may not be predicted accurately within the model.

The cloud prediction schemes in many models have been empirically derived from observational or 'climatological' data. The following examples of cloud prediction schemes devised for different types of climate models demonstrate how the definition of cloud varies between climate models.

The Zonal Atmospheric Model (ZAM), developed at the Lawrence Livermore National Laboratory, USA, was designed to test climatic sensitivity and has been used in the study of natural and anthropogenerated atmospheric changes. The horizontal resolution is $10°$ latitude, and there are nine vertical layers in the atmosphere. In each latitude zone, fractional cloud cover A is calculated at 850, 600, 400 and 200mb levels as a function of relative humidity r and a model coefficient c

$$A = \text{minimum } (0 \cdot 6, c(\phi,p)r^2) \qquad (3.6)$$

where ϕ is latitude; and p the pressure level. This relationship has been calibrated using London's data (see Chapter 1). Recently, an additional cloud layer at 1000 mb has been incorporated into the predictive scheme. This permits cloud formation at 1000 mb, A_{1000}, as follows

$$A_{1000} = \text{maximum } (0 \cdot 0, \ 2 \cdot 6(r_{1000} - 0 \cdot 65)) \tag{3.7}$$

This formulation was derived in an attempt to improve the National Center for Atmospheric Research's (NCAR) GCM performance in predicting trade-wind cumulus and high-latitude stratus cloud. The perceived need to include this further parameterization in at least two separate models underlines the current concern among climate modelling groups to simulate cloud cover correctly.

The 11-layer GCM at the United Kingdom Meteorological Office uses a cloud prediction scheme developed from the GARP Atlantic Tropical Experiment (GATE) data. The prediction scheme depends on relative humidity and, for low cloud amounts, the lapse rate, such that:

For high cloud amount, A_H
when $r \geqslant 80\%$, $A_H = (r - 80)^2/400$ $\qquad\qquad$ (3.8)
when $r < 80\%$, $A_H = 0$

For medium cloud amount, A_M
when $r \geqslant 65\%$, $A_M = (r - 65)^2/1225$ $\qquad\qquad$ (3.9)
when $r < 65\%$, $A_M = 0$

For low cloud amount, A_L, in the absence of a temperature inversion
when $r \geqslant 80\%$, $A_L = (r - 80)^2/400$ $\qquad\qquad$ (3.10)
when $r < 80\%$, $A_L = 0$

when $\dfrac{\Delta\theta}{\Delta P_{min}} \leqslant -0 \cdot 07$,
$\qquad\qquad A_L = -16 \cdot 67(\Delta\theta/\Delta P_{min}) + \dfrac{\delta(r - 80)^2}{400} - 1 \cdot 167$ $\quad$ (3.11)

where $\delta = 1$ for $r > 80\%$, otherwise $\delta = 0$; and $\Delta\theta/\Delta P_{min}$ is the lapse rate in the most stable layer.

For convective cloud, A_C

$$A_C = aM_C \tag{3.12}$$

where M_C is the convective mass flux; and a is an empirical constant. The scheme has been reasonably successful in predicting the cloudiness associated with the ITCZ and north-east and south-east trade winds.

The cloud prediction scheme used in the Geophysical Fluid Dynamics Laboratory model assigned 80% cloud cover wherever condensation occurred. At all grid points where condensation does not occur, the cloud amount is assumed to be zero. The value of 80% was chosen so that the model atmosphere would equilibrate at a realistic temperature by maintaining a realistic areal mean cloud amount.

These schemes may be contrasted with the cloud prediction scheme used in the latest version of the GCM at the Goddard Institute for Space Studies (GISS) (Hansen *et al.*, 1983), which depends on sub-grid scale variations of temperature. The temperature deviation within each grid box is calculated from the model-resolved temperatures around the latitude zone. Large-scale cloud cover is the fraction of the grid square which is saturated, assuming that absolute humidity is constant, but temperature has the calculated sub-grid square variation. Convective activity is specified as the fraction of mass which can ascend to a higher layer as a function of the calculated sub-grid square temperature deviation. These cloud distributions have been compared (in Hansen *et al.*, 1983) with data from Landsberg for verification (see Chapter 1).

These cloud prediction schemes define cloud amount in a very different manner to cloud observations, either surface or satellite. Validation of model-predicted cloudiness against cloud observations requires an examination of the relationship between model-generated and observed cloud amounts. It is likely that the different cloud prediction methods and different resolutions of the various models will lead to diverse predictions of cloud amount in simulation experiments.

The spatial and also temporal averaging that would precede use of any new cloud climatology as a data set for a climate model would be different because of the varying horizontal and vertical resolutions, time-step procedures and cloud prediction schemes of climate models. Model predicted cloud climatologies seem to be at least as inherently different from one another as climatologies derived from observations (Hughes, 1983; see Chapter 1). As model climates may be very sensitive to local and regional cloudiness variations, this is a very disturbing result.

Climate model sensitivity to cloud specification

The model experiments described below examine model response to the specification of low, middle and high cloud amount. Few global cloud climatologies provide cloud height data. The most widely used climatology which does is largely based on data of London (see Chapter 1). Different estimates of global cloud amount at three layers are likely to be even more variable than the total cloud amount estimates described earlier in Chapter 1.

Recent experiments with the GFDL general circulation climate model use only zonally averaged and time-invariant cloud distributions. Meleshko and Wetherald (1981) examined further the effect of a geographical cloud distribution rather than zonally averaged cloud

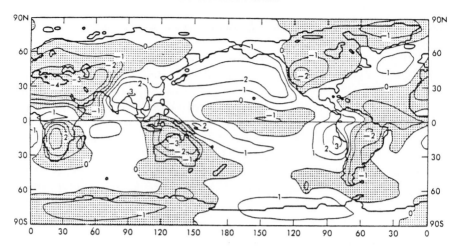

Figure 3.16. Geographical distribution of the differences in total cloud cover between the GFDL experiment using zonally averaged cloudiness and the experiment using geographical cloudiness. Contour interval in tenths (from Meleshko and Wetherald, 1981).

amount on the simulated climate of the GFDL model for July. Figure 3.16 shows the geographical distribution of the differences in total cloud cover between the experiments using geographical and zonally averaged cloudiness. The zonally averaged cloud is generally smaller than the geographical cloud cover over continents and generally larger over the oceans. Meleshko and Wetherald (1981) found that using a geographical cloud distribution increased the continental surface temperature by 2–4 K, decreased the surface pressure over continents and increased the surface pressure over oceans. The largest changes in surface pressure, ±12 mb, occur in the mid-latitudes. The largest differences in precipitation are observed in the tropics and the coastal regions of North and South America. Arid areas in the sub-tropical belt become more pronounced with a geographical distribution of cloud. This experiment demonstrated how the use of a geographical cloud distribution, rather than zonally averaged cloud amount, considerably improved the simulation capability of the GCM (Meleshko and Wetherald, 1981).

These results emphasize that the correct specification of cloud amount is an important parameter for modelling accurately the Earth's radiation budget and hence its climate. Climate models are hampered by the lack of adequate observations with which to derive cloud prediction schemes and verify results. Most climate models appear to be sensitive to the specification of cloud. An agreed and accurate cloud data set is required urgently by the climate modelling community.

3.6. The need for Earth radiation budget data—a summary

Significant advances have been made in recent years in the use of numerical climate models to simulate the Earth's climate and its sensitivity to changes in the external forcing. To validate the simulations, the distributions of the computed climatic variables are compared to the observed distributions of these elements. Due to the complexity of these models and the possibility of compensating errors in the treatment of various physical processes, good simulations of the observed climate are not a complete test of a climate model. Satellite observations of the radiation budget offer a means of validating model radiation calculations and of diagnosing possible causes of error in the simulations. An understanding of the present climate of the Earth and the calculation of climatic sensitivity to perturbations can only be achieved using consistent and well validated radiation budget data derived from satellites.

Acknowledgements

I am grateful to Drs A. Gruber, G. Ohring, N. A. Hughes, K. P. Shine and G. Kukla for allowing me to use material from preprints of manuscripts. I thank D. Robinson for the preparation of figure 3.10 and for details of the investigations using DMSP imagery in polar regions.

References

Briegleb, B. P. and Ramanathan, V., 1982, Spectral and diurnal variations in clear sky planetary albedo. *Journal of Applied Meteorology* **21**, 1160–71.

Budyko, M. I., 1974, *Climate and Life* (New York: Academic Press).

Cess, R. D., Briegleb, B. P. and Lian, M. S., 1982, Low-latitude cloudiness and climate feedback: comparative estimates from satellite data. *Journal of the Atmospheric Sciences* **39**, 53–59.

Hansen, J., Russell, G., Rind, D., Stone, P., Lacis, A., Lebedeff, S., Ruedy, R. and Travis. L., 1983, Efficient three dimensional global models for climate studies: Models I and II. *Monthly Weather Review* **11**, 609–662.

Hartmann, D. L. and Short, D. A., 1980, On the use of earth radiation budget statistics for studies of clouds and climate. *Journal of the Atmospheric Sciences* **37**, 1233–50.

Hughes, N. A. and Henderson-Sellers, A., 1983, A preliminary global oceanic cloud climatology from satellite albedo observations. *Journal of Geophysical Research* **88**, 1475–83.

Meleshko, V. P. and Wetherald, R. T., 1981, The effect of a geographical cloud distribution on climate. A numerical experiment with an atmospheric general circulation model. *Journal of Geophysical Research* **86**, 11995–12014.

Ohring, G. and Clapp, P. F., 1980, The effect of changes in cloud amount on the net radiation at the top of the atmosphere. *Journal of the Atmospheric Sciences* **37**, 447–54.

Ohring, G. and Gruber, A., 1983, Satellite radiance observations and climate theory. *Advances in Geophysics* **25**, 237–304.

Shine, K. P., Robinson, D., Henderson-Sellers, A. and Kukla, G. J., 1984, Evidence of Arctic-wide atmospheric aerosols from DMSP visible imagery. *Journal of Climate and Applied Meteorology* **23**, (in press).

Stephens, G. L., Campbell, G. G. and Vonder Haar, T. H., 1981, Earth radiation budgets. *Journal of Geophysical Research* **86**, 9739–9760.

Winston, J. S., Gruber, A., Gray, T. I., Varnadore, M. S., Earnest, C. L. and Manello, L. P., 1979, *Earth Atmosphere Radiation Budget Analyses Derived from NOAA Satellite Data June 1974–February 1978*, Volumes 1 and 2 (Washington D.C.: NOAA-NESS, Meteorological Satellite Laboratory).

4

Water and the photochemistry of the troposphere

Joel S. Levine
NASA Langley Research Center
Hampton, VA, USA

4.1. Introduction

The gaseous envelope that surrounds our planet extends several thousand kilometres above the surface, where it eventually merges with the interplanetary medium. The atmosphere is divided into four distinct regions defined by the temperature gradient within the region. The four regions of the atmosphere—the troposphere, the stratosphere, the mesosphere and the thermosphere—are shown in figure 1.7, p. 8. The troposphere extends from the surface to the base of the stratosphere and contains 80–85% of the total mass of the atmosphere. All but trace amounts of atmospheric water are found in the troposphere. Tropospheric species control cloud formation and precipitation, regulate climate, affect human health and the productivity of crops, and control the amount of ozone (O_3) in the stratosphere, which shields the surface of the Earth from lethal ultraviolet radiation from the Sun. The composition of the troposphere is summarized in table 4.1.

Over its history, the composition of the troposphere has been controlled by a complex series of biogeochemical cycles that transfer oxygen, nitrogen, carbon, hydrogen, sulphur and halogen species between the troposphere and the solid Earth, the oceans and the biosphere. On a shorter time scale (ranging from seconds to years), the composition of the troposphere is controlled by a series of atmospheric processes, including photochemical and chemical reactions, lightning and precipitation. These processes transform atmospheric species and return them to the solid Earth, the oceans and the biosphere to complete the biogeochemical cycle (e.g., Logan *et al.*, 1981; Levine and Allario, 1982). Over the last few decades, various anthropogenic activities have been identified as significant sources of tropospheric species (although this review will deal primarily with the chemistry of the 'natural' or unpolluted troposphere)

123

Table 4.1. The composition of the troposphere.

Species	Concentration[a]	Source[b]
Major and Minor Gases		
Nitrogen (N_2)	78·08%	Volcanic, biogenic
Oxygen (O_2)	20·95%	Biogenic
Argon (Ar)	0·93%	Radiogenic
Water vapour (H_2O)	Variable—up to 4%	Volcanic, evaporation
Carbon dioxide (CO_2)	0·032%	Volcanic, biogenic, anthropogenic
Trace Gases		
Oxygen species		
Ozone (O_3)	10–100 ppbv	Photochemical
Atomic oxygen (O) (ground state)	$10^3 \, cm^{-3}$	Photochemical
Atomic oxygen ($O(^1D)$) (excited state)	$10^{-2} \, cm^{-3}$	Photochemical
Hydrogen species		
Hydrogen (H_2)	0·5 ppmv	Photochemical, biogenic
Hydrogen peroxide (H_2O_2)	$10^9 \, cm^{-3}$	Photochemical
Hydroperoxyl radical (HO_2)	$10^8 \, cm^{-3}$	Photochemical
Hydroxyl radical (OH)	$10^6 \, cm^{-3}$	Photochemical
Atomic hydrogen (H)	$1 \, cm^{-3}$	Photochemical
Nitrogen species		
Nitrous oxide (N_2O)	330 ppbv	Biogenic
Ammonia (NH_3)	0·1–1 ppbv	Biogenic, anthropogenic
Nitric acid (HNO_3)	50–1000 pptv	Photochemical
Hydrogen cyanide (HCN)	~200 pptv	Anthropogenic (?)
Nitrogen dioxide (NO_2)	10–300 pptv	Photochemical
Nitric oxide (NO)	5–100 pptv	Anthropogenic, biogenic, lightning, photochemical
Nitrogen trioxide (NO_3)	100 pptv	Photochemical
PAN ($CH_3CO_3NO_2$)	50 pptv	Photochemical
Dinitrogen pentoxide (N_2O_5)	1 pptv	Photochemical
Pernitric acid (HO_2NO_2)	0·5 pptv	Photochemical
Nitrous acid (HNO_2)	0·1 pptv	Photochemical
Nitrogen aerosols:		
ammonium nitrate (NH_4NO_3)	10 pptv	Photochemical
ammonium chloride (NH_4Cl)	0·1 pptv	Photochemical
Carbon species		
Methane (CH_4)	1·7 ppmv	Biogenic, anthropogenic
Carbon monoxide (CO)	70–200 ppbv (N. Hemis.) 40–60 ppbv (S. Hemis.)	Anthropogenic, biogenic, photochemical
Formaldehyde (H_2CO)	0·1 ppbv	Photochemical
Methylhydroperoxyl radical (CH_3OOH)	$10^{11} \, cm^{-3}$	Photochemical
Methylperoxyl radical (CH_3O_2)	$10^8 \, cm^{-3}$	Photochemical
Methyl radical (CH_3)	$10^{-1} \, cm^{-3}$	Photochemical

Table 4.1 continued

Species	Concentration[a]	Source[b]
Sulphur species		
Carbonyl sulphide (COS)	$0 \cdot 5$ ppbv	Volcanic, anthropogenic
Dimethyl sulphide ((CH$_3$)$_2$S)	$0 \cdot 4$ ppbv	Biogenic
Hydrogen sulphide (H$_2$S)	$0 \cdot 2$ ppbv	Biogenic, anthropogenic
Sulphur dioxide (SO$_2$)	$0 \cdot 2$ ppbv	Volcanic, anthropogenic, photochemical
Dimethyl disulphide ((CH$_3$)$_2$S$_2$)	100 pptv	Biogenic
Carbon disulphide (CS$_2$)	50 pptv	Volcanic, anthropogenic
Sulphuric acid (H$_2$SO$_4$)	20 pptv	Photochemical
Sulphurous acid (H$_2$SO$_3$)	20 pptv	Photochemical
Sulphoxyl radical (SO)	10^3 cm^{-3}	Photochemical
Thiohydroxyl radical (HS)	1 cm^{-3}	Photochemical
Sulphur trioxide (SO$_3$)	10^{-2} cm^{-3}	Photochemical
Halogen species		
Hydrogen chloride (HCl)	1 ppbv	Sea salt, volcanic
Methyl chloride (CH$_3$Cl)	$0 \cdot 5$ ppbv	Biogenic, anthropogenic
Methyl bromide (CH$_3$Br)	10 pptv	Biogenic, anthropogenic
Methyl iodide (CH$_3$I)	1 pptv	Biogenic, anthropogenic
Noble gases (chemically inert)		
Neon (Ne)	18 ppmv	Radiogenic
Helium (He)	$5 \cdot 2$ ppmv	Radiogenic
Krypton (Kr)	1 ppmv	Radiogenic
Xenon (Xe)	90 ppbv	Radiogenic

[a]Species concentrations are given in % by volume, in terms of surface mixing ratio (parts per million by volume (ppmv = 10^{-6}); parts per billion by volume (ppbv = 10^{-9}); parts per trillion by volume (pptv = 10^{-12})), or in terms of surface number density (cm^{-3}). The species mixing ratio is defined as the ratio of the number density of the species to the total atmospheric number density ($2 \cdot 55 \times 10^{19}$ molecules cm^{-3}). There is some uncertainty in the concentrations of species at the ppbv level or less. The species concentrations given in molecules cm^{-3} are generally based on photochemical calculations and species concentrations in mixing ratios are generally based on measurements.
[b]Species major sources are divided into the broad category of volcanic, radiogenic, biogenic, anthropogenic and photochemical. This Chapter deals primarily with the photochemical production of tropospheric species.

(e.g., Levine and Graedel, 1982). Water in both gaseous and liquid state plays a major role in determining the photochemistry/chemistry and composition of the troposphere. The photochemistry/chemistry of the troposphere is controlled by several different processes initiated by and involving H$_2$O:

1. The hydroxyl radical (OH), produced by the reaction of H$_2$O with excited oxygen atoms (O(^{1}D))†, is a pivotal species involved in the photochemical/chemical transformation of almost every tropospheric species.

---

†The excited state of atomic oxygen is designated as O(^{1}D), as opposed to the ground state which is designated as O(^{3}P) or simply as O.

2. Atmospheric lightning generated by electric fields in clouds is a major natural source of nitric oxide (NO).

3. Cloud and rain droplets and water-covered aerosols are very efficient scavengers of water-soluble tropospheric species and lead to the formation of acid precipitation.

4. In addition to serving as an important sink for soluble tropospheric species, cloud and rain droplets and water-covered aerosols are sites for considerable chemical activity involving aqueous gas-to-gas and gas-to-ion reactions.

5. Due to their scattering, absorption and reflection of incoming solar radiation, clouds affect the photolysis rates and, hence, the photochemistry of tropospheric species.

This chapter deals with the role of water in its liquid and gaseous states in the photochemistry and chemistry of the global natural troposphere. Photochemical processes in the troposphere, the formation of the hydroxyl radical, the oxidation chains involving methane, carbon monoxide and ammonia, the chemistry of the sulphur, hydrogen, halogen and nitrogen species, and lightning as a source of tropospheric species are discussed in Section 4.2. Rainout, washout and the aqueous chemistry in cloud and rain droplets and water-covered aerosols are described in Section 4.3. Scientific questions concerning the chemistry and composition of the troposphere, particularly the impact of anthropogenic activities on the chemistry and composition of the Earth's atmosphere are presented in Section 4.4, which also includes a summary of the remote sensor technology and its application to answer the scientific questions posed earlier in the section.

4.2. *Water, clouds, lightning, precipitation and chemistry*

Photochemical processes in the troposphere

Tropospheric gases are transformed from one species to another as a result of both photochemical and chemical processes. The photochemical or photolytic process involves the dissociation of a molecule into simpler molecules or atoms as a consequence of the absorption of ultraviolet and/or visible radiation from the Sun. The photolysis process can be represented by

$$XY + h\nu \rightarrow X + Y \tag{4.1}$$

where h is Planck's constant; and ν is the frequency of the incoming solar radiation supplying the energy needed to dissociate species XY into species X and Y. Solar radiation down to about 300 nm can penetrate the atmosphere and reach the Earth's surface. Solar radiation between about 200 and 300 nm is absorbed by ozone in the stratosphere, and radiation less than 200 nm is absorbed by oxygen (O_2) and nitrogen (N_2) molecules in the thermosphere (see figure 2.1, p. 46). As it traverses the atmosphere to the top of the troposphere the transfer of incoming solar radiation is controlled by absorption and scattering due to molecules and aerosols. Once in the troposphere, the transfer of solar radiation is affected by the absorption, scattering and reflection properties of clouds, as well as molecules and aerosols. The transfer of solar radiation through the cloud-free troposphere has been discussed in detail in Chapter 2; see also Anderson and Meier (1979). The effects of clear-sky absorption and scattering on the photochemistry of the troposphere have been investigated by Augustsson and Levine (1982), and the effects of cloud absorption and scattering on photochemistry have been studied by Logan *et al.* (1981). Hence, clouds modify the transfer and distribution of incoming solar radiation in the troposphere and, therefore, affect the photolysis of tropospheric species. Many of the chemical transformations in the troposphere are initiated by reactions with the hydroxyl radical produced chemically from water vapour in the troposphere. Hence, to a very large extent photochemical and chemical transformations in the troposphere are controlled by water—in both its liquid and gaseous states.

The formation of the hydroxyl radical

With few exceptions, notably carbon dioxide (CO_2) and nitrous oxide (N_2O), almost all of the species supplied to the troposphere by the biosphere, such as methane (CH_4), ammonia (NH_3) and hydrogen sulphide (H_2S), are in reduced oxidation states. (CO_2 and N_2O are both very chemically inert in the troposphere, although by virtue of their strong infra-red absorption properties they are important greenhouse absorbers and, hence, affect the Earth's climate.) Once in the atmosphere, these reduced species are photochemically/chemically transformed and eventually returned to the solid Earth, the oceans and the biosphere by rainout, washout or dry deposition as oxidized species such as carbon dioxide (CO_2), nitrogen dioxide (NO_2), nitric acid (HNO_3), sulphur dioxide (SO_2) and sulphuric acid (H_2SO_4). The link between the reduced species emitted from the Earth's surface and the oxidized species returned to the surface is the hydroxyl radical (OH). The hydroxyl radical is responsible for the chemical transformation of reduced species to oxidized species in the atmosphere.

The hydroxyl radical is formed by the reaction of H_2O with excited atomic oxygen ($O(^1D)$). Excited oxygen is produced by the photolysis of tropospheric ozone by solar radiation with wavelengths equal to, or less

128   J. S. Levine

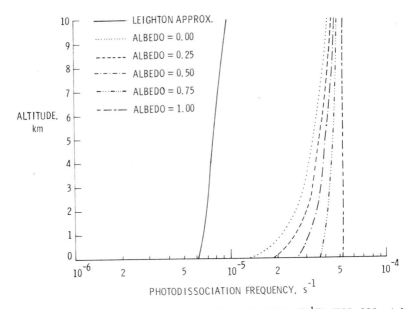

Figure 4.1. Photodissociation frequencies for ozone photolysis yielding O(^{1}D) (295–320nm) for the pure absorption case (Leighton approximation) and for absorption plus multiple scattering for various surface albedos (Augustsson and Levine, 1982).

than 320nm. While the bulk of the ozone in the atmosphere is located in the stratosphere (between about 15 and 50 km), about 10 % of the total atmospheric burden is found in the troposphere, where its concentration ranges from 10 to 100 parts per billion by volume (ppbv)†. Sources of tropospheric ozone include *in situ* photochemical production, as well as the transport of stratospheric ozone into the troposphere. The reactions leading to the production of OH in the troposphere can be summarized as

$$O_3 + h\upsilon \rightarrow O(^1D) + O_2, \quad \lambda \leqslant 320nm \quad (4.2)$$

$$H_2O + O(^1D) \rightarrow 2OH \quad (4.3)$$

Photochemical and kinetic rates for the reactions discussed in this chapter can be found in Logan *et al.*, 1981; Augustsson and Levine, 1982; and Baulch *et al.*, 1982.

The vertical distribution of the photodissociation rate or frequency

†Parts per billion by volume, ppbv = 10^{-9} is the species mixing ratio, which is the ratio of the number density of the particular species, ozone in this case, to the total atmospheric number density, which is $2\cdot55 \times 10^{19}$ molecules cm^{-3} at the surface. To determine the number density of the species, multiply its mixing ratio by the total atmospheric number density.

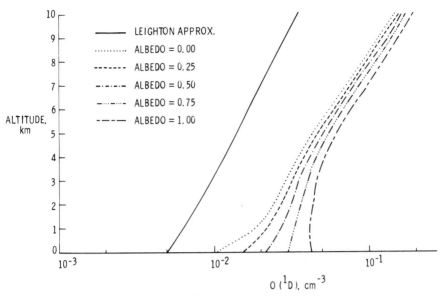

Figure 4.2. The vertical distribution of $O(^1D)$ *for the pure absorption case (Leighton approximation) and for absorption plus multiple scattering for various surface albedos (Augustsson and Levine, 1982).*

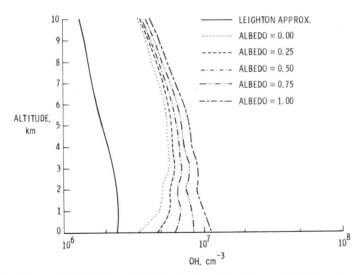

Figure 4.3. The vertical distribution of OH *for the pure absorption case (Leighton approximation) and for absorption plus multiple scattering for various surface albedos (Augustsson and Levine, 1982).*

(s^{-1}) for the photolysis of O_3 yielding $O(^1D)$ in the troposphere (reaction 4.2) is shown in figure 4.1. The vertical distributions of $O(^1D)$ and OH representing the balance between the production (reactions 4.2 and 4.3) and the various loss processes of these important species in the troposphere are given in figures 4.2 and 4.3, respectively. These figures are based on photochemical calculations of Augustsson and Levine (1982) using the clear-sky radiative transfer model of Anderson and Meier (1979). Calculations are given for six different cases of the attenuation of the incoming solar radiation through the troposphere: the case of molecular absorption only, the so-called 'Leighton approximation', which does not include effects of multiple scattering by molecules and aerosols and surface albedo, and the inclusion of multiple scattering for five different values of surface albedo (surface albedos = $0 \cdot 00, 0 \cdot 25, 0 \cdot 50, 0 \cdot 75$ and $1 \cdot 00$). All of the calculations shown in figures 4.1–4.3 were performed for a fixed solar zenith angle of 45°. The water vapour profile for the OH calculations was the average of the January and July profiles in the US Standard Atmosphere for a latitude of 30°N.

Ozone can also be photolysed by solar radiation of wavelengths between 320 and 1200 nm. This photolysis reaction leads to the production of groundstate atomic oxygen, $O(^3P)$ or simply O

$$O_3 + h\nu \rightarrow O + O_2, \quad 320 < \lambda < 1200 \text{nm} \quad (4.4)$$

However, the O atom rapidly reforms O_3 by recombining with O_2 in a three-body reaction

$$O + O_2 + M \rightarrow O_3 + M \quad (4.5)$$

where M is any third body, usually N_2 or O_2. Thus, reactions 4.4 and 4.5 form a null cycle with no net chemical effect.

The methane oxidation chain

The primary sources of CH_4 are the anaerobic microbiological decomposition of organic matter in rice paddy fields and wetlands, the enteric fermentation in ruminants (mostly cattle), biomass burning and natural gas leakage (see table 4.2 for estimates of source strength, removal processes, atmospheric lifetime and transport distances). The methane oxidation chain, initiated by the reaction of CH_4 with OH, leads to the photochemical production of methyl (CH_3), methylperoxyl (CH_3O_2), methoxyl (CH_3O) and hydroperoxyl (HO_2) radicals, nitric oxide (NO), nitrogen dioxide (NO_2), formaldehyde (H_2CO), carbon monoxide (CO), molecular hydrogen (H_2) and ozone. This reaction scheme is expressed as

$$CH_4 + OH \rightarrow CH_3 + H_2O \quad (4.6)$$

$$CH_3 + O_2 + M \rightarrow CH_3O_2 + M \quad (4.7)$$

$$CH_3O_2 + NO \rightarrow CH_3O + NO_2 \quad (4.8)$$

Table 4.2. Budgets of carbon species (Baulch *et al.*, 1982).

Gas	Source	Source strength (10^{14} g gas year^{-1})	Removed by	Atmospheric lifetime	Transport distances[a] Δx, Δy, Δz (km)
CO	Biomass burning	4–16	OH	2 months	4000, 2500, 10
	Industry	6·4			
	Oxidation of CH_4	6			
	Oxidation of C_5H_8 and $C_{10}H_{16}$	4–13			
CH_4	Rice paddy fields	0·7–1·2	OH	7 years	Complete
	Natural wetlands	0·3–2·2			
	Ruminants	0·6			
	Biomass burning	0·3–1·1			
	Gas leakage	0·5			
C_5H_8 $C_{10}H_{16}$	Trees	Unknown	OH	10 hours	400, 200, 1

[a]Diffusion distances in east–west (Δx), north–south (Δy), and vertical directions (Δz) (in km) over which concentrations are reduced to 30 % by chemical reactions; calculated with $[OH] = 7 \times 10^5$ molecules cm^{-3}.

$$NO_2 + h\nu \rightarrow NO + O, \quad \lambda \leqslant 400\,nm \qquad (4.9)$$

$$O + O_2 + M \rightarrow O_3 + M \qquad (4.5)$$

$$CH_3 + O_2 \rightarrow H_2CO + HO_2 \qquad (4.10)$$

$$HO_2 + NO \rightarrow NO_2 + OH \qquad (4.11)$$

$$NO_2 + h\nu \rightarrow NO + O, \quad \lambda \leqslant 400\,nm \qquad (4.9)$$

$$O + O_2 + M \rightarrow O_3 + M \qquad (4.5)$$

$$H_2CO + h\nu \rightarrow CO + H_2, \quad or \quad HCO + H, \quad \lambda \leqslant 340\,nm \qquad (4.12)$$

The net cycle for the methane oxidation chain can be represented as

$$CH_4 + 4O_2 \rightarrow H_2O + CO + H_2 + 2O_3 \qquad (4.13)$$

Non-methane hydrocarbons

Perhaps the greatest deficiency in our understanding of the reactive carbon cycle in the troposphere concerns the photochemistry and chemistry and the global emission rates of non-methane hydrocarbon (NMHC) species. NMHCs include alkanes, such as ethane (C_2H_6), propane (C_3H_8) and butane (C_4H_{10}); olefins, such as ethylene (C_2H_4) and propylene (C_3H_6); alkynes, such as acetylene (C_2H_2); aldehydes, such as formaldehyde (H_2CO) and acetaldehyde (CH_3CHO); peroxyacetylnitrate or PAN ($CH_3CO_3NO_2$); isoprene (C_5H_8); and terpenes ($C_{10}H_{16}$). Alkanes, olefins and alkynes are emitted from anthropogenic sources, while isoprene and terpenes are naturally emitted by vegetation. The other NMHC species are produced chemically from the oxidation of the emitted species by OH, O_3 and O. Oxidation of the simpler NMHCs (ethane, ethylene and acetylene) produces the same products that are produced from the oxidation of CH_4, i.e., H_2CO, CO, CH_3OOH, CH_3O_2 and CH_3O; in addition, the peroxyacetyl radical (CH_3CO_3), the radical precursor to peroxyacetylnitrate, is produced. The longer chain hydrocarbons are sources of longer chain radicals (such as ethylperoxyl ($CH_3CH_2O_2$) and ethoxy (CH_3CH_2O) radicals), which undergo reactions that parallel those in the CH_4 oxidation chain. As much as 90 % of the total flux of NMHCs may be due to the natural emission of isoprene and terpenes, with anthropogenic emissions accounting for the remaining 10 % of total NMHC emissions.

The carbon monoxide oxidation chain

In addition to the methane oxidation chain, other sources of CO are biomass burning, industry and the oxidation of non-methane hydrocarbons, i.e., C_5H_8 and $C_{10}H_{16}$ (see table 4.2 for estimates of source strength). The carbon monoxide oxidation chain leads to the photochemical production of carbon dioxide (CO_2), atomic hydrogen (H) and other species: OH, HO_2, NO_2, NO and O_3. This reaction scheme is expressed as

$$CO + OH \rightarrow CO_2 + H \qquad (4.14)$$

$$H + O_2 + M \rightarrow HO_2 + M \qquad (4.15)$$

$$HO_2 + NO \rightarrow NO_2 + OH \qquad (4.11)$$

$$NO_2 + h\nu \rightarrow NO + O, \quad \lambda \leqslant 400\,nm \qquad (4.9)$$

$$O + O_2 + M \rightarrow O_3 + M \qquad (4.5)$$

The net cycle for the carbon monoxide oxidation chain can be represented as

$$CO + 2O_2 \rightarrow CO_2 + O_3 \qquad (4.16)$$

The ammonia oxidation chain

Sources of ammonia include biomass burning, emissions from natural (non-fertilized) and fertilized fields, domestic and wild animals and coal burning (see table 4.3 for estimates of source strength). The details of the ammonia oxidation chain are known with less certainty than the CH_4 and CO oxidation chains. The oxidation of NH_3 by OH leads to the formation of the amine radical (NH_2)

$$NH_3 + OH \rightarrow NH_2 + H_2O \qquad (4.17)$$

The amine radical may lead to either the production or destruction of nitrogen oxides ($NO_x = NO + NO_2$). The fate of the amine radical depends on the background concentration of NO_x. For NO_x concentrations below about 70 parts per trillion by volume (pptv)†, the amine radical may lead to the production of NO_x, while for NO_x concentrations

---

†Parts per trillion by volume, pptv = 10^{-12}.

Table 4.3. Budgets of nitrogen species (Baulch *et al.*, 1982).

Gas	Source	Source strength (10^{12} g N year^{-1})	Removed by	Atmospheric lifetime	Transport distances^a Δx, Δy, Δz (km)
NO$_x$ (NO + NO$_2$)	Biomass burning	10–40	OH	1·5 days	1500, 400, 1
	Industry	8·2–18·5			
	Lightning	3–4			
	Soils	0–15			
	Oxidation of N$_2$O (in stratosphere)	0·5–1·5			
	Jet aircraft	0·25			
HNO$_3$	OH + NO$_2$	22–77	Rain	3 days	300, 600, 1·5
N$_2$O	Oceans	4–10	Stratospheric photolysis	100 years	Global
	Loss of organic matter	2–6			
	Fertilized fields	<3			
	Biomass burning	1–2			
	Fossil fuel burning	1·8			
NH$_3$	Biomass burning	<60	Rain	<9 days	<9000, 1000, 3
	Natural fields	<30			
	Domestic animals	10–20			
	Coal burning	4–12			
	Wild animals	2–6			
	Fertilized fields	<3			

^{a}*Diffusion distances in east–west* (Δx), *north–south* (Δy), and vertical directions (Δz) (in km) over which concentrations are reduced to 30% by chemical reactions; calculated with [OH] = 7×10^5 molecules cm^{-3}.

greater than about 70pptv, the amine radical may lead to the destruction of NO_x. These NH_2-NO_x reactions are summarized below

$$NH_2 + O_3 \rightarrow NO_x + \text{products} \tag{4.18}$$

$$NH_2 + NO \rightarrow N_2 + H_2O \tag{4.19}$$

$$NH_2 + NO_2 \rightarrow N_2O + H_2O \tag{4.20}$$

Sulphur species chemistry

The hydroxyl radical is also a key species in the photochemistry of sulphur in the troposphere. The reaction of OH with hydrogen sulphide (H_2S) leads to the production of sulphur dioxide (SO_2). The main source of H_2S is microbiological activity, with several industrial sources, such as wood pulping and sewage treatment. Reaction of OH with SO_2 leads to the production of sulphuric acid (H_2SO_4), the primary constituent of acid rain. In addition to its production via the oxidation of H_2S, SO_2 is produced by volcanic emissions and is a product of fossil fuel combustion. These reactions are summarized below (HS is the thiohydroxyl radical, SO is the sulphoxyl radical and HSO_3 is the sulphuric acid radical)

$$H_2S + OH \rightarrow HS + H_2O \tag{4.21}$$

$$HS + O_2 \rightarrow SO + OH \tag{4.22}$$

$$SO + O_2 \rightarrow SO_2 + O \tag{4.23}$$

$$SO_2 + OH \rightarrow HSO_3 \tag{4.24}$$

$$HSO_3 + OH \rightarrow H_2SO_4 \tag{4.25}$$

Other tropospheric sulphur species include carbonyl sulphide (COS), carbon disulphide (CS_2), dimethyl sulphide ($(CH_3)_2S$) and dimethyl disulphide ($(CH_3)_2S_2$). The source of COS and CS_2 is probably volcanic emissions, and $(CH_3)_2S$ and $(CH_3)_2S_2$ are produced by anaerobic bacteria in marshlands and wetlands. Both COS and CS_2 react with OH to produce oxidized sulphur species

$$COS + OH \rightarrow HOS + CO \tag{4.26}$$

$$CS_2 + OH \rightarrow HOS + CS \tag{4.27}$$

$$HOS + O_2 \rightarrow HSO_3 \qquad\qquad (4.28)$$

$$CS + O_2 \rightarrow SO + CO \qquad\qquad (4.29)$$

$$SO + O_2 \rightarrow SO_2 + O \qquad\qquad (4.30)$$

The details of the reactions of $(CH_3)_2S$ and $(CH_3)_2S_2$ with OH are less certain, but SO_2 is again the expected result.

Halogen species chemistry

The concentrations of the halogen species in the troposphere are much less than the species so far discussed. The most abundant naturally produced halogen species are hydrogen chloride (HCl) and methyl chloride (CH_3Cl). HCl results from volcanic emissions and from sea salt spray, while CH_3Cl is biologically produced in the oceans. Both HCl and CH_3Cl begin their chemical cycles in the troposphere by reaction with OH

$$HCl + OH \rightarrow Cl + H_2O \qquad\qquad (4.31)$$

$$CH_3Cl + OH \rightarrow CH_2Cl + H_2O \qquad\qquad (4.32)$$

The chlorine atom (Cl) reacts with several tropospheric species

$$Cl + CH_4 \rightarrow CH_3 + HCl \qquad\qquad (4.33)$$

$$Cl + H_2 \rightarrow H + HCl \qquad\qquad (4.34)$$

$$Cl + HO_2 \rightarrow O_2 + HCl \qquad\qquad (4.35)$$

$$Cl + O_3 \rightarrow ClO + O_2 \qquad\qquad (4.36)$$

The reaction of CH_3Cl with OH initiates a reaction similar to the methane oxidation chain leading to the production of HCl at the end of the cycle. Bromine and iodine species are much less abundant than the chlorine species in the troposphere and will not be considered.

Hydrogen species chemistry

OH is also involved in reactions that lead to the photochemical production of the hydroperoxyl radical (HO_2) and hydrogen peroxide (H_2O_2).

The reactions linking the oxides of hydrogen are summarized below

$$OH + O_3 \rightarrow HO_2 + O_2 \qquad (4.37)$$

$$H + O_2 + M \rightarrow HO_2 + M \qquad (4.38)$$

$$HO_2 + NO \rightarrow OH + NO_2 \qquad (4.39)$$

$$HO_2 + O_3 \rightarrow OH + 2O_2 \qquad (4.40)$$

$$HO_2 + HO_2 \rightarrow H_2O_2 + O_2 \qquad (4.41)$$

$$H_2O_2 + OH \rightarrow H_2O + HO_2 \qquad (4.42)$$

$$H_2O_2 + h\nu \rightarrow 2OH \quad \lambda \leqslant 350 \, nm \qquad (4.43)$$

Lightning as a source of tropospheric species

In addition to the OH-initiated reactions involving tropospheric species described earlier, the excitation, ionization and dissociation due to atmospheric lightning plays an important role in the transformation and production of tropospheric species. The lightning flash generated by charge separation in clouds consists of several individual strokes. Each of the estimated 500 lightning strokes occurring around the globe every second deposits about $10^5 \, Jm^{-1}$ of energy into the lightning channel having a length of up to 10km and a radius of $10^{-3}-10^{-2}$m. The high pressures and temperatures produced within this lightning channel lead to a radially propagating cylindrical shock wave. As the shock front moves radially outwards, the surrounding air is heated to temperatures as high as 30000K in the immediate vicinity of the channel and to lower temperatures at greater distances. Within the heated shock wave, atmospheric species, such as N_2, O_2, H_2O and CO_2, are ionized, excited and dissociated into atoms, molecules and radicals, such as N, O, OH, H and CO. Eventually, the heated air is cooled to atmospheric temperatures by radiation, expansion and entrainment of cooler ambient air. During the cooling stage certain atmospheric species 'freeze out' and form stable species. A theoretical calculation of the production of tropospheric species based on the atmospheric lightning fast-cooling shock wave/thermochemical model of W. L. Chameides (Georgia Institute of Technology) is shown in figure 4.4 (Levine *et al.*, 1979). In this figure the theoretically calculated equilibrium volume mixing ratio f° of various species is given as a function of the cooling temperature T of a parcel of tropospheric air initially heated by a lightning discharge.

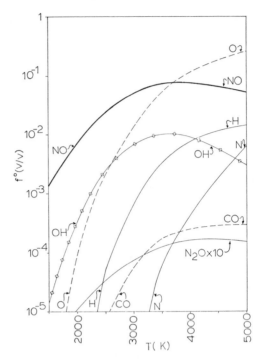

Figure 4.4 Theoretically calculated equilibrium volume mixing ratio, f° of N_2O, CO, OH, N, H, NO and O as a function of temperature T for heated tropospheric air, with a mass density of $1\cdot2$ $kg\,m^{-3}$ (calculation by W. L. Chameides, Georgia Institute of Technology; see Levine et al., 1979).

Nitrogen species chemistry

Atmospheric lightning is a major natural global source of nitric oxide (NO) (it has also been identified as a localized source of N_2O and CO). Other natural sources of NO include biomass burning, biogenic activity in soils and the oxidation of N_2O in the stratosphere (see table 4.3 for estimates of source strength). The largest global source of NO is probably an anthropogenic source: high-temperature combustion. Once formed, NO is quickly converted to NO_2 by reactions with either O_3 or HO_2

$$NO + O_3 \rightarrow NO_2 + O_2 \qquad (4.44)$$

$$NO + HO_2 \rightarrow NO_2 + OH \qquad (4.11)$$

NO can also be converted to nitrous acid (HNO_2) by reacting with OH

$$NO + OH + M \rightarrow HNO_2 + M \qquad (4.45)$$

The NO_2 formed in these reactions (4.44 and 4.11) is then transformed to nitric acid by reaction with OH

$$NO_2 + OH + M \rightarrow HNO_3 + M \qquad (4.46)$$

Nitric acid, the fastest increasing component of acid precipitation, at present accounts for about one-third of its acidity (sulphuric acid accounts for the remaining two-thirds). Nitric acid is the dominant active nitrogen species in the troposphere. Active nitrogen, NO_y, is defined as the sum of NO, NO_2, HNO_3, HNO_2, nitrogen trioxide (NO_3) and dinitrogen pentoxide (N_2O_5). The two minor active nitrogen species, NO_3 and N_2O_5, are produced via the following reactions

$$NO_2 + O_3 \rightarrow NO_3 + O_2 \qquad (4.47)$$

$$HNO_3 + OH \rightarrow NO_3 + H_2O \qquad (4.48)$$

$$NO_3 + NO_2 \rightarrow N_2O_5 \qquad (4.49)$$

$$N_2O_5 + M \rightarrow NO_3 + NO_2 + M \qquad (4.50)$$

The loss of active nitrogen in the troposphere is controlled by the rainout of HNO_3. The dissolution of water-soluble species in cloud and rain droplets, such as HNO_3, HNO_2, H_2SO_4, H_2O_2, H_2CO and NH_3, constitutes a major loss mechanism for these species.

Aerosol formation is another sink for gaseous NH_3, HNO_3 and H_2SO_4. NH_3 reacts with HNO_3 and H_2SO_4 to form ammonium nitrate (NH_4NO_3) and ammonium sulphate (($NH_4)_2SO_4$) aerosols, respectively

$$NH_3 + HNO_3 \rightarrow NH_4NO_3(\text{solid}) \qquad (4.51)$$

$$2NH_3 + H_2SO_4 \rightarrow (NH_4)_2SO_4(\text{solid}) \qquad (4.52)$$

Rainout and washout are also sinks for nitrate and sulphate aerosols (see Section 4.3). Rainout and washout return tropospheric species in the form of nitrates, nitrites, sulphates and oxides to the biosphere, where they are transformed by microbiological metabolic processes (such as nitrification and denitrification) and are eventually recycled back into the troposphere. Microbiological recycling completes the biogeochemical cycle of tropospheric species and is of major importance in controlling the composition of the troposphere. In the absence of microbiological recycling the Earth's atmosphere would contain only trace amounts of molecular nitrogen, its major constituent. Molecular oxygen, the second most abundant constituent of the atmosphere, is also produced by biological activity. Oxygen is produced as a by-product in the photosynthetic process, where H_2O and CO_2 are converted to carbohydrates ($C_m(H_2O)_n$) in the presence of sunlight by chlorophyll in green plant cells. This important process is represented by

$$nH_2O + mCO_2 + h\nu \xrightarrow{\text{chlorophyll}} C_m(H_2O)_n + mO_2 \qquad (4.53)$$

4.3. Aqueous chemistry in rain and aerosols

The collection of liquid water in cloud and rain droplets and around solid aerosol particles is an important sink for tropospheric species. The loss of tropospheric species in both gaseous and particulate form is due to rainout and washout. Rainout involves the scavenging of water-soluble gases within the cloud or condensation of cloud droplets around particulate matter, while washout involves the scavenging of aerosol particles below the cloud as the precipitation falls to the surface. The dissolution of a soluble gas on to a cloud or rain droplet or on to a water-covered aerosol can be calculated based on the concentration of the soluble gaseous species, the aqueous solubility of the gas (the Henry's Law equilibrium constant) and the liquid water content of the cloud and precipitation. The aqueous solubility of selected tropospheric gases at one atmospheric pressure as a function of temperature compiled by Graedel and Weschler (1981) is summarized in table 4.4. Typical concentrations of soluble tropospheric gases in the aqueous phase of atmospheric aerosols compiled by Graedel and Weschler (1981) are summarized in table 4.5.

Table 4.4. Aqueous solubility of selected gases at one atmosphere pressure (Graedel and Weschler, 1981). Solubilities are given as cm^3 of gas per cm^3 of water.

Compound	0°C	10°C	20°C	25°C	30°C
O_2	0.0489	0.0380	0.0310	0.0283	0.0261
O_3	0.0184	0.0140	0.0098	0.0065	0.0036
H_2	0.0214	0.0193	0.0178	0.0171	0.0163
CO_2	1.71	1.18	0.860	0.738	0.642
CO	0.0352	0.0278	0.0227	0.0208	0.0192
NH_3	5234.0	3369.0	2235.0	1840.0	1524.0
NO	0.0738	0.0571	0.0471	0.0430	0.0400
NO_2	—	—	—	0.269	—
H_2S	4.37	3.59	2.91	2.61	2.33
SO_2	79.79	56.65	39.37	32.79	27.16
CS_2	0.601	0.572	0.528	0.498	0.457
OCS	1.33	0.836	0.561	0.468	0.403
HCl	505.0	—	—	430.0	413.0
HCHO	—	—	—	180.0	—

Table 4.5. Typical concentrations of selected gases in the atmosphere and in the aqueous phase of atmospheric aerosols at 25°C (Graedel and Weschler, 1981).

Species	Urban Gas conc. (ppb)	Urban Aqueous conc. (M)	Rural Gas conc. (ppb)	Rural Aqueous conc. (M)
O_2	2×10^8	3×10^{-4}	2×10^8	3×10^{-4}
O_3	70	2×10^{-11}	30	9×10^{-12}
H_2	2×10^3	2×10^{-9}	600	5×10^{-10}
H_2O_2	20	3×10^{-6}	1	3×10^{-7}
CO	8×10^3	7×10^{-9}	120	1×10^{-10}
CO_2	5×10^5	2×10^{-5}	$3 \cdot 3 \times 10^5$	1×10^{-5}
NO	10	2×10^{-11}	$0 \cdot 1$	2×10^{-13}
HNO_3	1	5×10^{-5}	$0 \cdot 3$	1×10^{-6}
NH_3	25	2×10^{-6}	$1 \cdot 5$	1×10^{-7}
H_2S	2	2×10^{-10}	$0 \cdot 1$	1×10^{-11}
CS_2	$0 \cdot 15$	3×10^{-12}	$0 \cdot 04$	9×10^{-13}
OCS	$0 \cdot 6$	1×10^{-11}	$0 \cdot 6$	1×10^{-11}
SO_2	50	7×10^{-8}	1	2×10^{-9}
HCl	6	1×10^{-7}	$0 \cdot 7$	1×10^{-8}
HCHO	6	5×10^{-8}	$0 \cdot 2$	2×10^{-9}

Since many tropospheric gases become acids when dissolved in water, their solubilities become a function of the acidity of the solution. The acidity of the cloud or rain droplet is set initially by equilibrium with gaseous CO_2, which produces a weak carbonic acid solution (H_2CO_3) of pH ∼ 5·6, in the absence of any other ions in solution. However, rain is considerably more acidic in some regions of the world, with pH values of about 4 common, and acidities below pH = 2 occasionally measured. This enhanced acidity of rain results from the presence of H_2SO_4 and HNO_3 and other acids in the droplets and leads to the acid rain phenomenon. Washout expressions for sulphate and nitrate aerosols have been parameterized in terms of the aerosol concentrations in and below the precipitating cloud. A washout relationship can be represented by an expression in the general form of

$$W = cR_x \qquad (4.54)$$

where W is the washout ratio, which is the fraction of the sulphate or nitrate aerosol removed; R is the rate of precipitation; and c and x are empirically derived parameters for a particular precipitation event.

Not only are cloud and rain droplets and water-covered aerosols sites for the dissolution of soluble tropospheric species, but they are an active medium for considerable chemical activity. In addition to the gas-phase reactions between the dissolved species in solution, discussed earlier in Section 4.2, gas–ion and ion–ion reactions are also possible (Graedel and

Weschler, 1981). In bulk water, water molecules establish an equilibrium relationship with their ions

$$2H_2O \rightleftharpoons H_3O^+ + OH^- \qquad (4.55)$$

In addition, other dissolved gases, including H_2O_2, H_2SO_4, SO_2, HNO_3, HNO_2, NH_3, HCl, H_2CO_3, H_2S, $HCOOH$, CH_3COOH and CO_2, establish equilibrium relationships with their ions in solution. The equilibrium relationships and the ionization constants for these soluble gases in aqueous atmospheric aerosol systems compiled by Graedel and Weschler (1981) are summarized in table 4.6. This table summarizes the ionization reaction in water solution, the ionization constant K (the degree to which a species is ionized in water solution) and the rate constants for the forward (k_f) and reverse (k_r) reactions. In addition to the ionization constants, the hydrolysis constants for a number of important reactions involving SO_2 and CO_2 are included in the table. The formation of ions in water leads to a number of gas–ion and ion–ion reactions in aqueous solution, in addition to the gas–gas reactions. These reactions discussed by Graedel and Weschler (1981) are summarized in table 4.7 (this table should be used in conjunction with table 4.6).

The chemistry of aqueous aerosol systems is a relatively new area of atmospheric chemistry. The most comprehensive study of this subject is

Table 4.6. Ionization constants in aqueous atmospheric aerosol systems (Graedel and Weschler, 1981).

Reaction	K (25°C)	k_f	k_r
$2H_2O \rightleftharpoons H_3O^+ + OH^-$	$1 \cdot 008 \times 10^{-14} M^2$	$1 \cdot 3 \times 10^{-3} Ms^{-1}$	$1 \cdot 3 \times 10^{11} M^{-1}s^{-1}$
$H_2O_2 \rightleftharpoons H^+ + HO_2^-$	$1 \cdot 5 \times 10^{-12} M$	$\sim 7 \cdot 5 \times 10^{-2} s^{-1}$	$\sim 5 \times 10^{10} M^{-1}s^{-1}$
$H_2SO_4 \rightleftharpoons H^+ + HSO_4^-$	$> 1 M$	$> 5 \times 10^{10} s^{-1}$	$\sim 5 \times 10^{10} M^{-1}s^{-1}$
$HSO_4^- \rightleftharpoons H^+ + SO_4^{2-}$	$\sim 10^{-2} M$	$1 \times 10^9 s^{-1}$	$1 \times 10^{11} M^{-1}s^{-1}$
$SO_2 \cdot xH_2O \rightleftharpoons H^+ + HSO_3^-$	$1 \cdot 7 \times 10^{-2} M$	$3 \cdot 4 \times 10^6 s^{-1}$	$2 \cdot 0 \times 10^8 M^{-1}s^{-1}$
$HSO_3^- \rightleftharpoons H^+ + SO_3^{2-}$	$6 \cdot 2 \times 10^{-8} M$	$3 \times 10^3 s^{-1}$	$5 \times 10^{10} M^{-1}s^{-1}$
$HNO_3 \rightleftharpoons H^+ + NO_3^-$	$> 1 M$	$> 5 \times 10^{10} s^{-1}$	$\sim 5 \times 10^{10} M^{-1}s^{-1}$
$HNO_2 \rightleftharpoons H^+ + NO_2^-$	$6 \cdot 0 \times 10^{-6} M$	$\sim 3 \times 10^5 s^{-1}$	$\sim 5 \times 10^{10} M^{-1}s^{-1}$
$NH_3 \cdot H_2O \rightleftharpoons NH_4^+ + OH^-$	$1 \cdot 8 \times 10^{-5} M$	$6 \times 10^5 s^{-1}$	$3 \cdot 4 \times 10^{10} M^{-1}s^{-1}$
$HCl \rightleftharpoons H^+ + Cl^-$	$> 1 M$	$> 5 \times 10^{10} s^{-1}$	$\sim 5 \times 10^{10} M^{-1}s^{-1}$
$H_2CO_3 \rightleftharpoons H^+ + HCO_3^-$	$2 \times 10^{-4} M$	$\sim 1 \times 10^7 s^{-1}$	$\sim 5 \times 10^{10} M^{-1}s^{-1}$
$HCO_3^- \rightleftharpoons H^+ + CO_3^{2-}$	$4 \cdot 84 \times 10^{-11} M$	$\sim 2 \cdot 5 s^{-1}$	$\sim 5 \times 10^{10} M^{-1}s^{-1}$
$H_2S \rightleftharpoons H^+ + HS^-$	$5 \cdot 7 \times 10^{-8} M$	$4 \cdot 3 \times 10^3 s^{-1}$	$7 \cdot 5 \times 10^{10} M^{-1}s^{-1}$
$HS^- \rightleftharpoons H^+ + S^{2-}$	$1 \cdot 3 \times 10^{-13} M$	$\sim 6 \cdot 5 \times 10^{-3} s^{-1}$	$\sim 5 \times 10^{10} M^{-1}s^{-1}$
$HCOOH \rightleftharpoons H^+ + HCOO^-$	$1 \cdot 7 \times 10^{-4} M$	$8 \cdot 6 \times 10^{-6} s^{-1}$	$\sim 5 \times 10^{10} M^{-1}s^{-1}$
$CH_3COOH \rightleftharpoons H^+ + CH_3COO^-$	$1 \cdot 6 \times 10^{-5} M$	$7 \cdot 8 \times 10^5 s^{-1}$	$\sim 5 \times 10^{10} M^{-1}s^{-1}$
$H_2SO_3 \rightleftharpoons SO_2 + H_2O$	$1 \cdot 9 \times 10^1$	$6 \cdot 3 \times 10^7 s^{-1}$	$3 \cdot 4 \times 10^6 s^{-1}$
$SO_2 + H_2O \rightleftharpoons HSO_3^- + H^+$	$1 \cdot 7 \times 10^{-2} M$	$3 \cdot 4 \times 10^6 s^{-1}$	$2 \times 10^8 M^{-1}s^{-1}$
$H_2CO_3 \rightleftharpoons CO_2 + H_2O$	$3 \cdot 5 \times 10^2$	$15 s^{-1}$	$4 \cdot 3 \times 10^{-2} s^{-1}$
$CO_2 + H_2O \rightleftharpoons HCO_3^- + H^+$	$7 \cdot 7 \times 10^{-7} M$	$0 \cdot 043 s^{-1}$	$5 \cdot 6 \times 10^4 M^{-1}s^{-1}$
$HCO_3^- \rightleftharpoons CO_2 + OH^-$	$7 \cdot 1 \times 10^{-9} M$	$1 \times 10^{-4} s^{-1}$	$1 \cdot 4 \times 10^4 M^{-1}s^{-1}$

Table 4.7. Summary of gas–ion and ion–ion reactions in the aqueous atmospheric aerosol system (Graedel and Weschler, 1981).

Oxygen–hydrogen reactions

$O^- + O_2 \rightarrow O_3^-$

$O_3^- \rightarrow O^- + O_2$

$O^- + H_2O \rightarrow OH^- + OH$

$O^- + HO_2^- \rightarrow OH^- + O_2^-$

$O^- + H_2O_2 \rightarrow O_2^- + H_2O$

$O_2^- + OH \rightarrow OH^- + O_2$

$O_2^- + HO_2 \rightarrow HO_2^- + O_2$

$O_2^- + H_2O_2 \rightarrow OH^- + OH + O_2$

$OH + OH^- \rightarrow O^- + H_2O$

Inorganic carbon reactions

$O_2^- + HCO_3^- \rightarrow CO_3^- + HO_2^-$

$OH + HCO_3^- \rightarrow CO_3^- + H_2O$

$OH + CO_3^{2-} \rightarrow CO_3^- + OH^-$

Nitrogen reactions

$O^- + NO_2^- \xrightarrow{H_2O} 2OH^- + NO_2$

$Cl^- + NO_3 \rightarrow products$

$OH + NO_2^- \rightarrow OH^- + NO_2$

$OH + NO_3^- \rightarrow OH^- + NO_3$

$N_2O_4 \xrightarrow{3H_2O} 2H_3O^+ + NO_2^- + NO_3^-$

$OH^- + HNO_2 \rightarrow NO_2^- + H_2O$

Sulphur reactions

$O^- + SO_3^{2+} \xrightarrow{H_2O} SO_3^- + 2OH^-$

$OH + HS^- \rightarrow OH^- + HS$

$OH + CH_3S^- \rightarrow OH^- + CH_3S$

$OH + CH_3SSCH_3 \rightarrow OH^- + CH_3SSCH_3^+$

$OH + HSO_3^- \rightarrow SO_3^- + H_2O$

$OH + HSO_4^- \rightarrow SO_4^- + H_2O$

$OH + SO_3^{2-} \rightarrow SO_3^- + OH^-$ or $SO_4^{2-} + H$

$O_3 + HSO_3^- \rightarrow HSO_4^- + O_2$

$O_3 + SO_3^{2-} \rightarrow SO_4^{2-} + O_2$

$H_2O_2 + HSO_3^- \xrightarrow[2H_2O]{H_3O^+} H_2SO_4$

$SO_3^- + O_2 \rightarrow SO_5^-$

Halogen reactions

$OH + Cl^- \rightarrow ClOH^-$

$ClOH^- \rightarrow OH^- + Cl$

$OH + Br^- \rightarrow BrOH^-$

$BrOH^- \rightarrow OH + Br^-$

---

that of Graedel and Weschler (1981). The following is a brief summary of the major findings of this study.

Water serves as a solvent, as an acid and as a base. O_2 is among the primary oxidizing agents with a typical aqueous phase concentration of $\sim 3 \times 10^{-4}$ M. Hydrogen peroxide, with an aqueous phase concentration of $\sim 3 \times 10^{-6}$ M, has perhaps an even greater role in aerosol chemistry, since it serves as both an oxidant and as a source of hydroxyl radicals. Ozone dissolved in water is a source of HO_x radicals; this source is potentially significant in situations where H_2O_2 concentrations are low. The initial pH of the aqueous phase of the aerosol is determined by the CO_2–HCO_3^-–CO_3^{2-} system. The chemical reactions of these ions do not appear to be significant. Ammonia (and the ammonium ion) and nitric acid (and the nitrate ion) are the most important inorganic nitrogen compounds in atmospheric aerosols. Ammonia is the principal species that reacts with strong acids in the atmosphere, as evidenced by the large concentrations of ammonium salts found in aerosols. Although nitric acid can serve as an oxidizing agent, its chief action is to reduce the pH of the aqueous aerosol. The most common fate for the NO_x compounds (NO and NO_2) is conversion to either nitric acid or nitrous acid. Since nitrous acid (or nitrite ion) will eventually be converted to nitrate ion, the net effect is for the majority of the NO_x molecules in atmospheric aerosols to be oxidized to nitrate. Sulphate is the dominant sulphur-containing species in atmospheric aerosols. The high aqueous solubility of the submicron fraction of atmospheric aerosols (often greater than 50 % by weight) reflects the high sulphate content. It also follows that the reduced pH of the submicron fraction is due primarily to sulphuric acid and acid ammonium sulphate. Sulphur dioxide has the highest aqueous phase concentration of the sulphur-containing gases. It is the most important reducing agent in atmospheric aerosols, and its own eventual fate is oxidation to sulphate. The aqueous phase concentrations of hydrogen sulphide, carbonyl sulphide and carbon disulphide are several orders of magnitude less than that of sulphur dioxide. Carbonyl sulphide hydrolyses to yield H_2S, carbon disulphide can yield OCS, H_2S or SO_2, and hydrogen sulphide is oxidized to sulphate. Hence, sulphate is the eventual product of reduced sulphur gases dissolved in the aqueous phase of the aerosol. Although chloride salts are major constituents of atmospheric aerosols, their role in aerosol chemistry appears to be disproportionately small. The conversion of chloride ions to chlorine atoms leads to a number of radical processes in solution, and acidification of chloride-containing aerosols can generate HCl.

4.4. Remote sensing of tropospheric species

A series of questions concerning the composition and chemistry of the global troposphere, including an assessment of the impact of anthropogenic activities on the composition and chemistry of the global troposphere, have been identified in a series of NASA workshops and conferences. Some of these scientific questions are summarized in table 4.8. *In situ* sampling methods for studying the problems of the global troposphere provide only spatially and temporally limited sets of data. In the long run, remote sensor measurements from orbiting spacecraft may be the only technique to answer the questions listed in table 4.8. At the present time NASA is developing a series of remote sensors to study the composition and chemistry of the global troposphere. Some of these remote sensors operate from the ground, others from aircraft, and one

Table 4.8. Summary of scientific questions related to tropospheric pollution.[a]

Question	Question
What are the relative roles of stratospheric–tropospheric exchange, photochemistry and surface sinks in the global tropospheric ozone budget?	What is the relative importance of homogeneous versus heterogeneous processes in the SO_2 gas-to-particle conversion cycle?
Will variations in global CO and NO_x levels shift the ozone balance?	What are the major sources of tropospheric OCS? And, if man's activities are a major source, what might be the climatic implications of a COS build-up in the atmosphere?
What is the detailed degradation chain for methane and what are its consequences on ozone and HO_x?	
What is the role of non-methane hydrocarbons in the ozone budget?	What is the reliability of present photochemical theory in predicting H_xO_y distributions in both urban and remote atmospheric air parcels?
What are the detailed exchange mechanisms between the troposphere and stratosphere?	Can the asymmetry in global CO levels lead to asymmetry in global OH distributions?
	What is the lifetime of H_2O_2 in the atmosphere, and what are the major sinks for this compound?

Table 4.8 continued.

Question	Question

What are the tropospheric concentrations of the active nitrogen compounds?

What are the natural sources of active nitrogen?

What are the major sinks of atmospheric NO_x and total active nitrogen?

Can increased levels of NO_x influence the tropospheric ozone balance?

What are the global concentrations, sources and sinks of ammonia?

What are the natural sources and sinks of N_2O and what are the implications of increased use of nitrogenous fertilizers on the tropospheric N_2O budget?

Are higher carbon monoxide levels in the Northern Hemisphere compared with the Southern Hemisphere due to natural or anthropogenic processes and will continued CO emissions perturb atmospheric OH levels, especially the relative abundance in the two hemispheres?

Do soils always act as a sink of atmospheric carbon monoxide?

What is the role of non-methane hydrocarbons in the CO budget?

What are the exchange mechanisms between the two hemispheres, and how important are they in controlling the CO abundance?

Is the deforestation and wood-burning hypothesis plausible for the increase of CO_2 global concentration?

What are the climatic implications of changes in the atmospheric composition of trace gases, such as CO_2, CH_4, N_2O, fluorocarbons, and aerosols?

What is the yearly quantity of sulphur now being released from the natural biosphere, and what are the major chemical forms in which this sulphur is being released?

What is the global distribution of SO_2?

What are the short-term and long-term impacts of anthropogenic emissions of SO_2 on the tropospheric sulphur budget and on the budgets of other trace gases such as NH_3?

What importance (in terms of a global average concentration value for OH) can be assigned to the inter-hemispheric gradient in methyl chloroform?

What are the major sources of atmospheric HCl and how important is this in determining rain acidity in both remote and urban areas?

What are the stratospheric implications of anthropogenic releases of halocarbons (such as CH_3CCl_3, $CHCl_2F$ and $CHClF_2$) in terms of their OH-controlled chemical lifetimes in the troposphere?

What are the optical properties of tropospheric aerosols?

What are the physical and chemical properties of tropospheric aerosols?

How do aerosols influence the tropospheric radiation balance?

What is the origin and fate of large-scale tropospheric haze?

What are the meteorological processes governing the long-range transport of pollutants?

What is the relative importance of the various removal processes on the long-range transport of pollutants?

What chemical processes influence O_3 formation in point-source and urban plumes?

What chemical processes govern the conversion of SO_2 to sulphates in point-source and urban plumes?

What chemical processes govern the conversion of NO_x to nitrates in point-source and urban plumes?

What are the contributions of photochemistry involving man-made and natural hydrocarbons, long-range transport of O_3 and precursors, and stratospheric intrusions in producing elevated O_3 concentrations in rural areas?

What is the source and composition of acid rain?

*a*From Seinfeld, J.H., Allario. F., Bandeen, W.R., Chameides, W. L., Davis, D.D., Hinkley, E.D. and Stewart, R.W., 1981, *Report of the NASA Working Group on Tropospheric Program Planning.* NASA Reference Publication RP–1062, National Technical Information Service. Springfield, Va. 22161, 161 pp.

J. S. Levine

Table 4.9. Summary of Earth observation satellite systems (Levine and Allario, 1982).

Category and instruments	Spectral coverage (μm)		Spatial resolution (km)	Satellite system
VISSR	IR images	(10·5–12·6)	8	SMS-1/2
(Visible Spin Scan Radiometer)	VIS images	(0·55–0·75)	4	GOES-1/2/3
	VIS images	(0·55–0·75)	2	
	VIS images	(0·55–0·75)	1 and 2	
VAS	IR and VIS imaging			GOES-4/5/6
(VISSR Atmospheric Sounder)	Identical to VISSR with			
	additional sounding channels at			
	IR	(3·73–8·0)	16	
	IR	(8·0–14·7)		
AVHRR	VIS images	(0·55–0·70)	1·1	TIROS-N
(Advanced Very High	Near IR images	(0·735–0·90)	1·1	NOAA-6/7/8
Resolution Radiometer)	IR images	(3·55–3·93)	1·1	
	IR images	(10·5–11·5)	1·1	
MSS	VIS	(0·5–0·6)		LANDSAT-1/2
(Multi-Spectral Scanner)	IR	(0·6–0·7)		
RBV	IR	(0·7–0·8)	0·08	
(Return Beam Vidicon Camera)	IR	(0·8–1·1)		
	Blue-green	(0·475–0·575)		
	Orange-red	(0·580–0·680)		
	Near IR	(0·690–0·830)		
MSS	Same as LANDSAT-1 and -2		0·08	LANDSAT-3
(5-Band Multi-Spectral Scanner)	with additional MS channel			
RBV	IR	(10·4–12·6)	0·237	
(Return Beam Vidicon Camera)				
Thematic mapper	VIS	(0·45–0·52)	0·03	LANDSAT-4
	VIS	(0·52–0·60)	0·03	
	VIS	(0·63–0·69)	0·03	
	IR	(0·79–0·90)	0·03	
	IR	(1·55–1·75)	0·03	
	IR	(10·4–12·5)	0·120	

sensor has already obtained measurements of a tropospheric species from space aboard the Space Shuttle. The current ground-based and aircraft remote sensors may also one day obtain global measurements of tropospheric species from orbit. The development and application of remote sensor technology within NASA has been reviewed by Levine and Allario (1982). Three classes of remote sensors have demonstrated unique capabilities in meeting some of the measurement needs identified in table 4.8. The first class includes imaging spectroradiometers currently utilized in meteorological and Earth resource measurements, the so-called Earth observation satellite systems. These sensors have demonstrated the ability to detect and track regions of elevated haze layers and aerosol loadings

Orbit	Launch schedule			Application/objective
Geosynchronous	SMS-1 5/74	SMS-2 2/75		Cloud images, sea surface temperature, high-resolution images
	GOES-1 10/75	GOES-2 6/77	GOES-3 6/78	
Geosynchronous	4 9/80	5 5/81	6 4/83	Atmospheric sounding, sea surface temperature, cloud-top temperatures, meteorological high resolution, Earth images
Sun-synchronous	TIROS-N 10/78 NOAA-6 6/79	NOAA-7 6/81	NOAA-8 3/83	Cloud images, sea surface temperature, high-resolution images
Sun-synchronous	1 7/72 (ERTS-1)	2 1/75		Solar-reflected radiation, land use, water resources, agricultural survey
Sun-synchronous	3/78			Solar-reflected radiation, land use, water resources, agricultural survey
Sun-synchronous	7/82			Solar-reflected radiation, land use, water resources, agricultural survey

in the troposphere. Earth observation satellite systems are summarized in table 4.9. A second class of instruments includes passive remote sensors which measure spectral emission or absorption of atmospheric molecules using external sources of radiation, i.e., direct incoming solar radiation, reflected solar radiation or Earth-emitted thermal infra-red radiation. Vertical distributions of molecular species, pressure and temperature throughout the troposphere can be inferred using inversion algorithms. Passive remote sensors developed or under development at NASA are summarized in table 4.10. A third class of instruments includes active remote sensors in which lasers in the ultraviolet, visible and infra-red portions of the spectrum are used in a similar mode as an active radar

Table 4.10. Summary of passive remote sensors (NASA) (Levine and Allario, 1982).

Category and instruments	Species	Spectral coverage (μm)	Field of view	Vertical resolution (km)	Viewing mode	Platform	Current status
Gas Filter Correlation (GFC)							
A/C MAPS	CO, CH_4; extendable to other gases	4·52-4·8	7·5°	<5	Thermal IR	A/C	A/C operational
OFT MAPS (Shuttle)	CO; extendable to other gases	4·52-4·8	4·5°	<5	Thermal IR	Shuttle	CY81 flight test
PMR	CH_4, CO, N_2O, NH_3	4·52-4·8 8·00-11·0	4·5°	<5	Thermal IR	A/C	Proposed engineering model development
DCR	CH_4, NH_3, CO, N_2O, HCl, NO_2, SO_2 (NMHC)	2·0-9·0	2·7°	Total burden	Solar-reflected	A/C	A/C operational
Interferometry							
COPE	CO, CH_4	2·35	7°	Total burden	Solar-reflected	A/C	A/C operational
CIMATS	NH_3, H_2O, N_2O, CO, CH_4, CO, SO_2	2·0-2·4 4·0-9·0	2-7°	<5	Solar-reflected Thermal IR	A/C	Engineering model developed
HSI	Spectrally scanning	2·0-5·0	7°	Stratospheric profiles	Direct solar	Balloon ground-based	Balloon/ground-based operational
ATMOS	Spectrally scanning	2·0-15·0	2°	Stratospheric profiles	Direct solar	Spacelab	Scheduled for Spacelab flight
Infra-red heterodyne radiometry							
IHR	O_3, NH_3	9-12 (discrete steps with CO_2 laser)	0·25mr	<5	Direct solar	A/C	A/C, Ground-based operational
LHS	Spectral coverage (continuous)	7·5-13	0·25mr	<5	Direct solar	Ground-based	A/C engineering model
Spectroradiometers							
SBUV-TOMS	O_3; 12 selected channels	(0·25-0·34)	3°	Total burden	Solar-reflected (UV)	Nimbus-3	Satellite

system. Through a combination of scattering by aerosols and molecules and selective absorption by molecules, these sensors can provide range-resolved measurements of molecules and aerosol loading in the troposphere. Active remote sensors developed or under development at NASA are summarized in table 4.11. The remote sensor instrumentation and technology listed in tables 4.9–4.11 are discussed in more detail in Levine and Allario (1982).

In general, remote sensors for atmospheric applications can be developed to measure changes in the three basic properties of electro-magnetic waves including energy (absorption or emission), wavelength (frequency shifts) and polarization. Passive spectroscopic remote sensors correspond to the class of sensors which measure one of these properties (absorption or emission) using external sources of radiation. For tropospheric measurements from space or aircraft observing platforms, the downward viewing mode (side-scanning from nadir) provides the widest vertical and horizontal coverage. Solar occultation† measurements from space and airborne platforms have been used extensively in stratospheric applications, but due to the extent of global cloud cover, geographical coverage of the troposphere is severely limited in this opera-tional mode. External sources of radiation for nadir viewing experiments include upwelling thermal radiance of the Earth–atmosphere system, reflected radiation from the surface of the Earth and scattered radiation from molecules and aerosols in the atmosphere – see Chapters 2 and 3. Compared to detection of direct solar radiation through the atmosphere, as in stratospheric solar occultation measurements, these sources of radiation are relatively weak and require more sensitive detection instruments.

Before discussing available passive remote sensing techniques, some basic properties of atmospheric radiation (see Chapter 2) for the various external sources of radiation should be recalled. In viewing the upwelling thermal radiance of the Earth–atmosphere system, passive instruments detect radiation which is a composite of energy transmitted through the atmosphere and that absorbed and re-emitted by atmospheric molecules at all altitudes between the source and the sensor. The thermal emission of this radiation is primarily governed by the temperature of the lower atmosphere, which ranges from approximately $300\,K$ near the ground to $220\,K$ in the lower stratosphere. Since the Planck function of a blackbody radiator peaks at approximately $10\,\mu m$ for a $300\,K$ blackbody, the desired wavelength region varies from $4\cdot5$ to $15\cdot0\,\mu m$. Also, in this wavelength range the contribution to atmospheric radiance from scattered solar

---

†Observing a bright object (a star or the Sun) as it is eclipsed by a dark object (in this case the Earth) permits derivation of information about the atmosphere of the Earth, since the radiation source is viewed through the limb of the Earth's atmosphere.

J. S. Levine

Table 4.11. Summary of active remote sensors (NASA) (Levine and Allario, 1982).

Category and instrument	Species	Spectral range (μm)	Transmitting laser
Fixed wavelength Large telescope lidar system	Stratospheric aerosols	0·693	Ruby
A/C stratospheric aerosol lidar system	Stratospheric aerosols	0·693 1·06	Ruby Nd: YAG
Ground-based ruby lidar system	Tropospheric aerosols	0·693 1·06	Ruby Nd: YAG
High spectral resolution lidar (HSRL)	Aerosol extinction	0·4–0·45	N_2 pumped dye CuCl laser
Laser absorption spectrometer (LAS)	O_3, NH_3	9–12 (discretely tunable)	CO_2 waveguide laser
CO_2 lidar (pulsed)	Aerosols, O_3	9–12 (discretely tunable)	CO_2 TEA laser
Tunable wavelength UV DIAL	SO_2, O_3, NO_2, aerosols	0·28–0·31	Nd:YAG pumped dye
Near IR DIAL	H_2O, pressure, temperature profiles	0·70–0·75	Ruby pumped dye
A/C DIAL system	SO_2, O_3, NO_2, H_2O, aerosols, temperature, pressure	0·28–0·9	Nd: YAG pumped dye
Mid-IR DIAL	CO, HC, CH_4, CO_2, H_2O, N_2O, NH_3, H_2S (NMHC)	1·4–3·7	Optical parametric oscillator
Shuttle lidar	Multiple species, clouds, aerosols, met. parameters	0·30–13·0	Evolutionary lidar system
CO_2 lidar (pulse)	Not extendable to other gases	9–12 (discretely tunable) Doubled frequencies	CO_2 TEA laser

radiation is small and can be considered a second-order contribution. Because of the pressure broadening of the spectroscopic absorption or emission lines, the energy received in the wings of the atmospheric emission lines reflects the presence of molecules at high pressures (i.e. lower altitudes), while energy received near line centres is representative of molecules at lower pressures (i.e., higher radiation altitudes). This gives rise to the possibility that by measuring radiation selectively at

Receiver telescope diameter (cm)	Viewing mode	Platform	Current status	Field applications
122	Atmos. backscatter	Mobile, ground	Operational	Ground-truth for sateilites (SAM-II, SAGE), volcanic eruptions
36	Atmos. backscatter	A/C	Operational	Ground-truth for satellites (SAM-II, SAGE)
31	Atmos.backscatter	Mobile, ground	Operational	Power plant plume studies
36	Atmos. backscatter	A/C	Operational	Regional aerosol extinction profiles
15	Reflected ground (molecular absorption)	A/C	Operational	Total burden
30	Atmos. backscatter	Ground	Operational	Tropospheric aerosol/O_3
31	Differential absorption from atmos. backscatter	Mobile, ground	Operational	Regional/urban plume studies
51	Differential absorption from atmos. backscatter	Ground/vertical profile	Operational	Tropospheric meteorological parameters
36	Differential absorption from atmos. backscatter	A/C	Operational	Regional study of O_3, SO_2, aerosols, tropical H_2O profiles
36	DIAL. Ground reflection	A/C	Under development	Regional HC and tropospheric species survey
125	Atmos. backscatter fluorescence DIAL	Spacelab	Study complete	Global tropospheric stratosphere, mesosphere, thermosphere
	Atmos. backscatter and retros.	Ground	Operational	Elevated pollution levels

various spectral regions from line centre, one could in principle generate vertical profiles of gas concentrations. This is analogous to the inference of temperature profiles by remote sensing methods using thermal infrared wavelengths, where wavelengths in various parts of the emission line of a uniformly mixed gas (e.g., carbon dioxide and nitrous oxide) are used to obtain altitude discrimination (see Chapter 5). In order to invert radiances detected at the top of the atmosphere to concentration profiles, inversion algorithms must be developed which take into account emission by the Earth–atmosphere system, governed by the temperature lapse rate. In inverting the radiances to obtain useful concentration accuracies, temperature profiles must be simultaneously measured with the radiance to a desired accuracy of approximately $\pm 2\,K$. For measurements of minor trace gases in the troposphere, thermal emissions are weak due to a combination of low concentrations and low temperatures. Also, the temperature lapse rate and the relatively low pressure gradient in the atmosphere limit the degree of vertical discrimination possible, independently of the spectral resolving power of the instrument beyond a resolution of approximately $0\cdot01\,cm^{-1}$. Furthermore, radiances emitted and absorbed near the ground (within the first 3 km) are faced with the limitation of a small temperature gradient between the Earth and the layer of atmosphere near the Earth, making fine discrimination of layers near the ground difficult.

In the free troposphere, however, distinct layers of the atmosphere can be identified when simulating the inversion of the Earth–atmosphere radiance in the thermal infra-red and weighted averages of gas concentrations at specific altitudes can be obtained. Since this method of establishing concentrations of gaseous species from remote platforms provides both global coverage and synoptic information, it can be used in studies of (*a*) global budgets and distributions of gas concentrations; and (*b*) the relative changes which occur over time of some of the well-mixed gases such as methane, nitrous oxide and carbon dioxide, especially in the free troposphere.

The other two external sources of radiation which can be used in the tropospheric sounding of atmospheric trace gases are reflected radiation from the Earth's surface and scattered radiation from the Earth's atmosphere. In the latter mode, scattered radiation from aerosols and molecules is predominant in the ultraviolet and visible portions of the spectrum and can be used to infer integrated molecular concentrations by interpretation of atmospheric absorption spectra. In the ultraviolet and visible portions of the spectrum, the dependence of line width on pressure is smaller than in the infra-red, and vertical layering of the atmosphere using the physical process is more difficult to infer than in the thermal infra-red. However, inversion of radiance data in the ultraviolet and visible is less sensitive than the thermal infra-red to the knowledge of atmospheric temperature profiles. Relevant tropospheric molecules in this spectral region included ozone, sulphur dioxide and nitrogen dioxide.

In viewing reflected solar radiation from the Earth's surface, one should restrict observations approximately to the $1\cdot0$–$3\cdot5\,\mu m$ region, since in this range the radiance at the top of the atmosphere is primarily due to reflected solar radiation and molecular absorption can be used as the physical process to infer molecular concentrations. Some information on vertical layering of molecular concentrations can be achieved through interpretation of the effects of pressure broadening of the molecular absorption lines in the atmosphere, and integrated measurements to the ground are possible. The accuracy of the retrieved concentrations has a smaller functional dependence on the temperature profile than in the upwelling thermal infra-red region. In the intermediate spectral band ($3 \cdot 5 - 4 \cdot 5 \mu$m) the radiance at the top of the atmosphere is composed of comparable values of the upwelling thermal radiance and reflected solar radiation—see figure 1.6, p. 7. Although interesting absorption and emission lines of major atmospheric molecules lie in this region, inversion of the radiance measurements to molecular concentrations is complicated by the complexity of the radiative transfer equation, which makes it difficult to invert radiances quantitatively to derive molecular concentrations.

The use of active remote sensors, i.e., laser techniques for measuring range-resolved concentrations of aerosols, molecular constituents and meteorological parameters in the stratosphere and troposphere has represented a major research activity over the last decade. In considering some of the fundamental limitations in using external sources of radiation for measuring tropospheric molecules, following the discussion in Chapter 2 it will be clear that the use of powerful and tunable, monochromatic sources in the ultraviolet, visible and infra-red portions of the spectrum has the potential to remove some of the limitations of passive remote sensing imposed by the physical processes of the atmosphere. For example, in probing the troposphere from the top of the atmosphere in the nadir mode, the laser beam has the potential to probe down to the surface, and, through the process of molecular absorption and range gating the vertical distribution of molecular concentrations, aerosols and meteorological parameters can in principle be obtained to a spatial resolution $\leq 0 \cdot 1$ km. The ability to achieve such accuracy is dependent on the energy of the laser and the magnitude of the differential absorption cross-section of the species to be measured.

In general, the ability of lidar measurements to obtain individual measurement parameters to a given accuracy depends on the magnitude of optical scattering coefficients and molecular absorption cross-sections, given sufficient flexibility in systems parameters such as telescope size, detector quantum efficiency and energy of the laser transmitter. For example, the Differential Absorption Lidar (DIAL) instrument uses two laser pulses which are backscattered to the receiver telescope by atmospheric aerosols and molecules, so that the return signal represents a time history of the scattering and absorption properties of the atmosphere for each laser pulse. This time history is related to a spatial profile of these scattering and absorption properties of the atmosphere, through the equation $(c\Delta t)/2 = \Delta R$, where Δt represents a time gating interval which can be selected in time to correspond to a range gate interval

ΔR. In the data processing mode, the ratio of return signals at the 'on' wavelength, P_{on}, to the 'off' wavelength, P_{off}, are measured and, through the lidar equation, can be related to the molecular concentration of the gas in the atmosphere as a function of range from the transmitting telescope. Another useful mode for a DIAL experiment is to employ a continuous wave (CW) laser which uses reflection from the ground to return the two laser wavelengths back to the receiving telescope. In the infra-red, where pressure broadening of atmospheric spectral lines as a function of altitude is large relative to the ultraviolet and visible portion of the spectrum, it is possible to probe selectively various segments of the absorption line by tuning the wavelength of the transmitting laser from the central peak into the wings to obtain information on vertical layering of the selected tropospheric molecule.

The following discussion will concentrate on three remote sensors listed in tables 4.10–4.11. All three instruments were developed at the NASA Langley Research Center, Hampton, Virginia, as part of NASA's Office of Space Science and Applications' Tropospheric Air Quality Program. All three instruments are operational and have been used to obtain measurements of tropospheric species from different platforms—from the ground, from an aircraft and from space.

There are primarily three types of radiation sources which can be used in measurements designed to determine tropospheric composition and chemistry. These are the Sun, the Earth and laser sources on board the platform itself. The latter two of these sources have already been used successfully on remote platforms, i.e., MAPS and DIAL. To date, however, remotely derived information about tropospheric species using the Sun as the energy source has only been possible using ground-based instrumentation, i.e., IHR. But as technology advances there is no reason why incident and reflected solar radiation should not be monitored from space in order to retrieve similar profiles from a satellite platform. This technology is progressing.

Each of the three instruments described below is based on different remote sensing principles and, hence, is representative of a whole class of remote sensors. The IHR (Infra-red Heterodyne Radiometer) is a passive ground-based instrument, using the Sun as a radiation source, routinely used for measurements of the vertical distribution of ammonia in the troposphere and lower stratosphere (see entry 9 in table 4.10). The airborne DIAL system is an active instrument used for measurements of tropospheric ozone and aerosols (entry 9 in table 4.11). MAPS (Measurement of Air Pollution from Satellites) is a passive gas-filter correlation instrument that uses Earth-emitted infra-red radiation as a source. This instrument flew aboard the second orbital flight test (OFT-2) of the Space Shuttle and obtained near-global measurements of tropospheric carbon monoxide (see entry 2 in table 4.10).

The Infra-red Heterodyne Radiometer

The IHR system is routinely used in a tropospheric measurement programme. The following discussion is from the instrument description given by Hoell *et al.* (1982). The high spectral resolution available from optical heterodyne instruments is typically achieved through the use of one or more narrow bandpass intermediate-frequency (IF) filters. These filters are generally selected to coincide with particular portions of the infra-red spectrum that have been shifted by the heterodyne process into the microwave frequency region. This is illustrated in figure 4.5(*a*), where a synthetic atmospheric spectrum of a selected absorption line of ammonia is shown along with the relative position of the laser local oscillator (LO) and six IF channels. Note that because of the image property of heterodyne radiometers, the output from each IF channel is the sum of the radiation over the spectral regions $(\bar{\nu}_{LO} - \nu_{IF})/c \pm \beta_{IF}/2c$ and $(\bar{\nu}_{LO} + \nu_{IF})/c \pm \beta_{IF}/2c$, where $\bar{\nu}_{LO}$ is the LO wave number; ν_{IF} the centre frequency of the IF filter, and β_{IF} the IF filter bandpass. Thus, analysis of data from a heterodyne radiometer must account for radiation from both of these sidebands.

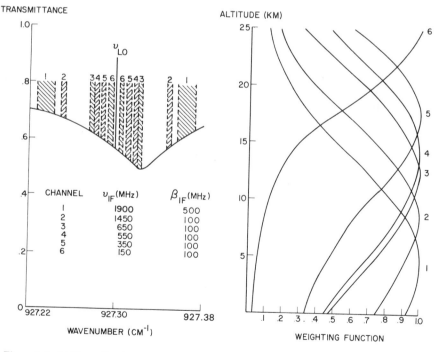

Figure 4.5. (a)Channelization of the $927\cdot32323\,cm^{-1}$ ammonia absorption line; and (b) weighting functions corresponding to each IF channel (Hoell et al., *1982).*

Ammonia profiles are obtained from solar absorption measurements at the six IF channels shown in figure 4.5(a) along with a seventh channel (described below) which provides a normalizing reference signal. The average solar radiance I_j detected in the jth IF channel can be expressed as

$$I_j = I_j^0 \tau_{jc} \tau_{jg} \qquad (4.56)$$

where I_j^0 is the unattenuated solar radiance; τ_{jc} the atmospheric continuum transmittance related to molecular (i.e., water vapour) and aerosol absorption and scattering; and τ_{jg} the non-continuum molecular transmittance. Note that each parameter identified in equation 4.56 is considered to be averaged over the upper and lower sideband of the given channel. Normalization of the solar radiance detected in each IF channel by that observed in the reference channel is required to minimize errors associated with the continuum interference effects τ_{jc} and the fluctuations in the source intensity I_j^0. The desired quantity is the atmospheric transmission ratio, given by

$$R_j = \tau_{jg}/\tau_g^{\text{ref}} \qquad (4.57)$$

where τ_g^{ref} is the atmospheric transmissivity due to molecular line absorption at the reference channel wavelength. Equation 4.57 is only obtained from equation 4.56 if the source radiation and atmospheric continuum transmittance are equal at the wavelengths for the reference and jth signal channel.

The IHR provides for a reference and the six signal channels through the use of two $^{13}C^{16}O_2$ laser LOs. The reference and signal LO are the $R(8)$ transition at $920 \cdot 2194 \text{cm}^{-1}$ and $R(18)$ transition at $927 \cdot 3004 \text{cm}^{-1}$ respectively. Effects on the measured R_j due to the more abundant isotope of CO_2 are minimized by using $^{13}C^{16}O_2$ LOs; and for these LO transitions, the contribution to R_j due to the combined effects of water vapour continuum, solar radiation and atmospheric $^{13}C^{16}O_2$ is negligible. Moreover, analysis of lower resolution atmospheric absorption spectra and synthetic spectra calculated using available molecular line parameters indicates that contributions to equation 4.57 from other interfering species are also negligible.

Vertical profiles of ammonia are inferred from a set of R_js measured simultaneously at different positions on an absorption feature. Collisional broadening of the absorption feature provides the mechanism by which the spectrally resolved transmission ratios are coupled to an altitude region. The altitude dependence of the absorption coefficient for each channel serves as the weighting function and a plot of altitude vs. the normalized absorption coefficient for a given IF channel (figure 4.5(b))

indicates the vertical sensitivity that can be obtained. The upper altitude limit available from this technique is governed by the decrease in the collisional broadening effects as the pressure decreases with altitude. For ammonia, the upper limit is around 30 km, where the collisional and Doppler broadened half-widths are approximately equal.

The density profile $\varrho(z)$ is inferred from R_j through the use of an iterative inversion technique, with successive approximations to $\varrho(z)$ at altitude z given by

$$\varrho^{k+1}(z) = \varrho^k(z)\frac{\sum_j K_j^*(z)\ln R_j/\ln R_j^k}{\sum_j K_j^*(z)} \quad (4.58)$$

where K_j^* is the normalized absorption coefficient for ammonia averaged over channel j; and R_j and R_j^k are the measured and the kth approximation to the normalized transmission ratio respectively. The inversion process starts with an initial ammonia profile $\varrho^0(z)$. The transmission ratio R_j^k is calculated and compared with the measured R_j. If the rms difference between R_j^k and R_j is greater than the system noise, a new approximation to $\varrho(z)$ is calculated using equation 4.58.

For each iteration of $\varrho^k(z)$, the calculated R_j^k is obtained using a line-by-line computer model for all transmittance and absorption calculations. This model considers both Lorentz and Voigt line shapes and includes temperature and pressure effects on spectral line strengths and half-widths. The data for the observed ammonia line are given as: line centre $= 927 \cdot 32323 \, \text{cm}^{-1}$; line strength $= 4 \cdot 09 \times 10^{-19} \, \text{cm}^{-1}\text{molecule}^{-1}\text{cm}^{-2}$; foreign gas broadened half-width $= 0 \cdot 08 \, \text{cm}^{-1}\text{atm}^{-1}$. The atmospheric model in the inversion algorithm includes H_2O, CO_2 and NH_3 profiles along with a pressure and temperature profile. The H_2O, pressure and temperature profiles were obtained from the US Standard Atmosphere Supplements, for 30°N latitude and the appropriate climatological period.

The IHR is shown schematically in figure 4.6. The optical package consists of two Dicke-switched radiometer sections which share a common input lens, reference blackbody source and calibration source. The solar radiance is collected using a 5 cm (actually 2 in) diameter ($F \# 6 \cdot 5$) zinc selenide lens which focusses the incoming radiance on to two high-speed photovoltaic HgCdTe photomixers. These two photomixers and their corresponding CO_2 laser LOs provide the reference and ammonia absorption channels. A long-pass optical filter, which cuts in at 8 μm, limits the amount of solar radiance reaching the photomixers. The infrared Dicke switch alternately switches the field of view of each photomixer between the solar radiance and the reference blackbody. Approximately

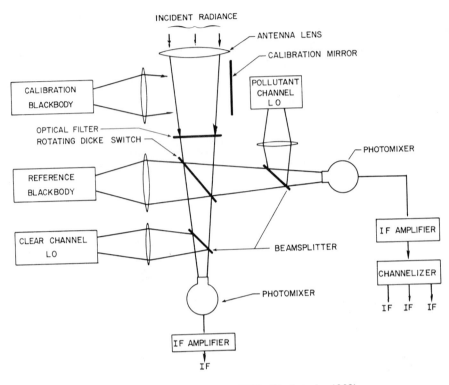

Figure 4.6. The Infra-red Heterodyne Radiometer (IHR) (Hoell et al., 1982).

10 % of the radiation from each LO is directed on to its respective photomixer. During alternate half-cycles of the Dicke switch, radiation from the Sun or reference blackbody is combined with the LO radiation via the beamsplitter. The calibration mirror is used to insert a variable 300–1300 K calibration blackbody into the optical path.

The measurement programme began at Langley Research Center during the latter part of March 1979. The initial measurements, performed with the IHR, were supplemented with ground-level *in situ* measurements beginning in August 1979. The IHR was located in a laboratory area with a 20 cm heliostat on the roof to track and direct solar radiation into the laboratory. Throughout the year the solar viewing angle was generally limited to about 80° from the zenith for both sunrise and sunset. Results from the IHR illustrating the seasonal variability in atmospheric ammonia observed from March 1979 to April 1980 are shown in figure 4.7. The shaded areas contain approximately 90 % of the profiles that were obtained during the indicated time period. The sensitivity of the IHR is about 0·1 ppbv of ammonia. Each profile

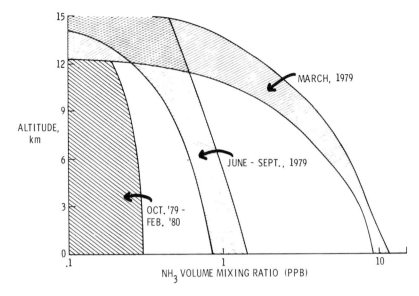

Figure 4.7. The range of the seasonal variation of the vertical distribution of atmospheric ammonia. The sensitivity of the instrument is about 0·1ppb (Hoell et al., 1982).

contained within the shaded area was inferred from four sets of the transmission ratios R_j measured over approximately 1h. For each set of data, two initial profiles—one a constant 25ppbv and the other a constant 0·3ppbv—were used to initiate the inversion algorithm. In general, the difference between the profiles resulting from each initial guess was less than 0·05ppbv at all altitudes. For each inversion profile, the rms residual between the calculated and measured transmission ratios was less than 0·01.

The airborne Differential Absorption Lidar (DIAL) system

The airborne lidar system uses the Differential Absorption Lidar (DIAL) technique for the remote measurement of gas profiles. The following discussion is from the instrument description given in Browell *et al.* (1983). The DIAL technique determines the average gas concentration over some selected range interval by analysing the difference in lidar backscatter signals for laser wavelengths tuned 'on' and 'off' a molecular absorption peak of the gas under investigation. The value of the average gas concentration N between range R_1 and R_2 can be determined from the ratio of the lidar signals at the on and off wavelengths. This relationship is given by

$$N = \frac{1}{2(R_2 - R_1)(\sigma_{on} - \sigma_{off})} \ln \frac{P_{off}(R_2) P_{on}(R_1)}{P_{off}(R_1) P_{on}(R_2)} \qquad (4.59$$

where $\sigma_{on} - \sigma_{off}$ is the difference between the absorption cross-sections a the on and off wavelengths; and $P_{on}(R)$ and $P_{off}(R)$ are the signal power received from range R at the on and off wavelengths respectively. Th analysis assumes that the aerosol and molecular optical properties ar equal at the on and off DIAL wavelengths. If there is an interfering ga which does not have the same absorption coefficient at these wavelengths the concentration of this gas must be known or determined by a separat measurement.

A block diagram of the DIAL system is shown in figure 4.8. Tw frequency doubled Nd: YAG lasers are used to pump two high conversio efficiency tunable dye lasers. All four lasers are mounted on a rigi support structure which also contains the transmitting and receivin optics. The dye laser on and off wavelengths that are used in the DIA measurement are produced in sequential laser pulses with a time separa tion of 100 μs or less. This close spacing ensures that the same atmospheri scattering volume is sampled at both wavelengths during the DIA measurement. The output beams are separated and steered usin

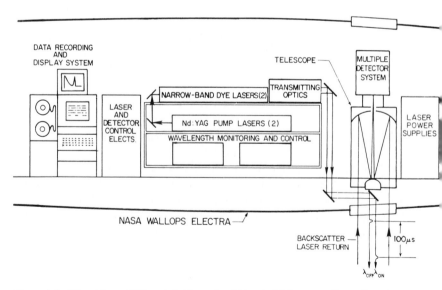

Figure 4.8. The airborne Differential Absorption Lidar (DIAL) system (Browell et al., 1983).

dielectric coated optics. They are transmitted out of the aircraft through a 40 cm diameter quartz window, coaxially with the receiver telescope. The wavelength of the two dye lasers is determined using a one-metre monochromator and a spectral reference lamp. The monochromator output is displayed in real time by an optical multichannel analyser (OMA). Simultaneous operation with the spectral reference provides the laser wavelength to an accuracy less than ± 10 pm.† When more accurate wavelength control is needed, such as for H_2O DIAL measurements, it is accomplished using a closed-loop wavelength control system. This system uses a stepping motor to control the dye laser grating angle and it provides wavelength control to better than $\pm 0 \cdot 3$ pm.

The receiver system consists of a 35 cm diameter Cassegrain telescope with optics to direct the received signals on to the detectors, which are gateable photomultiplier tubes. As many as three photomultiplier tubes can be accommodated. Normally, only one or two are used. When the system is operating in the visible or near infra-red, only one tube is needed for the on and off lines, with the off line also providing an aerosol measurement. Frequency doubling crystals are used to double the visible radiation into the UV when making measurements in this spectral region, and the residual off-line visible wavelength is transmitted and used to measure atmospheric aerosols. Two photomultiplier tubes are used, one optimized for the UV wavelength region and the other optimized for the visible wavelength. Three 10-bit transient digitizers, operating at a 10 MHz conversion rate, digitize sequentially the on- and off-line DIAL and aerosol return signals. The data are then stored on a 1600 bpi high-speed magnetic tape unit by means of a PDP 11/34 minicomputer. Gas concentration profiles can be calculated in real time and displayed on a video system or hardcopy printer for real-time operator experiment control.

The multipurpose airborne DIAL system can operate from 280 to 1064 nm for measurements of ozone, sulphur dioxide, nitrogen dioxide, water vapour, temperature, pressure and aerosol backscattering. Only ozone measurements will be discussed here. Measurements of ozone obtained with the airborne DIAL system near Wallops Island, Virginia, on the morning of 29 May 1980 are shown in figure 4.9. The mixed layer had a height of about 500 m above mean sea level with an average ozone concentration of less than 115 ppbv. An ozone-enriched layer was found above the mixed layer which had concentrations exceeding 130 ppbv. This stable layer extended up to 1500 m with lower ozone values towards the top of the layer. The DIAL and *in situ* measurements (based on the

---

† pm = picometre = 10^{-12} m.

J. S. Levine

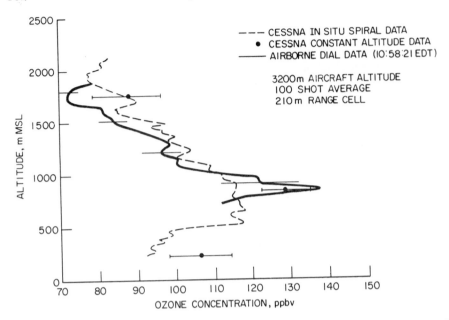

Figure 4.9. Ozone profile comparison of DIAL and in situ measurements on 29 May 1980 (Browell et al., 1983).

ozone–ethylene chemiluminescent technique) of ozone show agreement to within the uncertainties of the two techniques, which is about 10 %. The magnitude of the DIAL measurement uncertainty is due primarily to photon statistical errors which can be reduced by averaging over a larger horizontal or vertical extent. However, the natural spatial variability of ozone also contributes to the standard deviation of the average DIAL ozone profile. The spatial variability of ozone can be seen in figure 4.10, where a sequence of five DIAL profiles is shown over a horizontal distance of 10km. Very low values of ozone (25–30ppbv) persisted at 2700m, where the air was clean and dry. The ozone concentration increased to 60–65ppbv in the mixed layer, which increased in height from about 1800 to 2250m over the 10km measurement leg.

The airborne DIAL system has the flexibility to operate in the ultraviolet for measurements of ozone or sulphur dioxide, in the visible for nitrogen dioxide and in the near infra-red for water vapour, atmospheric temperature (using water vapour or oxygen absorption lines) and pressure (using oxygen lines). Aerosol backscatter investigations in the ultraviolet, visible and near infra-red can be conducted simultaneously with the DIAL measurements. The capabilities of the airborne DIAL system are functionally the same as those proposed for an early

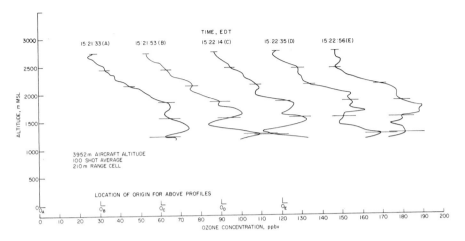

Figure 4.10. Sequential ozone profiles obtained by the airborne DIAL system on 5 June 1980. The zero ozone concentration level is shifted by 30ppbv for each subsequent ozone profile. The subscript identifies the appropriate profile. EDT is Eastern Daylight Time—1300 EDT is 1700 GMT) (Browell et al., 1983).

phase of the NASA Shuttle lidar program. The experience gained with the airborne DIAL system will be useful for a preliminary evaluation of several potential Shuttle lidar investigations.

4.5. Measurement of Air Pollution from Satellites (MAPS)

MAPS is a gas-filter correlation radiometer. The following discussion is from the instrument description given by Reichle and Hesketh (1976). Energy emitted by or reflected from the surface of the Earth propagates through the atmosphere and is gathered by the downward looking instrument. As the radiation passes through the atmosphere it undergoes selective absorption and re-emission that varies depending on the composition, temperature and pressure of the atmosphere. After entering the instrument, the radiation is directed through either of two cells, one containing the gas of interest and one being evacuated. The energy then passes to a detection system (figure 4.11). As the gas in one of the cells acts as a highly selective filter, there is generally a difference in the energy transmitted to the detection system through the two cells. This signal is related to the amount of gas with absorption features that correlate positively with those of the atmospheric gas of interest. In the MAPS

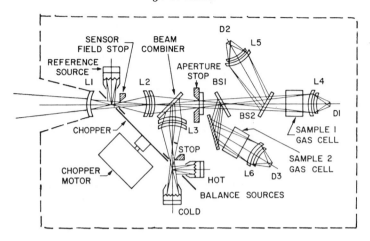

Figure 4.11. Schematic diagram of the MAPS instrument (Reichle and Hesketh, 1976).

instrument the energy is divided by a beam splitter and is directed simultaneously through the two cells to two different detectors. The outputs of these two detectors are then electrically differenced to determine the difference in the absorption of the two cells. A difference signal would exist even if no gas were present between the instrument and the energy source simply because the gas-filled cell removes energy from the beam while the evacuated cell does not, assuming that the optical properties of the elements in the two legs are identically matched. To reduce the sensitivity of the output to changes in the source temperature, the instrument is 'balanced'; that is, the gain (either optical or electrical) of one leg of the system is adjusted such that the output of the instrument does not vary as it views sequentially either of two blackbodies whose temperatures bracket the expected scene temperature. One of the crucial problems in operating the instrument is the requirement that the balance be established within the range from one part in 10^3 to one part in 10^4 of the total incoming radiance. The error effect is therefore reduced, provided that the temperatures of the balance blackbodies are well chosen and that the spectral characteristics of the optical elements are as wavelength independent as possible.

Scene energy enters the instrument through an aperture (left-hand side of figure 4.11) and is chopped against the reference blackbody. The energy passes through the field stop that sets the field of view at $4 \cdot 5°$. The scene energy then continues through lens L2, the beam combiner, the aperture stop and the bandpass filter. At the first beamsplitter BS1, part of the energy is diverted through gas cell two and field lens L6 to detector D3. The energy passing through BS1 is further divided at BS2, the reflected component passing to detector D2 through field lens L5 (this constitutes the vacuum path), while the transmitted component passes through a gas cell and a field lens to detector D1. All detectors are PbSe, thermoelectrically cooled to 195 K. Two gas cell paths were used because analytical studies had indicated that placing different CO partial pressures in the two cells would improve the data recovery. The signals from the two detectors D1 and D2 are electrically differenced as are the signals from detectors D2 and D3. This produces two separate outputs, called ΔV and $\Delta V'$. This instrument as described above comprises the actual data-collecting optical system.

Balancing of this instrument is achieved as follows. Two balance blackbodies are chopped (at a frequency equal to twice that of the scene energy) such that radiation from them passes through lens L3 to the beam combiner where it is introduced into the path traversed by the scene energy. The signals from these balance blackbodies are measured and the electrical gain of the gas cell leg is adjusted such that the difference between the balance signals from D1 and D2 and from D2 and D3 is zero. It can be shown that this achieves balance, as previously defined, as long as the frequency response of the system does not change.

The MAPS experiment flew as part of the OSTA-1 (NASA's Office of Space and Terrestrial Applications) payload on the second flight of the Space Shuttle (STS-2) during early November 1981. The experiment was designed to remotely measure the mixing ratio of carbon monoxide in the middle (centred around 8 km) and upper (centred around 12 km) troposphere. The instrument performed excellently and acquired approximately 32 h of data between 38°N and 38°S. The MAPS measurements of carbon monoxide represented the first measurement set of a tropospheric trace species obtained from an orbiting spacecraft.

In summary, water is an important species in the photochemistry/chemistry of the troposphere. Water vapour is the photochemical precursor of the hydroxyl radical which, as has been described in the early part of this chapter, is the key species involved in the photochemistry of the troposphere. Liquid water cloud droplets modify the transfer of solar radiation through the troposphere, and hence affect the photolysis rates of tropospheric species. Cloud and rain droplets and water-coated aerosols absorb soluble tropospheric species and serve as sites for active aqueous phase chemistry. These processes, which involve water in both gaseous and liquid states, control the photochemistry and composition of the troposphere. Remote sensor technology is just now beginning to offer the potential to measure and monitor tropospheric species to assess the biogeochemical and atmospheric processes that control the photochemistry and composition of the global troposphere.

References

Anderson, D. E. and Meier, R. E., 1979, Effects of anisotropic multiple scattering on solar radiation in the troposphere and stratosphere. *Applied Optics* **18**, 1955–60.

Augustsson, T. R. and Levine, J. S., 1982, The effects of isotropic multiple scattering and surface albedo on the photochemistry of the troposphere. *Atmospheric Environment* **16**, 1373–80.

Baulch, D. L., Cox, R. A., Crutzen, P. J., Hampson, R. F., Jr., Kerr, J. A., Troe, J. and Watson, R. T., 1982, Evaluated kinetic and photochemical data for atmospheric chemistry. Supplement 1. *Journal of Physical and Chemical Reference Data* **11**, 327–496.

Browell, E. V., Carter, A. F., Shipley, S. T., Allen, R. J., Butler, C. F., Mayo, M. N., Siviter, J. H., Jr. and Hall, W. M., 1983, The NASA multipurpose airborne DIAL system and measurements of ozone and aerosol profiles. *Applied Optics* **22**, 522–34.

Graedel, T. E. and Weschler, C. J., 1981, Chemistry within aqueous atmospheric aerosols and raindrops. *Reviews of Geophysics and Space Physics* **19**, 505–39.

Hoell, J. M., Jr., Levine, J. S., Augustsson, T. R. and Harward, C. N., 1982, Atmospheric ammonia: measurements and modeling. *AIAA Journal* **20**, 88–95.

Levine, J. S. and Allario, F., 1982, The global troposphere: biogeochemical cycles, chemistry, and remote sensing. *Environmental Monitoring and Assessment* **1**, 263–306.

Levine, J. S. and Graedel, T. E., 1981, Photochemistry in planetary atmospheres, *EOS, Transactions of the American Geophysical Union* **62**, 1177–81.

Levine, J. S., Hughes, R. E., Chameides, W. L. and Howell, W. E., 1979, N$_2$O and CO production by electric discharge: atmospheric implications. *Geophysical Research Letters* **6**, 557–59.

Logan, J. A., Prather, M. J., Wofsy, S. C. and McElroy, M. B., 1981, Tropospheric chemistry: a global perspective. *Journal of Geophysical Research* **86**, 7210–54.

Reichle, H. G., Jr. and Hesketh, W. D., 1976, A gas filter correlation instrument for atmospheric trace constituent monitoring. *Special Environmental Report No. 10, Air Pollution Measurement Techniques*, WMO No. 460.

5

Vertical temperature sounding of the atmosphere

Joel Susskind
Goddard Space Flight Center
Greenbelt, MD, USA

5.1. Introduction and basic principles

The concept of remote sensing of the atmospheric temperature structure
by monitoring the upwelling radiation in a number of spectral intervals
in the thermal infra-red (frequencies less than $3000\,\mathrm{cm}^{-1}$) and microwave
regions of the atmosphere was first introduced by Kaplan (1959). In these
spectral regions, scattering by the cloud-free atmosphere can be neglected
and the upwelling radiation leaving the atmosphere at frequency ν is
given primarily by the radiation emitted by the surface attenuated by the
atmosphere, plus the radiation emitted by the atmosphere attenuated by
the atmosphere above it, as described in Chapter 2. In addition, for a partially reflecting (non-black) surface, radiation reflected from the surface,
coming from both downwelling atmospheric and solar radiation, and attenuated by the atmosphere, represents a non-negligible contribution to
the signal. Specifically, the upwelling radiation at angle θ can be written

$$R_\nu(\theta) = \epsilon_\nu(\theta)B_\nu(T_s)\tau_\nu(P_s,\theta) + \int_{\ln P_s}^{\ln \overline{P}} B_\nu[T(P)]\frac{\partial \tau_\nu(P,\theta)}{\partial \ln P}\mathrm{d}\ln P + R_\nu'(\theta) \quad (5.1)$$

where $\epsilon_\nu(\theta)$ is the emissivity of the surface at frequency ν and angle of
observation θ; $B_\nu(T)$ is the Planck function for emitted radiance of a
blackbody at frequency ν, in cm^{-1}; and temperature T, in K, is given,
from equation 2.1b, p. 52 by

$$B_\nu(T) = 1\cdot191 \times 10^{-8}\nu^3[\exp(1.439\nu/T) - 1]^{-1} \quad (5.2)$$

$\tau_\nu(P,\theta)$ is the atmospheric transmittance from pressure P to the top of the

167

atmosphere, passing through the atmosphere at angle θ, given by

$$\tau_\nu(P,\theta) = \exp\left[- \int_0^P k_\nu(P)\sec\theta\,\mathrm{d}P \right] \qquad (5.3)$$

where $k_\nu(P)$ is the atmospheric absorption per unit pressure at frequenc
ν and at pressure P; and $R_\nu{}'(\theta)$ represents the contribution of reflecte
radiation. The atmospheric absorption is due primarily to absorptio
lines from CO_2, H_2O, O_3 and O_2, the strongest of which are in band
clustered about the band centres shown in table 5.1. Spectral region
called windows, in which the atmosphere is almost transparent, are als
indicated (cf., for instance, figure 6.1, p. 204) in the table, together wit
their primary absorption species.

Since the absorption coefficient $k_\nu(P)$ is non-negative, the transmittanc
function $\tau_\nu(P)$ starts at 1 at the top of the atmosphere and falls to a valu
$\tau_\nu(P_s)$ which is greater than or equal to zero. The derivative $\partial\tau_\nu/\partial\ln P$
which appears in equation 5.1, is called the weighting function and ha
the properties that it is approximately zero at the (top) part of the atmos
phere where $\tau\approx 1$, zero at the (bottom) part of the atmosphere wher
$\tau\approx 0$, and peaks in the middle, centred about a pressure of $\tau_\nu(P)\approx 1/e$ (se
Kaplan *et al.*, 1977 for a detailed discussion about the properties o
weighting functions). If it were possible to monitor outgoing radiation a
a number of frequencies in which the atmospheric transmittances $\tau_\nu(P$
reach unity at different levels in the atmosphere, with levels which ar
relatively profile independent, then the observed radiances would giv

Table 5.1. Atmospheric absorption characteristics.

Feature centre		Bandpass	
(cm^{-1})	(μm)	(cm^{-1})	Absorbing species
2·0	60[a]	±5[a]	O_2
4·0	120[a]	±2[a]	O_2
6·1	183[a]	±5[a]	H_2O
250	40	±250	H_2O
667	15·0	±80	CO_2
870	11·5	±110	Window (H_2O)
1040	9·6	±50	O_3
1200	8·3	±100	Window (H_2O)
1500	6·7	+·500 −200	H_2O
2325	4·3	+60 −140	CO_2, N_2O
2700	3·7	+300 −300	Window (N_2, H_2O)

[a]frequencies in GHz.

information about a number of integrals of the atmospheric temperature profile, each peaking at different levels in the atmosphere.

Analysis of these integrals to give estimates of the entire temperature profile is one of the major elements of the sounding problem. There is no unique solution to the problem without the addition of constraints on the solution (for a detailed discussion of the theory see Rodgers, 1976). Two major classes of methods to solve equation 5.1 exist: statistical methods, which rely primarily on regression relationships between observed radiances and temperature profiles as measured by colocated radiosondes; and physical methods, which try to find atmospheric and surface conditions which, when substituted in equation 5.1, match the observations to a desired accuracy. Clearly, in the latter case, the accuracy with which it is possible to compute expected radiances as a function of atmospheric conditions is also a key element in the retrieval of vertical temperature profiles. Perhaps the most significant element of all is accounting for the effects of clouds on the observations (cf. Chapter 1). Equation 5.1 can be interpreted as applying to either clear or cloudy skies, but generally τ is taken to be for the case of clear skies. Methods of determining atmospheric temperature profiles from the observations, as well as accounting for cloud effects and other factors affecting the observations, will be discussed in later sections.

In practice, one cannot measure monochromatic radiation leaving the atmosphere with any appreciable signal to noise. Instead, one measures outgoing radiation with a broad-band radiometer with channels having characteristic normalized response functions $f_i(\nu)$ where the radiance measured by channel i, R_i, is given by

$$R_i = \int_0^\infty f_i(\nu) R_\nu \, d\nu \qquad (5.4)$$

Provided $f_i(\nu)$ is non-negligible over only a moderate spectral interval, characterized by effective central frequency ν_i and band pass $\Delta\nu_i$, equations 5.4 and 5.1 can be combined to give

$$R_i(\theta) = \epsilon_i B_i(T_s)\tau_i(P_s,\theta) + \int_{\tau_i(P_s)}^1 B_i[T(P)] \frac{d\tau_i(P)}{d\ln P} \, d\ln P + R_i'(\theta) \quad (5.5)$$

where

$$B_i(T) = B_{\nu_i}(T)$$

$$\epsilon_i = \epsilon_{\nu i}$$

$$\tau_i(P,\theta) = \int_0^\infty f_i(\nu)\tau_\nu(P,\theta) \, d\nu$$

and R_i' will be treated later.

For the purpose of temperature sounding, one would like the atmospheric absorption to be due primarily to gases of fixed atmospheric distribution, and have only minor contributions from the variable gases H_2O and O_3. Therefore, from table 5.1, four spectral regions are available for sounding. In the microwave region, there are the 60 GHz and 120 GHz regions in which O_2 absorbs, while in the infra-red region there are the 15 μm and 4·3 μm regions where CO_2 and N_2O absorb. If, on the other hand, it is necessary and desirable to determine the distribution of absorbing gases such as H_2O and O_3, then such information can be obtained from measurements in spectral regions where these gases are the primary absorbers. Remote sensing of constituents has already been dealt with in Chapter 4. In addition, if radiation is measured in atmospheric windows, information about the surface can be obtained.

The current temperature sounding system on the TIROS series of polar orbiting meteorological satellites comprises HIRS2, a 20-channel infra-red sounder, and MSU, a four-channel microwave sounder. This instrument package is based on the considerations just discussed. In addition, the satellite carries SSU, a three-channel pressure-modulated stratospheric temperature sounding unit, which will be not be discussed in this chapter. The following sections will discuss (*a*) HIRS2/MSU, a multi-spectral sounder; (*b*) a comparison of statistical and physical retrieval systems; (*c*) a statistically constrained solution to the radiative transfer equation; (*d*) determination of cloud fields, surface temperature and surface emissivity; and (*e*) results for January 1979.

5.2. HIRS2/MSU—a multispectral sounder

Table 5.2 shows the channel centres of the HIRS2 and MSU instruments. Also shown for each temperature sounding channel is the peak of $d\tau/d\ln P$ computed using a US standard atmosphere at nadir and the peak of the radiance contribution function $Bd\tau/d\ln P$ computed using the same conditions. In addition, those channels used by the Goddard Laboratory for Atmospheric Sciences (GLAS) in their analysis of the HIRS2/MSU data, and their principal users, are indicated in the table. A description of the GLAS analysis procedure and results will be shown later.

Channels 1–7 on HIRS are in the 15 μm region and sound portions of the atmosphere ranging from the mid-stratosphere to the surface. Channels 13–17 are temperature-sounding channels in the 4·3 μm region. While similar portions of the atmosphere appear to be sounded in the 4·3 μm region as in the 15 μm region, the blackbody function is much more temperature sensitive at the higher frequency and the peak radiance contribution function is shifted much more to pressures with warmer temperatures and away from the tropopause. This has the property of greatly narrowing the portions of the atmosphere from which an appreciable signal comes in the mid–lower tropospheric sounding channels 13–15. On the other hand, it significantly broadens the regions sounded by $4 \cdot 3 \mu m$ channels whose weighting functions peak near the tropopause, such as channel 16. $4 \cdot 3 \mu m$ channels whose weighting functions peak in the stratosphere, such as channel 17, have radiance contribution functions shifted up to the stratopause and are highly perturbed by effects of non-local thermodynamic equilibrium in the upper stratosphere. Such channels are not useful for temperature sounding. HIRS2 also has an $11 \mu m$ window channel, two window channels with centres at $3 \cdot 7 \mu m$ and $4 \mu m$, three channels in the $6 \cdot 7 \mu m$ water vapour band and one channel in the $9 \cdot 6 \mu m$ ozone band. MSU has four channels in the 60 GHz oxygen band. The three highest frequency MSU channels are useful for atmospheric temperature sounding while the $50 \cdot 3$ GHz channel is primarily a window and is more sensitive to surface properties. Since the blackbody function is nearly linear in temperature in the microwave region, the radiance contribution functions are essentially unchanged from the weighting functions.

Two very important considerations with regard to the information content and utility of temperature sounding channels are: (a) the sensitivity of the observation to changes in atmospheric temperature; and (b) the sensitivity of the observation to other parameters such as water vapour, ozone, ground temperature, clouds, etc. The first factor represents the signal and the second represents sources of noise. These sources of noise are significant only to the extent that they cannot be accounted for. In general terms, one can summarize the strengths and weaknesses of temperature sounding in different spectral regions, using current instrumental techniques, as follows.

1. Microwave observations are relatively insensitive to clouds and trace constituents, have moderately sharp stratospheric and upper tropospheric temperature sensitivity functions, and slightly broader lower tropospheric temperature sensitivity functions which are also sensitive to surface effects and rain.

2. $4 \cdot 3 \mu m$ channels have sharp mid–lower tropospheric temperature sensitivity functions which are relatively insensitive to water vapour and ozone but are very cloud sensitive, and have very broad upper tropospheric and stratospheric sounding channels which are also affected by non-local thermodynamic equilibrium in the upper stratosphere.

3. $15 \mu m$ channels have moderate resolution mid–lower tropospheric sounding channels which are affected significantly by water vapour and O_3 as well as clouds, and have very broad upper tropospheric and stratospheric sounding channels which are also ozone and water vapour

sensitive. Based on these considerations and the desire to minimize redundant temperature information, especially if the channels are sensitive to effects of trace gases, the seven channels whose sensitivity functions are shown in figure 5.1 were chosen by GLAS for determining atmospheric temperature profiles from the HIRS2/MSU data. The sensitivity functions, which will be treated in detail later, represent the derivative of the brightness temperature (or equivalent blackbody temperature) with respect to changes in atmospheric temperature. As indicated in the above discussion, $4 \cdot 3 \mu m$ channels are best for sounding the mid–lower troposphere, microwave channels are best for sounding the upper troposphere–stratosphere, while $15 \mu m$ channels are less advantageous for any atmospheric region. The $15 \mu m$ channels 2 and 4 are included in the sounding set to augment the upper tropospheric and stratospheric data provided by the two microwave channels M3 and M4. Microwave channel 2 gives a good measure of mean tropospheric

Table 5.2. HIRS2 and MSU channels.

Channel	$\lambda(\mu m)$	$\nu(cm^{-1})$	Peak of $d\tau/d\ln P$(mb)	Peak of $B\,d\tau/d\ln P$(mb)
H1	14·96	668·40	30	20
H2[a]	14·72	679·20	60	50
H3	14·47	691·10	100	100
H4[a]	14·21	703·60	280	360
H5	13·95	716·10	475	575
H6[b]	13·65	732·40	725	875
H7[b]	13·36	748·30	Surface	Surface
H8[b]	11·14	897·70	Window, sensitive to water vapour	
H9	9·73	1027·90	Window, sensitive to O₃	
H10	8·22	1217·10	Lower tropospheric water vapour	
H11	7·33	1363·70	Middle tropospheric water vapour	
H12	6·74	1484·40	Upper tropospheric water vapour	
H13[a,c]	4·57	2190·40	Surface	Surface
H14[a]	4·52	2212·60	650	Surface
H15[a]	4·46	2240·10	340	675
H16	4·39	2276·30	170	425
H17	4·33	2310·70	15	2
H18[d]	3·98	2512·00	Window, sensitive to solar radiation	
H19[d]	3·74	2671·80	Window, sensitive to solar radiation	
M1[e]	0·596[f]	50·30[g]	Window, sensitive to surface emissivity	
M2[c]	0·558[f]	53·74[g]	500	
M3[a]	0·548[f]	54·96[g]	300	
M4[a]	0·518[f]	57·95[g]	70	

[a]Used by GLAS to compute temperature profiles.
[b]Used by GLAS to compute cloud fields.
[c]Used by GLAS in cloud correction.
[d]Used by GLAS to compute surface temperature.
[e]Used by GLAS to compute surface emissivity.
[f]λ in cm.
[g]ν in GHz.

temperature but is effectively redundant with the $4 \cdot 3 \mu m$ channels and is not used directly in the temperature sounding algorithm. It is perhaps the single most important channel in the whole processing system, however, and is used both in the algorithm to correct the infra-red channels for cloud effects and also as a final quality check on the retrieved profile.

The window channels in different spectral regions also have different properties. The most transparent and least humidity-sensitive windows are in the $4 \cdot 0$ and $3 \cdot 7 \mu m$ regions. In addition, as a result of the greater dependence of the blackbody function on temperature at high frequencies, radiances in these channels are much more sensitive to changes in temperature and the retrieved temperatures are less sensitive to given percentage noise errors, such as variability in surface emissivity. A negative attribute of the short-wave window channels is their sensitivity to reflected solar radiation during the day, and, like the $11 \mu m$ windows, to clouds. The $11 \mu m$ window is much more sensitive to the effects of water vapour. In the case of tropical profiles observed at large satellite zenith angles, the surface transmittance can be as low as $0 \cdot 2$. Under such conditions, it is better utilized as a low-level humidity sounding channel than a surface temperature sounding channel. The $50 \cdot 3 \, \text{GHz}$ channel on MSU is more sensitive to surface emissivity changes than surface temperature and is used to determine the emissivity of the surface, from which ice and snow cover can be inferred. The primary uses of the channels in the GLAS analysis system are indicated in table 5.2. All channels are actually used interactively. The two short-wave window channels are used to solve for both ground temperature and reflected solar radiation effects during the day.

5.3. Comparison of statistical and physical retrieval systems

Observations in temperature sounding and other auxiliary channels give information about the atmosphere through equation 5.5. As a result of the broadness of the sensitivity functions, it can be shown that no unique solution exists which when substituted into equation 5.5 can match the observation within noise levels. Instead, some form of constraint must be applied, including the use of additional information, to reach an unique solution. In terms of operational constraints, this information should be readily available at least 6 hours beforehand and should not include, for example, radiosonde information concurrent in time with the satellite observations.

A very powerful technique for determining atmospheric temperature profiles from satellite observations utilizes previously determined statistical relationships between satellite radiance observations and atmospheric temperature profiles given by radiosondes colocated in space and time. Such a procedure is used operationally by NOAA/NESS (Smith and Woolf, 1976). In general terms, the approximately linear relationship between change in brightness temperature (equivalent blackbody temperature) for a channel and change in temperature at a level is used

$$(T - \bar{T}) = B(T_B - \bar{T}_B) \tag{5.6}$$

where T, the solution, is a vector of atmospheric temperature at a set of discrete levels; $\bar{T}$ is the mean of the vectors of atmospheric temperature profiles used previously to determine the matrix B; T_B is the vector of observed brightness temperatures; and $\bar{T}_B$ is the mean of the vectors of observed brightness temperatures used to construct B. The matrix B contains not only the information of the atmospheric weighting functions but also additional information about properties of atmospheric temperature profiles, including correlations of deviations of atmospheric profiles from mean values in different parts of the atmosphere. In practice, separate B matrices and means are used in different latitude bands. The appropriate coefficients are also updated every other week to account for temporal changes in atmospheric structure and changes in instrument characteristics. In addition to the ability to incorporate statistical properties about the behaviour of the atmosphere into the solution, the regression method does not require the ability to compute accurately the radiative transfer equation 5.5.

The regression approach would be ideal if the radiances depended only on temperature profile. Other factors affect the radiances, however, such as zenith angle of observation, humidity and ozone profiles, surface properties (elevation, emissivity and temperature), reflected solar radiation and, most important of all, clouds. In addition, retrievals cannot be done using this method if there are no direct measurements of temperature such as in the upper stratosphere or on other planets. Operationally statistical relationships between microwave observations and infra-red observations under conditions determined to be clear are used to correct for cloudiness and also to adjust the observations at the observed zenith angle to what would have been observed at nadir. Other factors are treated implicitly to the extent that their effects are correlated with temperature profile.

The physical approach to a solution of equation 5.5 is totally different. The basic method is to find atmospheric and surface conditions which, when substituted in equation 5.5, match the observations to within a specified noise level. This noise level involves instrument noise, scene noise (variability in the field of view) and computational noise, that is, the accuracy with which equation 5.5 can be computed given the surface and atmospheric conditions. The combined effects have been shown (Susskind *et al.*, 1983) to be about $0 \cdot 75\,\mathrm{K}$ in the rms sense. The procedure used b

GLAS attempts to account for all relevant factors in an iterative manner. No zenith angle correction is required because all calculations are done at the angle of observation. The cloud correction, surface temperature, surface emissivity, reflected solar radiation calculation and temperature profile retrieval are found in an interactive manner. While no statistical information relating observed brightness temperatures to atmospheric temperature profiles is used, a first guess 6-hour forecast temperature–humidity profile is utilized as additional *a priori* information. If a solution to the radiative transfer equation which matches the observations to the specified accuracy cannot be found, or if it is estimated to be too cloudy to account for the cloud effects on the infra-red channels, no retrieval is performed.

It will be shown in the next section that the iterative relaxation equation is of the form

$$(T^{N+1} - T^N) = A(T^N - \bar{T}_G) + B^N(\hat{T}_B^N - T_B^N) \qquad (5.7)$$

where T^N is the vector of the Nth iterative temperature profile; $\bar{T}_G$ is the global mean temperature; $\hat{T}_B^N$ is the Nth iterative estimate of the clear-column brightness temperature; T_B^N is the vector of calculated brightness temperatures using the Nth iterative set of conditions; B^N is a profile dependent matrix computed using both statistical constraints and the weighting function for the particular sounding; and A is a matrix introduced to stabilize the solution. While equation 5.7 is very similar in form to the regression equation 5.6, equation 5.7 differs significantly in that: (*a*) it contains the full physics of the problem; (*b*) the cloud corrected radiances are part of the iterative scheme; (*c*) the interpolation matrix B is computed based on the weighting functions; and (*d*) initial guess information is incorporated via T^0. The next section describes the principles behind the relaxation method and its implementation in the GLAS retrieval scheme.

5.4. A statistically constrained relaxation solution to the radiative transfer equation

The relaxation method of finding a solution to a set of radiative transfer equations was originally developed by Chahine (1968). It differs from other methods in that it does not attempt, in any iteration, to find a best solution to the set of equations (observations) but only to provide a set of parameters giving better agreement of observed and calculated radiances than obtained in the previous iteration. The iterative method is computationally fast and stable. Moreover, in order to solve the inverse radiative transfer equation it is necessary first to put the equation in approximate linear form with coefficients which are profile-dependent. Therefore, an 'exact' solution must be iterative in any event.

In order to use the difference of reconstructed and computed brightness temperatures in the temperature sounding channels to estimate the error in the Nth iterative temperature profile, it is useful to look at the response of the brightness temperature of a channel to changes in atmospheric temperature profile.

From equation 5.5 it can be seen that, to a good approximation, brightness temperatures T_i computed for two closely related temperature profiles $T(P)$ and $T(P) + \delta(P)$ will differ by

$$T_i[\,T(P) + \delta(P)\,] - T_i[\,T(P)\,] = \int W_i(P)\,\delta(P)\mathrm{d}\ln(P) \qquad (5.8)$$

where

$$W_i(P) = \left(\frac{\mathrm{d}T_i}{\mathrm{d}R_i}\right)\bigg|_{R_i}\left(\frac{\mathrm{d}R_i}{\mathrm{d}T}\right)\bigg|_{T(P)}$$

$$= \left(\frac{\mathrm{d}T}{\mathrm{d}B_i}\right)\bigg|_{T_i}\left(\frac{\mathrm{d}B_i}{\mathrm{d}T}\right)\bigg|_{T(P)}\left(\frac{\mathrm{d}\tau_i}{\mathrm{d}\ln P}\right)\bigg|_P \qquad (5.9)$$

assuming all other parameters remain constant. These sensitivity weighting functions, as defined in equation 5.9, which relates the change in brightness temperature to the change in atmospheric temperature profile, are shown in figure 5.1 for the seven channels used in determining the atmospheric temperature profile.

It can be shown that

$$\int W_i(P)\mathrm{d}\ln P \approx 1 - \tau_i(P_s) \qquad (5.10)$$

Therefore, to a good approximation, for two profiles differing by a constant δ

$$T_i[\,T(P) + \delta\,] \approx T_i[\,T(P)\,] + \delta[\,1 - \tau_i(P_s)\,] \qquad (5.11)$$

that is, a constant shift of temperature profile throughout the atmosphere produces approximately the same change in brightness temperature, reduced by a small amount if the channel sees the surface. Moreover, if, for example, channel M4 in figure 5.1 is considered, the brightness temperature in that channel can be seen to be virtually independent of temperature above 10 mb and below 300 mb and is mostly dependent on

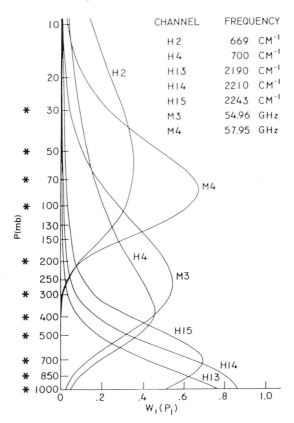

CHANNEL	FREQUENCY
H 2	669 CM^{-1}
H 4	700 CM^{-1}
H 13	2190 CM^{-1}
H 14	2210 CM^{-1}
H 15	2243 CM^{-1}
M 3	54.96 GHz
M 4	57.95 GHz

Figure 5.1. Weighting functions, for a US standard atmosphere at nadir viewing, for the seven HIRS2/ MSU channels used to determine the temperature profile. These channels are used to estimate mean temperatures for the layers bounded by *.

temperature between 40 mb and 130 mb. Therefore, it is possible to attribute a difference δ between observed clear-column brightness temperature T_{M4} and computed brightness temperature T_{M4}^{N} to a comparable difference in the true (130–40 mb) layer mean temperature, $\bar{T}_j = \int T(P) \, d\ln P$, and that of the Nth guess.

In order to reduce the effects of noise and stabilize the solution, it is desirable to average the estimates of mean layer temperature in layer j obtained from channel i

$$\bar{T}_j^{N+1} = \bar{T}_j^N + \sum_i \bar{W}_{ij} [\, \hat{T}_i^N - T_i^N \,] / \sum_i \bar{W}_{ij} \qquad (5.12)$$

where $\bar{W}_{ij}$ is the mean value of $W_i(P)$ in layer j. There is no need for a one-to-one relationship between the number of channels and number of layers.

In the analysis, mean layer temperatures for the ten layers shown in figure 5.1 are used to determine the total temperature profile $T^{N+1}(P)$. To ensure uniqueness, a constraint is put on the system that

$$T^{N+1} = T_G + \sum_{k=1}^{K} A_k^{N+1} F_k \qquad (5.13)$$

where $T_G(P)$ is a global mean temperature profile; $F_k(P)$ are empirical orthogonal functions at 52 pressure levels extending from 1000 mb to 30 mb, given by the first K eigenvectors, with largest eigenvalues, of the covariance matrix of a set of global radiosonde profiles; and A_k^{N+1} are iterative coefficients. The coefficients A_k uniquely define the solution.

The K coefficients, A_k^{N+1} can be determined for each iteration from the $N+1$th estimate of M mean layer temperature $\bar{T}^{N+1}$ and any other available information. If, for example, it is desired to establish the K coefficients that compute mean layer temperatures which agree most satisfactorily (in a least squares sense) with the M specified values then

$$A^{N+1} = (\bar{F}'\bar{F})^{-1} \bar{F}' \Delta^{N+1} \qquad (5.14)$$

where A is the vector of K coefficients, $\bar{F}$ is an M by K matrix with elements $\bar{F}_{jk}$ given by the mean value of $F_k(P)$ in layer j, and Δ_j^{N+1} is the difference between $\bar{T}_j^{N+1}$ and $\bar{T}_{G,j}$, the mean layer temperature of the global mean

$$\Delta_j^{N+1} = \bar{T}_j^{N} - \bar{T}_{G,j} \qquad (5.15)$$

As a result of uncertainties in the mean layer temperatures, increased stability is obtained by finding the coefficients which minimize a combination of the difference between estimated and computed mean layer temperatures on the one hand, and maximize the likelihood of the solution on the other. An appropriate equation is given by

$$A^{N+1} = (F'F + \sigma H)^{-1} F' \Delta^{N+1} \qquad (5.16)$$

where H is a diagonal M by M matrix with H_{kk} being the inverse of the fraction of total variance arising from eigenvector k; and σ is a constant. In practice, equation 5.16 is used with $K = 6$ and $\sigma = 5 \times 10^{-4}$.

It is seen from figure 5.1 that while very little detailed information is contained about temperatures above 50 mb, the observations in channels 2 and M4 are still quite sensitive to temperature changes above this level.

It was found that when the differences in the observed brightness for Channel 2 and that computed from the first guess were more than 5 K the retrieval was usually of poor quality at pressures less than 200 mb but satisfactory greater than 200 mb. These retrievals were flagged as good only in the troposphere. Occurrence is almost exclusively over land at the high latitudes.

The net result of equations 5.12, 5.13, 5.15 and 5.16 is the iterative equation

$$T^{N+1} = T^N + (B - I)(T^N - T_G) + BW^N(\hat{T}_B^N - T_B^N) \qquad (5.17)$$

where T_G is the 52-level global mean temperature profile; T^N is the Nth guess temperature profile; $(\hat{T}_B^N - T_B^N)$ is the difference between the reconstructed clear column brightness temperatures and those computed in the Nth iteration; W^N is the matrix of weighting functions, defined by equation 5.9 in the Nth iteration but normalized so that the sum of W over all channels equals 1 for any layer; I is the identity matrix; and B is given by

$$B = F(\bar{F}'\bar{F} + \sigma H)^{-1}\bar{F}'L \qquad (5.18)$$

where L is the matrix which produces layer average values from point values (e.g., $\bar{T} = LT$, $\bar{F} = LF$, $\bar{W} = LW$). The matrix BW is the fundamental interpolation matrix which produces a change in temperature profile given a difference between observed and computed brightness temperatures. It is composed of two elements: the profile dependent weighting functions which contain the atmospheric physics and the statistical B matrix which results from the constraints on the solution. The term $(B - I)(T^N - T_G)$ arises from the expansion of the solution about the global mean and tends to further stabilize the solution under high noise conditions. This term would drop out of equation 5.17 if T_G in equation 5.13 were replaced by T^N, i.e., if the iterative temperature profiles were constrained as an expansion about the Nth guess rather than about the global mean.

While the form of equation 5.18 is similar to that used in regression analysis, there are a number of significant differences. Foremost among these is that equation 5.18 contains the full physics of the problem, allows for the incorporation of initial guess information into the solution for the iterative treatment of the effects of clouds on the radiances, and for the ability to determine whether a solution can be found with satisfactory agreement of observed and computed radiances.

5.5. *Determination of cloud fields, surface temperature and surface emissivity*

Accounting for the effects of clouds on the infra-red observations

Accounting for the effects of clouds on the infra-red observations is the single most important step in the retrieval process. If one looks at an otherwise homogeneous but partially cloudy scene with cloud fraction α, then to a reasonable approximation, one can write the radiance observed in channel i to be

$$R_i = (1 - \alpha)R_{i,\text{clr}} + \alpha R_{i,\text{cld}} \tag{5.19}$$

where $R_{i,\text{clr}}$ is the clear-column radiance as computed from equation 5.5 and $R_{i,\text{cld}}$ is the radiance observed in the cloud-covered area. $R_{i,\text{cld}}$ depends not only on the variables already discussed but also on the detailed properties of the clouds. Rather than assume or attempt to determine the cloud properties simultaneously with the determination of atmospheric and surface properties, the method attempts to estimate, or 'reconstruct', from the observed radiances the clear-column radiances which would have been observed if no clouds had been present. These reconstructed clear-column radiances are used in determination of the atmospheric temperature profile, as discussed in the previous Section, as well as in the determination of ground temperature. The cloud field parameters are determined only after a complete atmospheric and surface solution has been obtained.

A two field-of-view approach, similar to that of Smith (1968), is used to extrapolate observed radiances to clear-column radiances. The clear-column radiance, $\hat{R}_i$, can be obtained from

$$\hat{R}_i = R_{i,1} + \eta(R_{i,1} - R_{i,2}) \tag{5.20}$$

where $R_{i,j}$ is the observation for channel i in the field of view j; and η is given by $\alpha_1/(\alpha_2 - \alpha_1)$, The fields of view are numbered in the sense that $R_{8,1} > R_{8,2}$, that is, the field of view with the higher radiance in the $11\,\mu m$ window is taken as the clearer field of view, with cloud fraction α_1. Once η is determined, clear-column radiances can be reconstructed from the observations by using equation 5.20. An estimate of η is obtained with each iteration. If $\alpha_1 \approx \alpha_2$, then either both fields of view are clear, or a large value of η will result.

It is seen from equation 5.20 that large values of η will tend to amplify noise in the observations and are, therefore, undesirable. Areas with large η are rejected, as described later. At the other extreme, $\eta = 0$ implies field of view 1 is clear. If both fields of view are clear η is indeterminate. If this occurs η is fixed at an artificial value of $\eta = -0.5$, which averages the observations in both fields of view.

As shown by Chahine (1974), η can be determined from the infra-red observations as part of an iterative scheme according to

$$\eta^N = \frac{R_7^N - R_{7,1}}{R_{7,1} - R_{7,2}} \tag{5.21}$$

where R_7^N is the computed clear-column radiance for the $15\mu m$ surface channel, using the Nth iterative temperature profile. In this case, the scheme will converge provided only $4.3\mu m$ infra-red channels are used for temperature sounding in the lower troposphere. The rate of convergence increases with the difference between the surface temperature and the cloud-top temperature. Under some high-noise, low-contrast conditions, divergent solutions can occur in the sense that an overestimate of η^N will cause an overestimate of the reconstructed $4.3\mu m$ clear-column radiances which, in turn, will yield an increased lower tropospheric temperature, produce an increased value of R_7^{N+1}, lead to an increased η^{N+1}, etc.

When a lower tropospheric-sounding microwave channel is available, such as Channel M2, a superior method for determining η can be used, making the estimate of η less sensitive to errors in the initial guess. This also alleviates the need for the use of $15\,\mu m$ channels, which are significantly affected by H_2O and O_3 absorption, in accounting for cloud effects. In the case of available MSU channels η is determined as in equation 5.21, except that the $4.3\,\mu m$ surface Channel 13 is used instead of Channel 7. The microwave channel is used to correct errors in R_{13}^N due to errors in the iterative temperature profile. The error in η^N determined from equation 5.21 is a result of either use of an incorrect temperature profile to estimate the clear-column radiance, computational uncertainties such as the effect of water vapour on the transmittance functions of Channel 13, observational errors in $R_{13,i}$ or errors in the assumption of only one degree of non-homogeneity in the combined fields of view. The error in R_{13}^N due to an incorrect temperature profile can be well accounted for by adjusting the computed brightness temperature for Channel 13 according to

$$T'_{13} - T_{13}^N = T_{M2} - T_{M2}^N \tag{5.22}$$

where T_{M2} and T_{M2}^N are the observed and calculated microwave brightness temperatures; T_{13}^N is the calculated clear-column brightness temperature for Channel 13; and T'_{13} is the corrected clear-column brightness temperature for Channel 13. This correction is based on the approximation that a bias in the iterative temperature profile in the mid–lower troposphere will produce approximately the same error in computed brightness temperature in the infra-red and microwave channels sounding that portion of the atmosphere. The corrected clear-column radiance for Channel 13 is then given by

$$R_{13}'^{N} = B_{13}(T_{13}'^{N} + T_{M2} - T_{M2}^{N}) \qquad (5.23)$$

and η is now computed according to

$$\eta^{N} = \frac{R_{13}'^{N} - R_{13,1}}{R_{13,1} - R_{13,2}} \qquad (5.24)$$

If the observations in the two fields of view are sufficiently close, it is most likely that both fields of view are either clear or overcast. It is possible to discriminate between these two cases by comparing T_{13}', the corrected clear-column brightness temperature for Channel 13, to $T_{13,1}$, the observed brightness temperatures for field of view 1. If $(T_{13}' - T_{13,1}) > 8 \, \text{K}$ and $\eta^{N} > 4$, the fields of view are considered too cloudy to undertake a retrieval. At the other limit, if $\eta^{N} \leq 0$ and $|T_{13,1} - T_{13,2}| \leq 1 \, \text{K}$, η is taken as $-0 \cdot 5$, that is, both fields of view are considered clear.

Utilization of the microwave sounding not only speeds up convergence under all conditions, but stabilizes the solution in the sense that a positive bias in the iterative temperature profile in the lower troposphere will not, to a first approximation, cause an increase in η.

Measurement of sea surface temperatures and ground temperatures

The main factors influencing the accuracy of retrieved sea surface temperatures from infra-red window observations are the effects of clouds and humidity on the observations. AVHRR, the operational infra-red sea surface temperature sounder (see Chapter 7), utilizes very fine spatial resolution observations of the order of $1 \, \text{km} \times 1 \, \text{km}$, in an attempt to find completely clear spots. The current analysis, using combined infra-red and microwave observations, requires neither high spatial resolution nor the existence of clear spots for the determination of accurate sea or land surface temperatures, which are determined as an integral part of the sounder processing system. The effects of clouds on the observations are accounted for by use of equations 5.20, 5.23 and 5.24, giving the clear-column radiances for all infra-red channels. The two $3 \cdot 7 \, \mu\text{m}$ window channels on HIRS2, whose brightness temperatures are relatively insensitive to humidity, are used to determine surface temperature rather than the $11 \, \mu\text{m}$ window, which has been used operationally on AVHRR.

In order to determine accurate surface temperatures from the reconstructed clear-column radiances $\hat{R}_i^N$ it is necessary to include the contribution of reflected radiation $R_i'(\theta)$ to equation 5.5. $R_i'(\theta)$ can be expressed as

$$R_i'(\theta) = (1 - \epsilon_i)R_i\!\downarrow \tau_i(P_s,\theta) + \varrho_i' H_i \tau_{is}(\theta,\theta_s) \qquad (5.25)$$

where the first term represents the contribution of downwelling atmospheric radiation, and the second term represents the contribution of reflected solar radiation. In equation 5.25, $R_i\!\downarrow$ is an effective atmospheric emission downward flux; ϱ_i' is the bi-directional reflectance of the surface of solar radiation from the Sun, with solar zenith angle θ_s to the satellite; H_i is the solar radiation at the top of the atmosphere; and $\tau_{is}(\theta,\theta_s)$ is the atmospheric transmittance along the entire path of incident and reflected solar radiation. At night, the reflected solar radiation term can be neglected in equation 5.25 and surface temperatures are easily obtained from each channel as

$$T_{s,i}^N = B_i^{-1} \frac{\hat{R}_i^N - [(1 - \epsilon_i)R_i^N\!\downarrow \tau_i^N(P_s)] - \displaystyle\int_0^{\tau_i(P_s)} B_i(T^N)\,d\tau}{\epsilon_i \tau_i^N(P_s)} \qquad (5.26)$$

The downward flux $R_i^N\!\downarrow$ is approximated as

$$R_i^N\!\downarrow = 2\cos\theta \int_0^{\tau_i(P_s)} B_i(T^N)\,d\tau \qquad (5.27)$$

This approximation is based on the assumption of an optically thin atmosphere and a Lambertian surface (see Chapter 2). In general, $T_{s,18}$ and $T_{s,19}$ are found to agree with each other to within $1\,\mathrm{K}$, even under partially cloudy conditions. The surface temperature, T_s^N, is taken as $0\cdot5(T_{s,18}^N + T_{s,19}^N)$.

During the day, the effects of solar radiation on the $3\cdot7\,\mu\mathrm{m}$ channels must be accounted for in obtaining accurate surface temperature retrievals from these channels. The solar radiation reflected off clouds in the field of view has already been accounted for by the clear-column radiance algorithm. If additional clouds are in the path of incident solar radiance with cloud fraction α, the solar radiation striking the ground will be attenuated by $(1 - \alpha)$. The solar radiation reflected off the clouds will not be seen by the instrument because of its narrow field of view. The net effect is to reduce the solar radiation by a factor of $(1 - \alpha)$.

One can attempt to account for reflected solar radiation directly by subtracting $\varrho_i' H_i \tau_{i,s}$ from $\hat{R}_i$ and substituting the result into equation 5.26.

In the case of $\alpha = 0$, $H_i \tau_{i,s}$, the mean solar radiation across the channel traversing the path from the Sun to the Earth and back to the satellite, can be well estimated as $2 \cdot 16\pi \times 10^{-5} B_i (5600\,\mathrm{K}) \cos\theta_H \tau_i (P_s, \theta_{\mathrm{EFF}})$ where θ_H is the solar zenith angle and the transmittance is computed at an effective zenith angle θ_{EFF}, whose secant is given by the sum of the secants of the solar and the satellite zenith angles. The case of $\alpha \neq 0$ is equivalent to an effective reflectivity $\varrho = \varrho'(1 - \alpha)$.

This procedure is impractical because of the uncertainty in ϱ_i, even if $\alpha = 0$. If the surface is Lambertian and the emissivity is known, ϱ_i', the directional reflectance, is equal to $(1 - \epsilon_i)/\pi$. Significant errors of up to a factor of 2 can be made in these estimations of ϱ_i', which may produce errors of up to $10\,\mathrm{K}$ in retrieved surface temperature. These errors arise from uncertainties in ϵ_i and the non-Lambertian character of the surface. The same uncertainties in ϵ_i, however, do not affect appreciably the calculated thermal radiation. Estimated values of $0 \cdot 85$ over land and $0 \cdot 96$ over ocean are used in the analysis. Rather than assume a value for ϱ, T_s and ϱ are solved for in an iterative manner, assuming only that ϱ is the same for both $3 \cdot 7\,\mu\mathrm{m}$ channels.

For the $3 \cdot 7\,\mu\mathrm{m}$ sounding channel i

$$\frac{\hat{R}_i^N - R_{\mathrm{atm},i}^N}{\epsilon_i \tau_i^N (P_s)} = B_i(T_s) + d_i H_i' = A_i^N \tag{5.28}$$

where $R_{\mathrm{atm},i}$ is the atmospheric contribution to the calculated clear-column radiance; $d_i = \bar{\varrho}/\epsilon_i$; and H_i' is given by $H_i' = H_i \tau_{is}/\tau_i(P_s)$. The left-hand side of equation 5.28, and consequently A_i, is known in a given iteration. Assuming ϱ_i and ϵ_i are the same for both $3 \cdot 7\,\mu\mathrm{m}$ channels, one obtains the equation

$$B_i(T_s) - aB_j(T_s) = A_i - aA_j = A \tag{5.29}$$

where $a = H_i'/H_j'$. This non-linear equation in one unknown, T_s, is solved iteratively according to

$$\frac{\exp(-h\nu/T_s^{M+1})}{\exp(-h\nu/T_s^M)} = \frac{A}{B_i(T_s^M) - aB_j(T_s^M)} \tag{5.30}$$

where $\nu = (\nu_i + \nu_j)/2$. This procedure converges rapidly. Once T_s is determined, d is calculated from equation 5.28. This provides a value of ϱ' which is used in equation 5.5 to correct the $4 \cdot 3\,\mu\mathrm{m}$ channels for reflected solar radiation effects.

The iterative ground temperature is used to compute the estimated clear-column radiances R_i^N for each channel. As shown in Section 5.4, temperature profiling utilizes a comparison of these radiances R_i^N, computed for the temperature sounding channels from T_s^N and $T^N(P)$, to the Nth reconstructed clear-column radiances. Over ocean, climatological sea surface temperatures have accuracies of better than $2\,K$ and can serve reasonably well for the computation of clear-column radiances. Under conditions when it is felt that it is too cloudy to retrieve sea surface temperatures of greater accuracy than climatology the sea surface temperature is fixed at its climatological value for the purpose of radiative transfer calculations. This decision is made only in the first iteration. Climatology is used if either $(T_{18} - T_{18,1}) > 20\,K$ or $|T_s - T_{CLIM}| > 5\,K$, or both $(T_{18} - T_{18,1}) > 10\,K$ and $|T_s - T_{CLIM}| > 3K$, that is, either the reconstructed brightness temperature is very far from the observed brightness temperature, indicating that a large cloud correction is necessary, or the retrieved sea surface temperature differs significantly from climatology, indicating a potential problem. If both indicators of a problem exist, the tolerance conditions made are more stringent. The sea surface temperature is fixed to climatology about $3\,\%$ of the time.

The accurate *a priori* knowledge of the sea surface temperature is also used to indicate low-level overcast which may have been missed by the cloud algorithm test for overcast described previously. If $(T_s - T_{CLIM}) < -3\,K$ and $|T_{18} - T_{18,1}| < 2\,|T_s - T_{CLIM}|$, that is, the retrieved sea surface temperature is more than $3K$ colder than climatology and the difference is greater than half the difference in reconstructed and observed brightness temperatures, the fields of view are considered to be fully overcast with low-level clouds. No retrieval is performed under these conditions. The retrieval is also rejected if the final retrieved sea surface temperature differs from climatology by more than $5\,K$.

Computation of surface emissivity for the microwave channels

The emissivity of the surface in the microwave region is much more variable than in the infra-red. At $50\,GHz$, the emissivity is typically $0\cdot45-0\cdot65$ for open ocean, and $0\cdot90-0\cdot95$ for land. Sea ice has an emissivity of the order of $0\cdot7$ or more, and snow has an emissivity of $0\cdot90$ or less, depending on the depth.

Given a temperature–humidity profile and surface temperature and pressure, the calculated microwave brightness temperature at a given zenith angle is much more sensitive to the surface emissivity than an equivalent infra-red channel, because of the greater dependence of radiance on temperature in the infra-red. To obtain an accurate calculation of brightness temperatures in microwave Channel 2, which has about a $10\,\%$ contribution from the surface, a precise knowledge of the microwave surface emissivity is needed. This parameter in turn also provides important information about the surface properties.

The microwave emissivity ϵ is calculated from the $50 \cdot 3$ GHz channel, as part of the iterative scheme, according to

$$\epsilon = \frac{R_i - \int T d\tau - R_i \downarrow \tau_i(P_s)}{(T_s - R_i \downarrow)\tau_i(P_s)} \qquad (5.31)$$

where R_i is the $50 \cdot 3$ GHz observed brightness temperature; $R_i \downarrow$ is the downwelling atmospheric flux; T_s is the iterative surface temperature; and $T(P)$ is the iterative atmospheric temperature profile used in the calculation of the upward and downward microwave fluxes emitted by the atmosphere.

Determination of cloud height and amount

Given a ground temperature and a temperature profile from which the clear-column radiance $R_{i,\text{clr}}$ can be calculated, an effective cloud height and percentage cloud cover which match the outgoing longwave radiation can be determined from equation 5.19, provided assumptions are made about the cloud properties. The cloud height is effective because of the possibility of the presence of multiple cloud layers in the field of view. The cloud fraction is effective both for the above reason and because it is the product of the true cloud fraction and the cloud emissivity. It represents the cloud opacity in the field of view. If a single cloud layer is assumed, whose emissivity is ϵ_{ic}, reflectivity is zero, transmissivity is $(1 - \epsilon_{ic})$, and whose top is at P_c with temperature $T(P_c)$, where $T(P)$ is the retrieved temperature profile, $R_{i,\text{cld}}$ can be evaluated at any assumed cloud-top pressure to give

$$R_{i,\text{cld}}(P_c) = \epsilon_{ic}\left(B_i[\,T(P_c)]\,\tau_i(P_c) + \int_{\tau_i(P_c)}^{1} B_i(\tau)\,d\tau\right) + (1 - \epsilon_{ic})R_{i,\text{clr}} \qquad (5.32)$$

Using equation 5.19, an effective cloud fraction $\alpha_i(P_c)$, consistent with the assumed cloud pressure, can be determined for channel i according to

$$\alpha_i(P_c) = \frac{R_{i,\text{clr}} - R_i}{R_{i,\text{clr}} - R_{i,\text{cld}}(P_c)} \qquad (5.33)$$

where R_i is the observation for channel i; and $R_{i,\text{clr}}$ is the calculated clear-column radiance using the retrieved temperature profile. For any set of channels, P_c and α can be determined which minimize the difference between the observed and computed radiances for the channels. In the current analysis, two channels were used and α and P_c were found such that $\alpha_i(P_c) = \alpha_j(P_c) = \alpha$.

In the global retrieval programme, one temperature retrieval is performed every $250 \times 250\,\text{km}$, and is localized in the $125 \times 125\,\text{km}$ quadrant having the field of view containing the warmest observed brightness temperature. This quadrant is chosen because it is assumed to be the single least cloud contaminated field of view. If cloud parameters are retrieved from the same quadrant as the temperature field, the general cloudiness would be systematically underestimated. To provide an estimate of cloudiness over the entire $250 \times 250\,\text{km}$ area, radiances from all four quadrants are averaged to be used in equation 5.33 for determination of cloud fraction, given a cloud height. The retrieved temperature profile from the clearest quadrant is assumed to be valid for the entire $250 \times 250\,\text{km}$ area and is used in equation 5.32 to estimate the cloud radiance as a function of cloud top pressure.

To maximize stability and minimize the effects of errors and uncertainties, the numerator and denominator of equation 5.33 should be maximized. Therefore, both the full overcast and the observed radiances should be as different as possible from the clear-column radiance. For this reason, Channels 6 and 7, the two $15\,\mu\text{m}$ channels sounding closest to the surface, are utilized for cloud height determination, and the observations used to determine cloud height are taken from the single coldest of the eight fields of view in the $250 \times 250\text{km}$ area. In addition, for very low clouds, for which $R_{i,\text{cld}}$ is not very different from $R_{i,\text{clr}}$, the numerator and the denominator of equation 5.33 become small and spurious values of α can be obtained. For this reason the cloud height is not allowed to have pressures greater than $850\,\text{mb}$ and is constrained to be no higher than the tropopause. Using this cloud height, the effective cloud fraction for the entire area is determined from the radiances of Channels 6 and 7 averaged over the entire field of view.

The cloud parameters obtained are effective in the sense that they reproduce the outgoing longwave radiation but not necessarily the detailed cloudiness. Under multiple cloud layers, for example, a single intermediate cloud height would be found and the total cloud fraction would be underestimated.

If the retrieval performed in the quadrant with the warmest observation has been rejected, cloud parameters can still be determined in an identical fashion, but the initial guess is used in equations 5.5 and 5.32 to compute clear and cloudy radiances rather than the solution. In the special case where overcast low level cloudiness has been found, 100% cloud cover is set at the pressure at which the guess temperature is equal to the retrieved surface temperature. If this pressure is greater than $600\,\text{mb}$, a second layer of clouds is looked for if the brightness temperatures in the coldest field of view are significantly lower than in the warmest field of view. The procedure is identical to that described previously, but it is assumed that there is complete overcast of the lower cloud deck throughout the entire 250×250 km field of view and the surface contribution to equation 5.5 is taken to come from the lower cloud level rather than the Earth.

This procedure can provide cloud parameters under almost all conditions. However, approximately 20 % of the time it is apparent from the observations that partial cloudiness exists and no consistent cloud height and cloud fraction can be determined using Channels 6 and 7 (equation 5.33). In such situations a cloud field is not returned. Thus, use of vertical sounder data is a new method of cloud field and characteristic retrieval, in addition to those described in detail in Chapter 6. At present a new procedure which will determine the amount and height of clouds by minimizing the residual of computed radiances for more than two tropospheric sounding channels to produce improved cloud parameters under all conditions is being tested.

5.6. Results for January 1979

440000 retrievals were run for the period 5 January–2 February 1979, with one retrieval every 250×250 km area. Of these, 60 % were acceptable and 40 % were rejected. For the class of acceptable retrievals, $36 \cdot 7$ % were found to be in clear cases, $21 \cdot 7$ % were in cases where no cloud field could be produced, $25 \cdot 6$ % were in cases with cloud fractions greater than zero but less than or equal to 40 %, 15 % were in cases of cloud fraction greater than 40 % but less than or equal to 70 %, and only 1 % were in cases of cloud fraction greater than 70 %. For the class of rejected retrievals, $29 \cdot 7$ % were clear, $18 \cdot 7$ % had no cloud field returned, $16 \cdot 5$ % were in cases of less than 40 % cloudiness, $14 \cdot 2$ % were in cases of between 40 % and 70 % cloudiness, and $20 \cdot 9$ % were in cases of cloudiness more than 70 %. Successful retrievals outnumbered rejected retrievals for all cloud fractions up to 60 % and were of comparable amount in the 60–70 % range. It is interesting to note also that for up to 70 % cloud cover the majority of rejected retrievals occurred because of non-convergence, and that the distribution of retrieved cloud amounts in non-convergent retrievals was similar to that in accepted retrievals. This indicates that successful HIRS2 retrievals can be performed in areas with a cloud fraction of up to 70 %. Retrievals indicating a cloud fraction of over 70 % in the 250×250 km area represent only 9 % of the total cases.

Figure 5.2 indicates the distribution of successful retrievals in terms of coverage per day on the 4° latitude $\times$ 5° longitude grid used in the GLAS general circulation model. Each grid point is counted 1 or 0 depending on whether a successful retrieval is found in a given six-hour period. The maximum possible yield is, in general, 2 retrievals per day because most

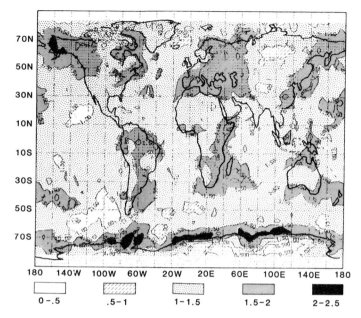

Figure 5.2. Average number of times per day at least one successful GLAS retrieval was performed per six-hour period in a 4° × 5° box in January 1979. 'Perfect' coverage is nominally twice per day.

grid points (except for high latitude points) are covered only twice daily. As shown in figure 5.2, retrieval coverage of at least one time per day is almost global, with the exception of a few areas where persistent cloudiness occurred for moderate periods of time. Conspicuous by their absence are features due to the ITCZ or mountain areas, indicating that these factors do not decrease the retrieval yield significantly.

Temperature retrieval accuracy

Accuracies of the retrieved temperature profiles for 5 January– 2 February 1979 are shown in figure 5.3, comparing mean layer temperatures of GLAS retrievals, in the nine pressure intervals shown, with mean layer temperatures reported by radiosondes colocated in space to 110 km and in time to 3 h. Retrievals flagged bad are not included in the statistics. In addition, retrievals flagged bad in the stratosphere are not included at 200 mb and above. Also shown are retrieval accuracies of the operational NESS retrievals for the same period of time†. In the case

---

†Subsequent to this period of time, NESS introduced changes to their operational clear-column radiance algorithm (McMillin and Dean, 1982), but this new processing system has not been applied to the January, 1979 data.

J. Susskind

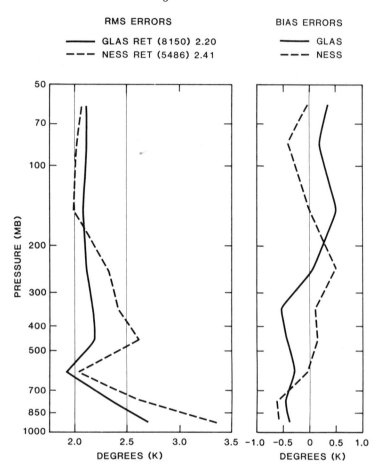

Figure 5.3. Errors of retrieved mean layer temperatures compared to colocated radiosonde for 5 January–2 February 1979 (± 3h, ± 110km).

of NESS retrievals, reported mean layer virtual temperatures are compared to virtual temperatures derived from the radiosonde temperature–humidity profiles. The GLAS retrievals are seen to be significantly more accurate in the troposphere, though slightly less accurate in the stratosphere. The total rms error of the 8150 colocated GLAS retrievals is $2 \cdot 20 \, \text{K}$, $0 \cdot 21 \, \text{K}$ lower than that of the 5486 colocated NESS retrievals. Also shown in figure 5.3 are the bias errors of the retrievals. The GLAS retrievals tend to have a $0 \cdot 4 \, \text{K}$ cold bias in the troposphere and a $0 \cdot 4 \, \text{K}$ warm bias in the stratosphere. The causes of these biases are currently under investigation. A tendency has been found for the bias to disappear in clear areas and be largest in tropical cloudy areas.

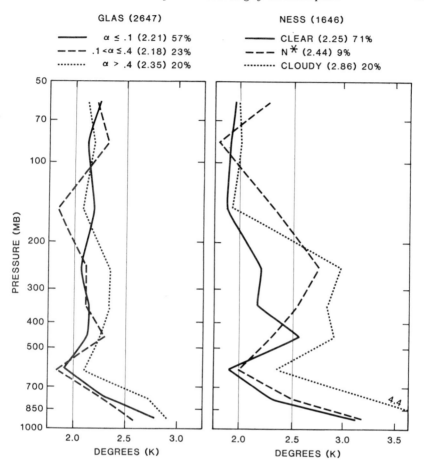

Figure 5.4. Rms mean layer temperature error compared to radiosondes (± 3 h, ± 110 km) vs. degree of cloudiness for 5 January–15 January 1979. GLAS retrievals are separated according to retrieved cloud fraction α. *NESS retrievals are separated according to reported retrieval type.*

Figure 5.4 shows a breakdown of the error statistics into categories of varying cloud cover for the period 5–15 January 1979. The NESS retrievals were stratified according to the reported retrieval type. In the retrievals marked clear, NESS treated the HIRS2 observations as not cloud contaminated and they applied no cloud correction to the HIRS2 radiances. In the retrievals marked N*, NESS performed a correction to account for cloud effects on the HIRS2 observations before the retrieval was performed. In the retrievals marked cloudy, the effects of clouds on the HIRS2 observations were considered by NESS to be too large to be accounted for accurately and only HIRS2 Channels 1–3, sounding the stratosphere, were used in the retrieval, together with the MSU and SSU observations. The GLAS retrievals were partitioned according to almost clear ($\alpha \leqslant 0 \cdot 1$), partially cloudy ($0 \cdot 1 < \alpha \leqslant 0 \cdot 4$), and highly cloudy ($\alpha > 0 \cdot 4$) conditions. The latter cut-off appears to be about the region where NESS began to apply their cloudy algorithm. NESS has a considerably higher percentage of clear retrievals than GLAS has. The main reason for this difference is that the GLAS cloud fraction refers to the entire 250×250 km area while NESS clear refers to the existence of clear 30 km spots.

The accuracy of the GLAS retrievals is seen to degrade much less with increasing cloudiness than that of the NESS retrievals. GLAS retrievals with cloud fractions between $0 \cdot 1$ and $0 \cdot 4$ are in fact comparable in accuracy to those obtained under clearer conditions. The NESS N* retrievals show a large degradation over the clear retrievals in the upper troposphere, possibly due to the effects of multiple cloud layers on the radiances. Such cases would hopefully be identified and flagged in the GLAS retrieval system. The NESS clear retrievals appear to degrade somewhat in the 400–500 mb layer, possibly also due to a residual cloud effect. The NESS cloudy retrievals, using only two microwave channels to sound the troposphere, are of significantly lower quality, primarily due to lack of data. The GLAS retrieval system shows that reasonably accurate retrievals, using both the HIRS and MSU channels, can be performed under almost all cloud conditions.

Sea surface temperatures (SST)

The sea/land surface temperatures produced by the GLAS retrievals can be used to produce global monthly mean fields of temperatures and their diurnal variations. In particular, monthly mean surface temperature fields are highly significant for climatological studies. Conventional *in situ* sea surface temperature measurements from ships and buoys are numerous in the Northern Hemisphere but coverage is sparse in space and time in the tropics and Southern Hemisphere. Sea or land surface temperatures, averaged over the 125×125 km area, are retrieved from each successful HIRS2/MSU retrieval except those in which the sea surface temperature was held fixed at climatology. At night, equation 5.26 is used to obtain the surface temperature while equations 5.27–5.29 are used during the day. Figure 5.5 shows the mean sea surface temperature field derived for the period 5 January–2 February 1979, obtained by averaging all sea surface temperature retrievals in $4° \times 5°$ latitude–longitude bins. No smoothing, rejection criteria other than those described in the text, or adjustments for bias removal, were applied to the

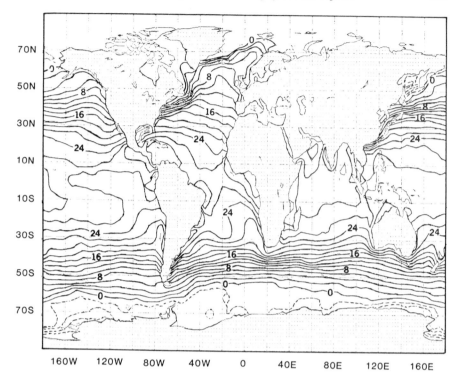

Figure 5.5. Monthly mean sea surface temperature field for January 1979 (from HIRS2/MSU), obtained by averaging all successful sea surface temperature retrievals in the 4° latitude × 5° longitude grid boxes and contouring the results. No empirical correction was undertaken on the data. The surface temperature is determined primarily by the $3 \cdot 7\mu m$ and $4 \cdot 0\mu m$ channels.

data. Differences between SST analyses using only night (3 a.m. local time) retrievals and using only day (3 p.m. local time) retrievals were very small and never more than 2 K in the open ocean. This indicates that the procedures used to account for the effects of solar radiation on the $3 \cdot 7 \mu m$ and $4 \cdot 0 \mu m$ channel observations can produce accurate daytime sea surface temperatures using these short-wave window channels. Both day and night retrievals are included in figure 5.5. The major climatological sea surface temperature features, such as the Gulf Stream, the Kuroshio, Humboldt and Benguela currents, and the sea surface temperature minimum at the equator in the eastern Pacific are readily observable.

Of particular interest is the sea surface temperature anomaly field for January 1979 shown in figure 5.6(a) obtained by subtracting the NCAR climatology, based on an average of data from 20 Januaries, from the GLAS January 1979 sea surface temperature field. The deviations from climatology are small, being less than 2 K in all cases, with the exception

194   *J. Susskind*

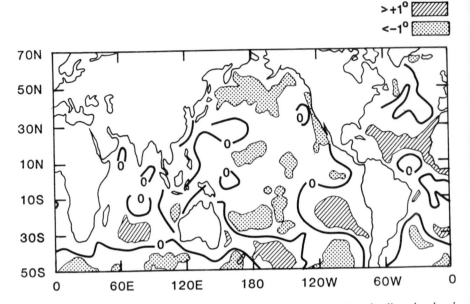

Figure 5.6(a). *Difference between the sea surface temperature field in figure 5.5 and a climatology based on an average of 20 Januaries compiled by NCAR.*

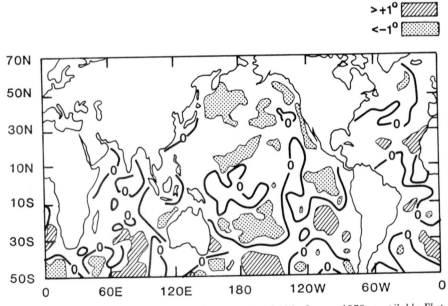

Figure 5.6(b). *Difference between the sea surface temperature field for January 1979, compiled by Fleet Numerical Weather Central from ship and buoy measurements, and the same NCAR January climatology.*

of the centres of the warm anomaly west of South America and the cold anomaly midway between South America and Australia.

Figure 5.6(*b*) shows the anomaly field for January 1979 computed by subtracting an analysis based on ship and buoy measurements, compiled by the Fleet Numerical Weather Center, from the same NCAR January climatology. The Fleet Numerical Analysis can be taken as a measure of ground truth in the areas of dense coverage. Agreement of the major anomaly features in the Northern Hemisphere, such as the cold Pacific areas north of 40° off the west coast of North America and centred at 180°, 10°N, as well as the warm Atlantic off the west coast of Africa, is excellent. When considering this map, it should be remembered that no bias errors were removed from the retrieved sea surface temperatures. Even small biases of a few tenths of a degree would have a significant effect on the location of the contour lines, especially the 0K bias line, which also matches extremely well. It can be concluded that the absolute accuracy of the climatological sea surface temperature data is quite high. Detailed comparison of the analyses gave rms differences of $0 \cdot 4$K in the North Atlantic Ocean and $0 \cdot 6$K in the North Pacific. Agreement in the Southern Hemisphere is also extremely good. Note, for example, the excellent agreement of the oscillating warm–cold–warm anomaly pattern in the latitude band from 10°S to 30°S. The largest difference in the fields occurs south of 40°S, where the conventional data field is noisy as a result of sparse data. The apparent large difference in the fields between 60°E–120°E and 40°S–50°S is in fact only a 1K difference in temperatures in this area. In general, the Southern Hemisphere anomaly field retrieved from the satellite data is less noisy and better defined than that from conventional data. Sea surface temperature retrieval is discussed further in Chapter 7.

Effective cloud cover and cloud height

Effective cloud heights and cloud fractions, consistent with upwelling longwave radiation, are derived for each 250×250km area through use of equations 5.32 and 5.33. In the special case of low-level overcast, 100 % cloudiness is assigned to the level whose temperature equals the retrieved surface temperature. The cloud-top pressure and cloud fraction are effective for two reasons. First, possible multiple-level cloud formations are assigned a single effective cloud-top pressure, which should lie somewhere in the range of the cloud-top pressures in the field of view, generally close to that of the highest (coldest) clouds. Second, the effective cloud fraction is a function of the computed effective cloud-top pressure for a given situation, with decreasing cloud-top pressure (increasing cloud-top height) corresponding to decreasing cloud fraction. Conse-

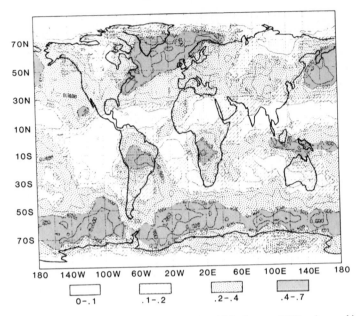

Figure 5.7. Average effective cloud fraction retrieved by GLAS for January 1979 using combined 3 a.m. and 3 p.m. soundings. The cloud parameters are determined primarily from Channels 6 and 7, the two 15 μm channels sounding lowest in the atmosphere.

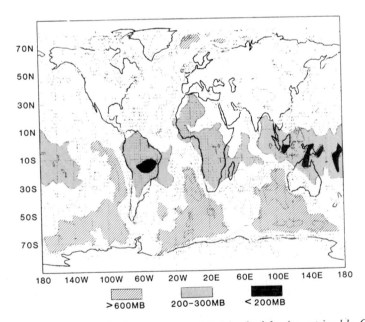

Figure. 5.8. Average effective cloud-top pressure, weighted by cloud fraction, retrieved by GLAS for January 1979. Only the areas of highest clouds ($P_c < 300$ mb) and lowest clouds ($P_c > 600$ mb) are shaded.

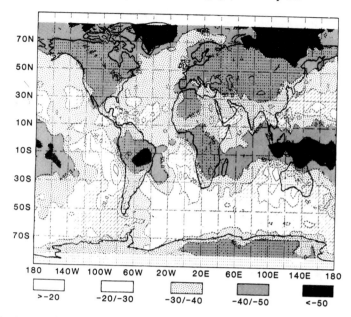

Figure 5.9. Average effective cloud-top temperature, weighted by cloud fraction, retrieved by GLAS for January 1979. Temperatures are in °C.

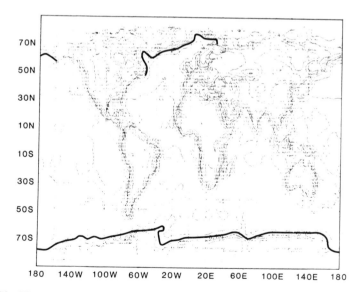

Figure 5.10. The average global surface emissivity at 50·3 GHz derived by GLAS for January 1979. Resolution is at the 4° × 5° grid. Emissivities for all the zenith angles of MSU have been averaged together without correction for angle. The 0·7 emissivity contour over oceans is indicated by the dark line. This contour is taken to be indicative of the extent of sea ice.

quently, cloud cover from low-level clouds may be significantly underestimated. Moreover, even for single-layer clouds, the effective cloud fraction corresponds to the cloud emissivity times the actual cloud fraction because an emissivity of one was assumed in generating the cloud fields. This means that cirrus clouds will be underestimated by up to a factor of 7 and mid-level clouds by up to 30 % (Paltridge and Platt, 1976).

Monthly mean fields of cloud fractions, cloud-top pressures and cloud-top temperatures were produced by averaging the retrieved cloud parameters in the 4° × 5° grid for the period 5 January–2 February 1979. The 20 % of the cases where no cloud field was retrieved were not included in the averages. The average cloud-top pressures and temperatures were taken as the average of appropriate quantities weighted by the corresponding cloud fractions. Cases in which no cloud field can be retrieved tend to have low–intermediate total cloud cover and their omission most likely does not affect the average cloud statistics significantly.

Figures 5.7–5.9 show contours of average cloud-top fractions, cloud-top pressure and cloud-top temperature for the period. The major features such as the Intertropical Convergence Zone, the storm tracks over the Atlantic and Pacific Oceans, the oceanic deserts and the region of the Siberian high are clearly visible in the cloud fraction map. In the cloud-top pressure map, in which only areas of high (< 300 mb) and low (> 600 mb) clouds have been indicated, the Intertropical Convergence Zone is again clearly visible, as well as the extensive areas of predominantly stratus cloud cover south of 50°S and off the west coast of the Southern Hemisphere continents south of 20°S. The global average effective cloud cover obtained for January 1979, sampled at 3 a.m. and 3 p.m. local time, is 25 %, a value considerably lower than the accepted value of 50 % global cloud cover based on ground observations—see Chapter 1 and especially comparisons of climatologies from surface and remote sites.

The cloud-top temperature field (figure 5.9) appears similar to the cloud pressure map (figure 5.8) in the tropics but very cold mid-level clouds are found in the Northern Hemisphere over land. It is interesting to note that the − 40°C contours, corresponding to the formation of ice clouds, closely follows the continental contours in the Northern Hemisphere.

Microwave surface emissivity—ice and snow cover

The emissivity of the surface in the microwave region is a strong function of surface conditions. At 50 GHz, open ocean has emissivity values ranging from 0·45 to 0·65; increasing with decreasing temperature and also increasing with increasing foam cover, which is a measure of wind speed. Land has emissivities typically greater than 0·9. Ice over ocean has an emissivity of the order of 0·7 or more depending on its history and snow over land has an emissivity of the order of 0·9 or less. Thus, the surface emissivity can give a measure of snow covered land, ice cover over ocean and possibly also boundary-layer wind speed over open ocean.

Passive microwave sounders designed to measure surface properties, such as ESMR and SMMR, use frequencies less than 40GHz in order to avoid attenuation by atmospheric oxygen, which mixes atmospheric effects into the signal (see Chapter 8 for a detailed discussion). Use of equation 5.31 accounts for atmospheric effects and also allows for the incorporation of surface temperatures, obtained from the infra-red 3·7µm channels, into the determination of surface emissivity from the HIRS2/MSU system. The spatial resolution of MSU is considerably lower than that of ESMR or SMMR, however, and this will degrade high-resolution features which could be significant for some applications.

Figure 5.10 shows the surface emissivity, averaged over the 4° × 5° grid for January 1979. As expected, the most obvious features are the continents, showing rapid gradients from 0·9 to 0·6. The intermediate emissivity values are partially due to contouring and partially due to soundings which include mixed fields of view at the coastlines. The

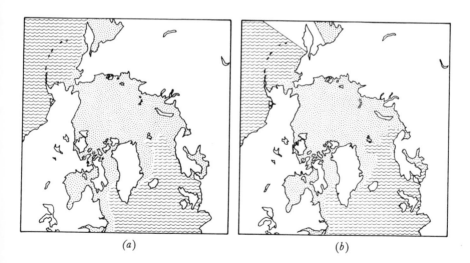

(a)   (b)

Figure 5.11. (a)North Polar projection showing average sea ice extent for January inferred from the 0·7 emissivity contour shown in figure 5.10. Sea ice is indicated by the dots; open ocean by the waves. The MSU spot size is 125 × 125km at nadir.
(b) Sea ice extent for January 1979 from analysis of SMMR data, with a 25 × 25km resolution. The ice/ocean line was taken as the line which represented 40% ice cover for the period (see discussion of the sea ice extent estimation in Chapter 8).

emissivity in South America, Africa and Australia is uniformly greater than 0·9. North America, Eurasia and Antarctica show complicated patterns containing lower emissivities, indicative of snow cover.

The oceans show generally expected features of emissivity between 0·45 and 0·65, with emissivity increasing with decreasing sea surface temperature. A distinct 0·7 contour, indicated by the solid line in figure 5.10, is observed in the Antarctic Ocean and in the Bering, Labrador, Greenland and Barents Seas. This contour can be interpreted as a measure of the sea-ice extent in these areas. Cavalieri has investigated sea ice and associated atmospheric interactions in both polar regions (see Chapter 8). Figure 5.11(a) shows the sea-ice line, as deduced from HIRS2/MSU, given by the 0·7 emissivity contour over water. Figure 5.11(b) shows the sea-ice line determined, by Cavalieri, from SMMR, with a 25 km resolution. The ice margin from SMMR data was taken as the contour of sea-ice coverage greater than 40%. The agreement is quite good, considering the difference in resolution of the instruments.

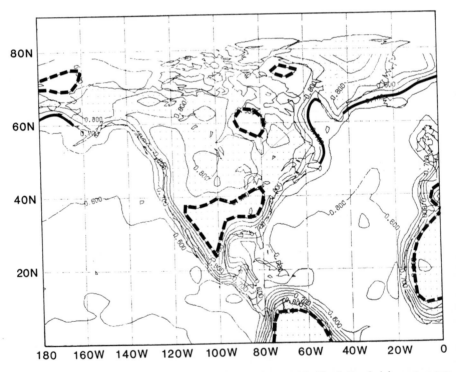

Figure 5.12. Enlargement of the north-west quadrant of figure 5.10. The 0·7 emissivity contour over ocean, indicative of sea ice extent, is shown as the solid line, as in figure 5.10. The 0·9 contour, indicative of land, is shown as the dashed line. Near oceans, it represents the land boundary. Inland, as in North America, it is indicative of snow-covered land. Further north, it is indicative of solid newly frozen sea ice.

The details of the emissivity over land may be indicative of snow cover. Figure 5.12 shows the averaged surface emissivity over North America. The $0 \cdot 7$ emissivity contours indicative of the onset of significant amounts of sea ice, and the $0 \cdot 9$ contour, indicative of typical snow-free land, are marked. Most of North America has land emissivities less than $0 \cdot 9$. Snow fields derived by averaging the weekly observations produced operationally by NOAA/NESS into the $4° \times 5°$ grid for the month of January indicate that the $0 \cdot 9$ emissivity contour lies completely in the snow-free area, while the $0 \cdot 85$ contour closely approximates the 60 % snow cover contour.

5.7. Summary and possible future developments

The physical principles affecting the determination of atmospheric temperature profiles from observation of outgoing infra-red and microwave radiation have been discussed. It was shown that simultaneous analysis of infra-red and microwave observations enhances the strength of observations in each spectral region. Infra-red soundings have higher sensitivity to lower tropospheric temperatures, ground temperatures and sea surface temperatures (see Chapter 7). Microwave observations are better for sounding the stratosphere, are sensitive to snow and ice cover (Chapter 8) and greatly aid correction of the infra-red observations for cloud effects (Chapter 6). In addition to atmospheric temperature profiles, sea and land surface temperatures, effective cloud parameters and snow and ice cover can be inferred from the observations. Thus 'temperature' sounders can also give information about a number of important climatic parameters.

Further improvements in instrumentation can be made. Kaplan *et al.* (1977) have shown that if outgoing radiation can be measured with higher precision at narrower spectral intervals, with $\Delta \nu \approx 0 \cdot 5 - 2 \, cm^{-1}$, then a set of $15 \mu m$ channels can be selected which are insensitive to H_2O and O_3 and have upper tropospheric and stratospheric sensitivity functions which are comparable to those found in the microwave region. In addition, $4 \cdot 3 \mu m$ channels can be selected which have much greater vertical resolution in the mid–lower troposphere than those of the HIRS2, and $3 \cdot 7 \mu m$ window channels can be found in which the atmosphere is almost completely transparent. Design studies for such an instrument are currently under way. In addition, an improved microwave sounder with higher spatial resolution and more sounding channels is under development. In combination, these instruments should greatly improve the quality of all products retrieved from the current sounding system.

References

Chahine, M. T., 1968, Determination of the temperature profile in an atmosphere from its outgoing radiance. *Journal of the Optical Society of America* **58**, 1634–37.

Chahine, M. T., 1974, Remote sounding of cloudy atmospheres. I. The single cloud layer. *Journal of the Atmospheric Sciences* **31**, 233–43.

Kaplan, L. D., 1959, Inference of atmospheric structure from remote radiation measurements. *Journal of the Optical Society of America* **49**, 1004–1007.

Kaplan, L. D., Chahine, M. T., Susskind, J. and Searl, J. E., 1977, Spectral band passes for a high precision satellite sounder. *Applied Optics* **16**, 322–25.

McMillin, L. M. and Dean, C., 1982, Evaluation of a new operational technique for producing clear radiances. *Journal of Applied Meteorology* **21**, 1005–1014.

Paltridge, G. W. and Platt, C. M. R., 1976, *Radiative Processes in Meteorology and Climatology* (Amsterdam: Elsevier Scientific Publishing Co.), pp. 3, 6, 206.

Rodgers, C. D., 1976, Retrieval of atmospheric temperature and composition from remote measurements of thermal radiation. *Reviews of Geophysics and Space Physics* **14**, 609–24.

Smith, W. L., 1968, An improved method for calculating tropospheric temperature and moisture from satellite radiometer measurements. *Monthly Weather Review* **96**, 387–96.

Smith, W. L. and Woolf, H. M., 1976, The use of eigenvectors of statistical covariance matrices for interpretation of satellite sounding radiometer observations. *Journal of the Atmospheric Sciences* **33**, 1127–40.

Susskind, J., Rosenfield, J. and Reuter, D., 1983, An accurate radiative transfer model for use in the direct physical inversion of HIRS2 and MSU sounding data. *Journal of Geophysical Research* **88**, 8550–68.

6
Cloud identification and characterization from satellites

James T. Bunting and Kenneth R. Hardy
US Air Force Geophysics Laboratory
Hanscom Air Force Base, MA 01731, USA

6.1. Atmospheric windows for cloud detection from satellites

This chapter summarizes several approaches to detecting clouds from satellites. In contrast to the previous chapter, emphasis is placed on *imaging* sensors responding to radiation in atmospheric *windows*, since these sensors provide the most information on cloud properties and their distribution in time and space. Imaging sensors record the considerable variability of clouds as viewed by satellites. Atmospheric windows are used so that low clouds as well as high clouds can be seen. For example, the top of a low cloud layer or fog might be only 100 m above sea level; in this case 99 % of the mass of the atmosphere would be between the cloud top and the satellite instrument so that the energy sent from the cloud towards the satellite will be very sensitive to attenuation by atmospheric molecules. For the highest clouds, such as in thunderstorms near the equator with tops above 17 km, almost 10 % of the atmospheric mass would still be between the cloud top and the satellite.

Figure 6.1 shows the spectral locations of atmospheric windows for visible and infra-red radiation with wavelengths from 0.1 to 28 μm. Atmospheric windows which have been used for cloud detection are in the approximate ranges of $0\cdot5-1\cdot3$, $1\cdot6-1\cdot7$, $2\cdot1-2\cdot3$, $3\cdot5-4\cdot0$ and $8-13\,\mu$m. The spectral response of satellite imagery is usually narrower than these bandwidths since it is desirable to avoid some molecular absorption bands within the windows, such as water vapour at $1\cdot14\,\mu$m or ozone at $9\cdot6\,\mu$m, and also to take advantage of variations in the reflectivity of Earth surfaces. For example, the reflectivity of most land surfaces tends to increase as a function of wavelength between $0\cdot5$ and $1\cdot0\mu$m, so that a spectral response from $0\cdot55$ to $0\cdot75\,\mu$m may give a sharper contrast for clouds over land than a response from $0\cdot5$ to $1\cdot0\,\mu$m. For the shorter-wavelength windows, sunlight is reflected so that they are useful

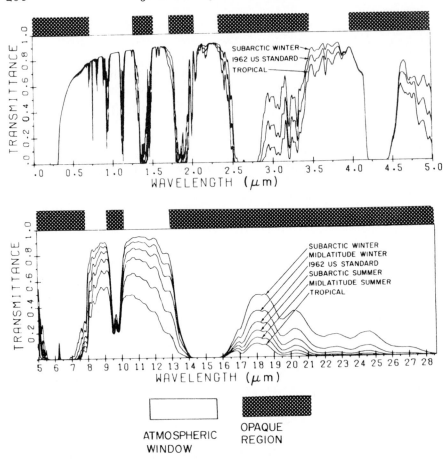

Figure 6.1. Atmospheric transmittance for a vertical path to space from sea level for six model atmospheres, calculated by Selby and McClatchey (1975). Atmospheric windows used for cloud detection are distinguished from more opaque wavelengths (available from Air Force Geophysics Laboratory Technical Report, *No. 75–0255, Hanscom, Mass. 01731, USA).*

only in daytime unless low-light-level technology is employed to sense reflected moonlight. The windows at longer wavelengths sense thermal energy originating from the clouds or their backgrounds plus smaller contributions from the atmosphere itself. They are used day or night. The $3 \cdot 5 - 4 \cdot 0\,\mu m$ window senses both reflected sunlight and thermal radiation and marks the transition from the reflected solar regime to the thermal infra-red region.

Thermal windows are also found at longer wavelengths known as millimetre or microwave frequencies. Cloud information is provided at wavelengths near $0 \cdot 32$, $0 \cdot 81$ and $1 \cdot 55$ cm, and the atmospheric transmittance for these windows is about 90 % from sea level to the satellite. These windows are more sensitive to the larger water particles and raindrops within clouds than they are to the entire cloud.

A common choice of bands for routine weather satellites is $10 \cdot 5 - 12 \cdot 5$ and $0 \cdot 55$ to $0 \cdot 75 \, \mu m$. Although most cloud information is provided by measurements within these windows, there are exceptions. The sounding channels discussed in Chapter 5 and the water vapour band centred at $6 \cdot 3 \, \mu m$ provide important information on high but thin clouds. Future satellites carrying active sensors, such as lasers, may be able to provide more information on cloud structure than the passive sensors currently in operation.

6.2. *Physical and optical properties of clouds*

Cloud optical properties such as reflectivity or emissivity depend on cloud physical properties as well as the wavelength of radiation observed by a satellite. For many applications, it is important to distinguish cloud particles from the larger precipitation particles and water particles from ice particles. Water-cloud particles tend to be much smaller than ice-cloud particles. A water cloud such as stratocumulus, covering broad areas of oceans, might contain cloud droplets with radii up to $30 \, \mu m$ in total number concentrations of about $300 \, cm^{-3}$ and with a total liquid mass density of $0 \cdot 3 \, g \, m^{-3}$. An ice cloud such as cirrostratus might contain ice particles shaped like hexagonal prisms with mass-equivalent radii of $10 - 80 \, \mu m$, number concentrations up to $0 \cdot 1 \, cm^{-3}$, and a total ice density up to $0 \cdot 05 \, g \, m^{-3}$. (These numbers are representative only. Real clouds have a splendid diversity which is difficult to summarize.) In short, water clouds have many small particles while ice clouds have relatively few but larger particles; however, both water and ice clouds have particle size distributions peaking at small sizes with fewer large particles.

Given the particle sizes and numbers, it is possible to calculate the radiation sent from the cloud towards the satellite. The calculations start with the amount of energy scattered, absorbed and emitted by individual particles. The results depend on the refractive index, which varies with the wavelength of the radiation and the water or ice phase of the particle. The results depend on the radius of the particle r and wavelength λ consistent with the relation

$$x = \frac{2 \pi r}{\lambda} \tag{6.1}$$

where x is known as the Mie size parameter and is dimensionless. Calculations are accurate for water droplets since their spherical shapes make Mie theory applicable. Calculations are less certain for ice particles since the particles often have irregular shapes.

Fine resolution weather-satellite imagery will have an individual field of view of about 1 km at cloud altitudes and will obviously sense billions of cloud particles for every cloudy picture element. Hence the radiative interaction among particles known as multiple scattering must be included in calculations to simulate satellite detection of clouds.

Results of various calculations of cloud optical properties are summarized in table 6.1. In this table, both water clouds and ice clouds are assumed to be horizontally homogeneous or stratiform and optically thick or poorly transmitting at short wavelengths. Both water and ice clouds have high reflectivities due to low emission/absorption at $0 \cdot 5 - 1 \cdot 3 \, \mu m$. Their high reflectivities make them important to the Earth's radiation budget since solar energy peaks in this bandwidth. The low absorption is directly observable since it can be seen that water and ice are fairly transparent. The high reflectivities of the clouds are due to sunlight refracted and scattered back to the satellite by cloud particles.

In the $1 \cdot 6 - 1 \cdot 7$, $2 \cdot 1 - 2 \cdot 3$ and $3 \cdot 5 - 4 \cdot 0 \, \mu m$ windows, water clouds tend to reflect more than ice clouds so that these bands can be used to distinguish water clouds from ice clouds during daytime. In the $8 - 13 \, \mu m$ window, both water clouds and ice clouds are poor reflectors but good emitters; therefore the radiance emitted by the clouds can be used to estimate the temperature of the clouds by means of the Planck function, which relates temperature and wavelength to emitted radiance (see Chapter 2). Some ice clouds are optically thin since they transmit a significant amount of energy from beneath the cloud and emit correspon-

Table 6.1. Representative cloud responses in atmospheric windows summarized from various sources. These numbers vary with cloud depth, width and particle size. Only typical values are given here. The radiation calculations assume that the clouds are horizontally extensive and that radiation sensors are looking straight down.

Response	Wavelength	Water cloud	Ice cloud	Ice cloud over rain
Reflectivity of sunlight when the Sun is overhead	$0 \cdot 7 \, \mu m$ $1 \cdot 7 \, \mu m$ $2 \cdot 1 \, \mu m$ $3 \cdot 7 \, \mu m$	$0 \cdot 80$ $0 \cdot 70$ $0 \cdot 50$ $0 \cdot 30$	$0 \cdot 80$ $0 \cdot 20$ $0 \cdot 05$ $0 \cdot 10$	$0 \cdot 80$ $0 \cdot 20$ $0 \cdot 05$ $0 \cdot 10$
Emissivity for thermal radiation	$3 \cdot 7 \, \mu m$ $11 \cdot 5 \, \mu m$	$0 \cdot 70$ $0 \cdot 95$	$0 \cdot 90$ $0 \cdot 95$	$0 \cdot 90$ $0 \cdot 95$
Temperature increase, due to clouds over an ocean (K)	$0 \cdot 32$ cm (94 GHz) $0 \cdot 81$ cm (37 GHz) $1 \cdot 55$ cm (19 GHz)	24 12 3	0 0 0	41 30 11

dingly less energy from the cloud. The energy sensed by a satellite is not a reliable indicator of the cloud temperature for these thin cirrus clouds and the clouds usually appear warmer than the true cloud temperature, due to warmer radiances transmitted through the thin ice clouds. Confirmation that many of these clouds are optically thin is provided by the fact that the Sun or Moon is often visible from the ground through these clouds.

At the microwave windows listed in table 6.1, water clouds also differ from ice clouds. At these frequencies, water clouds are partial absorbers of radiation while ice clouds are nearly transparent. Water clouds with raindrops and substantial water content stand out as warm regions against the cold background of oceans, which are poor emitters at these frequencies. Ice particles will scatter microwaves so that ice clouds may act as reflectors at the shorter wavelengths. However, looking down from a satellite there is practically no energy from space to be reflected back to the satellite and the Earth's surface or water clouds can usually be seen through the ice clouds. Land backgrounds have emissivities in the range of $0 \cdot 75$ to $0 \cdot 95$ in the microwave region; these are much higher than for oceans. In contrast, the emissivities for radiation at $8-13\,\mu$m are close to $1 \cdot 0$. Consequently, in the microwave region, neither water clouds nor ice clouds contrast with land backgrounds unless the water clouds contain ample raindrops.

The microwave windows also differ from the other windows since the radiation sensed at the satellite may be highly polarized. As a result, microwave imagers have been designed to scan conically so that the angle of incidence to the Earth's surface remains constant during the scan. The horizontal as well as vertical polarizations are measured for each window which provides additional information on the medium being observed.

6.3. Satellite instruments for cloud detection

Imaging sensors, often called imagers (see Chapter 2), are generally used for cloud identification and characterization from satellites. These sensors are designed to provide broad horizontal coverage of areas beneath the satellite at the finest horizontal resolution which can be sensed, stored, or transmitted from the satellite and used back on Earth. The fine resolution is required for the detection of the smallest clouds such as cumulus or scattered cirrus. The finest resolution available from weather satellites is about $0 \cdot 5-1$ km, and this resolution is still unable to detect some clouds which can be seen at the ~ 80 m resolution of Landsat. The finer resolutions of data are very useful for image analysis by meteorologists, but are not fully utilized in automated analysis due to the massive quantities of data involved. Moreover, sensing and spacecraft requirements limit the amount of high-resolution data available. Geostationary satellites are capable of 1 km visible and 5 km infra-red resolution directly under the satellite. These figures apply to the $0 \cdot 55 - 0 \cdot 75 \mu m$ band of GOES satellites and the $10 \cdot 5 - 12 \cdot 5 \mu m$ band of METEOSAT. Polar-orbiting satellites are capable of 1 km or better resolution in both visible and infra-red channels, but they can store only limited quantities of the data on the spacecraft, so that data must be smoothed on the spacecraft to 3 or 4 km resolution for most of the Earth.

The scanning part of the radiometer is usually a mirror which reflects energy into a Ritchey–Chrétien or a Cassegrainian telescope. Both visible and infra-red energy are handled through the same optics, although different detection elements are used for different spectral bands. A visible band is usually sensed by photomultiplier tubes or silicon diode detectors, while the $10-12 \mu m$ infra-red band is sensed with a HgCdTe detector. The spectral response of each band is determined by the sensor optics such as beam splitters or filters and by the detector properties; it can be measured precisely before launch. The thermal bands require extra attention since the sensor and spacecraft emit thermal energy which could contaminate the radiation from the Earth and clouds. Warm and cool reference temperatures are maintained and some parts of the sensor may be cooled by allowing them to radiate to space.

The spectral responses vary slightly from sensor to sensor even for two otherwise identical satellites from the same manufacturer. The sensor may degrade after launch, and inflight calibration checks are often limited since the sensor designs stress good horizontal coverage rather than radiometric accuracy. Consequently, calibration and intercomparison of the measurements from different satellites have been a recurring problem for quantitative applications of the imagery data, and particularly for the visible bands of imagers.

The fact that the Earth's surface is curved leads to four important effects when the surface is viewed from satellites. Firstly the degradation of the horizontal resolution of the imagery as a result of changing the nadir angle is accentuated since most sensors scan out for some distance from the satellite subpoint. Secondly, the data require corrections at oblique viewing angles for changing viewing geometry between the Sun, the cloud and the satellite for solar channels and for the increasing path through the atmosphere for infra-red channels. Thirdly, the satellite tends to see more clouds at oblique viewing angles since it looks through more atmosphere and has a higher probability of encountering a cloud. Moreover, it may confuse the sides of clouds with the tops of clouds. Finally, most Earth location procedures assign the pixel to a location at the Earth's surface based on spherical trigonometry. A high cloud may be mislocated horizontally by twice its altitude if it is viewed at an angle of 60° off local vertical. Although the oblique viewing leads to limitations in the scanned data, it is necessary in order to achieve adequate cloud coverage over space and time for the support of weather analysis.

Effects of the atmosphere and Earth's surface on microwave radiation are highly dependent on frequency, polarization and look angle. The atmospheric components of oxygen and water vapour contribute significantly to the total brightness temperature near the absorption lines at 22, 60, 118 and 183 GHz. Microwave channels are sometimes identified by frequency and sometimes by wavelength. The relation between λ and ν is $\lambda\nu = c$, where c is the speed of light. On the other hand, the effects of clouds and rain increase with increasing frequency and may form the dominant contributor to the total brightness temperature measured at high frequencies. Under conditions where the atmosphere is not opaque (a common occurrence at frequencies less than 20 GHz and between 25 and 40 GHz) the surface may be a significant contributor to the total radiation sensed. (These topics are more fully treated in Chapter 5).

The brightness temperature of the Earth's surface depends on surface type and condition. Water generally has a low emissivity while land has a high emissivity. Ocean emissivity, however, may increase with wind speed and land emissivity may decrease with increased soil moisture.

Channels selected for microwave measurements of clouds or precipitation are designed to maximize use of the differences in microwave responses to these geophysical parameters. For example, a channel near 19 GHz (1·6 cm wavelength) provides data on variations in surface parameters and was used in the Electrical Scanning Microwave Radiometer-5 (ESMR-5) by NASA to estimate mean rainfall amounts over the oceans. A channel near 37 GHz (0·8 cm wavelength) was used for ESMR-6 and also provided estimates of rainfall rates over the oceans. When the 19 and 37 GHz channels are used with a 22 GHz channel, which is highly sensitive to water vapour, then it becomes possible to estimate both total cloud water content and integrated water vapour over the oceans; both of these parameters are expected to be estimated using data from the Special Sensor Microwave Imager (SSM/I) which will be part of the DMSP in the mid-1980s. In addition to the 19, 22 and 37 GHz channels, the SSM/I will have an 85·5 GHz channel which will further enhance the instrument for obtaining cloud and precipitation information.

6.4. Interpretation of cloudy images

Using skills which are difficult and sometimes impossible to duplicate on computers, meteorologists with experience in the interpretation of satellite images can both identify and characterize clouds. The satellite images show spatial features of clouds and multispectral features when more than one spectral band is sensed. The human eye and brain have evolved to an exceedingly talented system for pattern recognition which has been applied to the interpretation of spatial features in images. As long as a clear/cloud contrast exists, people looking at cloudy images on photocopy or TV screens can readily adjust to different backgrounds, such as forest, farmland, mountains, oceans or coastlines, which have different reflectivities at solar wavelengths and often have different temperatures at thermal wavelengths; in an automated approach, these varying backgrounds may be confused with clouds. Recognizing geographical features such as lakes, river valleys or contrasts in vegetation tells the image analyst that clouds are not present. In fact, people looking at images can readily identify areas where clouds are often poorly detected by an automated analysis of the same image. The comparisons of automated cloud fields with the original satellite data in image format are important for the quality control of the automated cloud analysis.

Technology has evolved to support the image analyst in two important ways. Firstly, the spacecraft sensing, data transmission to Earth, assembly of scanlines to images and approximate Earth-location of the images are all automated so that the image analyst views the results of an automated process before he employs his personal skills of cloud interpretation. Secondly, this chapter will contain further references to interactive display and data access systems which are designed to display images on TV screens and allow the operator to specify enhancements or computations to be performed on a digitized version of the imagery data. These systems allow the personal skills of a meteorologist to direct the computer to areas where cloud information is most needed and to supervise computation for these areas. The operator may use the system to find critical reflectivities or temperatures separating clouds from clear areas, to identify cloud elements suitable for wind estimates, to write surface weather reports on the satellite images and to perform many other useful operations.

Elements of image interpretation

Aerial photography, which existed long before satellite meteorology, has identified the elements of image interpretation. For cloud analysis these elements should include the size, shape, tone, texture, shadow and pattern of individual clouds or cloudy regions. The context, which is the location of the clouds in relation to other weather or geographical information, and the cloud motion as revealed by successive images from geostationary satellites are also invaluable to the image analyst.

Cloud sizes vary considerably and give important information to the image analyst. Growing cumulus clouds may appear as a single bright picture element in the visible imagery at 1 km resolution. When the cumulus clouds develop to cumulonimbus in thunderstorms, they become larger and are evident in the visible or IR images with sizes greater than 10 km. If a number of cumulonimbus clouds are connected or merged into a mesoscale convective boundary, the cloudy region becomes even larger and will have a characteristic length of 100 km or longer. A central dense overcast region of high clouds over the tropical oceans may have a dimension of 300 km and may suggest an incipient tropical storm if other features are present. Mid-latitude low pressure areas or cyclonic storms and the fronts extending from them often have cloudy areas exceeding 1000 km. Meteorologists use these sizes to identify isolated clouds, mesoscale cloud lines, larger-scale fronts and storms.

Shape is used along with size as a means to characterize clouds. Cold air blowing over warm water often produces long lines of clouds known as cloud streets. Wave clouds, which form in the lee of mountains and which are due to the alternating rising and sinking of air, may be rather long, narrow and parallel to the mountains. Jet stream cirrus forms at high altitudes and has a characteristic shape on the poleward edge; this shape usually consists of a long smooth curve close to the maximum winds of the jet. Mature tropical storms called hurricanes or typhoons may appear nearly circular with a smaller circle or eye, lacking high clouds, in the centre. High clouds associated with cyclogenesis or storm formation at mid-latitudes may look first like an elongated leaf, smooth on the poleward side and ragged on the opposite, and later like a larger comma, the tail of the comma corresponding to clouds on a cold front.

The cloud tone represents how bright the cloud appears on the image. In solar channels, it relates to cloud reflectivity, brighter clouds being more reflective. In thermal channels, the greyscales for cloud pictures are usually reversed so that bright tones represent low thermal energy of cold clouds and dark tones represent high thermal energy of warmer clouds or clear areas. Figure 6.2 shows how cloud temperatures and reflectivities vary for different clear areas and cloud types. Each location represents the average of numerous cases which often overlap in the two-dimensional space of figure 6.2. All of the cases were identified and saved using an interactive display system. The low clouds (cumulus, stratocumulus and stratus) are water clouds and tend to appear warmer and brighter than the high ice clouds (cirrus, cirrostratus and cumulonimbus). Individual cloud elements for cumulus and cirrus are usually smaller than the field of view of the sensor, and consequently the measurements include background as well as cloud. These categories are relatively warmer and darker than the other categories, which are more likely to be overcast. When both visible and infra-red data are available, image analysts can

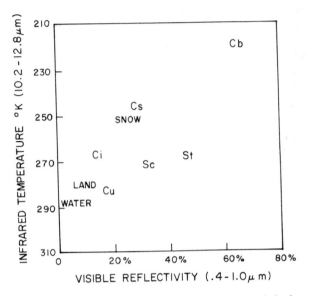

Figure 6.2. Mean temperatures and reflectivities for various clear areas and cloud types sensed by a DMSP satellite during December 1979 on passes over North and South America (after Air Force Geophysics Laboratory Technical Report, *No. 82–0027,* Bunting and d'Entremont, 1982, Hanscom, Mass. 01731, USA).

generally distinguish these cloud types using image tone along with size and shape. When only one image is available, fewer categories can be distinguished. When only visible is inspected, cirrus is most likely to be overlooked; when only infra-red is inspected, clouds with low tops are most likely to be overlooked.

Cloud texture can be defined as the visual impression of roughness or smoothness created by varying cloud elements, which may be too small to be discerned as individual objects. Cloud texture is easy to recognize in visible images. An area with small-scale clouds may appear rough in a visible image before the individual clouds or clear areas can be resolved. A solid area of clouds with a bumpy texture in the visible image may have embedded convection producing rain. Rough texture is useful for distinguishing cumuliform clouds from stratiform clouds. Infra-red images show fewer areas of rough texture even when the resolution is the same as a coincident visible image. The fields of thermal radiation emitted by clouds or the Earth's surface are smoother than the scattered sunlight at shorter wavelengths. Patchy cirrus clouds are an exception and often show rougher texture on infra-red images.

Cloud shadows appearing on visible imagery are also useful to the image analyst even though they may hinder an automated analysis. The length of the shadow can be used to estimate the height of clouds above a lower cloud deck or the Earth's surface. Jet-stream cirrus often casts shadows even though the clouds may be optically thin.

Cloud patterns are also used to characterize the clouds and the weather elements such as winds and humidities which generate the clouds. Patterns which are easy to recognize include wave clouds in the lee of mountains, cloud streets over the oceans or large lakes, and cellular cloud patterns. The cellular cloud patterns form as a result of convective mixing within a larger air mass. For example, the cold air behind a cold front may be heated as it passes over warmer water. A slow turnover process is initiated which may be confined to low altitudes beneath a temperature inversion. These cells with alternating cloudy and clear areas may extend over significant areas of the oceans. The cloud patterns are apparent to the analyst since the size and shape of individual clouds appear to be repeated many times.

When the satellite image analyst has access to other sources of data such as weather station reports of cloud bases and precipitation, radar echoes and temperature/humidity profiles from radiosondes, a more complete interpretation of the image can be made. In image analysis, this process is called association. These different sources of data can be readily handled and colocated on an interactive display and data access system. For example, precipitating clouds can be identified using radar echoes or weather station reports and distinguished from the other clouds in the image. Knowledge of the cloud background (land, water or ice), the local climate, time of year and time of day give clues as to what clouds may be expected. In many areas with tropical or summer weather, small and unresolved cumulus clouds are expected during the morning hours. In these situations, an image analyst may look out for other clues of their presence such as a general brightening in the area where these unresolved clouds might be present.

Examples of weather satellite images and applications

Examples of weather satellite images are shown in figures 6.3–6.6. Figures 6.3 and 6.4 are visible and infra-red images from the GOES-East satellite of an area centred over the eastern USA. The right side of the visible image shows a large mass of clouds which has the size and shape of a developing storm with a trailing cold front. The bumpy texture in the frontal clouds suggests convective activity producing rain. The cirro-stratus clouds along the north-western edge of the storm are casting shadows which provide information on their height. Cloud patterns are important for the mountain wave clouds near the centre of the image and also for cloud streets over the Great Lakes. Figure 6.4, the infra-red image, shows how high clouds stand out from low clouds or clear areas.

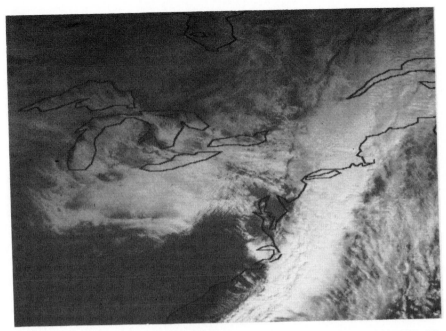

Figure 6.3. A visible image from the GOES-East satellite (13 November 1982, 1500 GMT). The image is centred on the Laurentian Great Lakes.

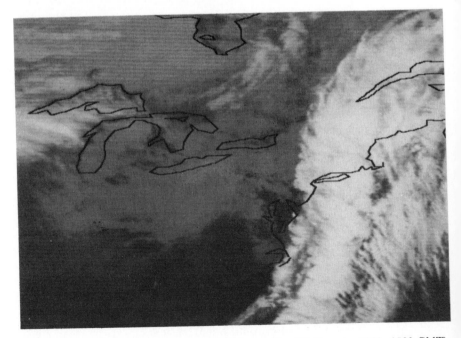

Figure 6.4. An infra-red image from the GOES-East satellite (13 November 1982, 1500 GMT). The image is centred on the Laurentian Great Lakes.

Figure. 6.5. A visible image from the DMSP satellite (18 July 1980, 1800 GMT). The image is centred on the Laurentian Great Lakes.

Figure 6.6. An infra-red image from the DMSP satellite (18 July 1980, 1800 GMT). The image is centred on the Laurentian Great Lakes.

The right-hand side of the picture has bright clouds with similar tops but varying thicknesses. The thickest clouds are near the centre of the storm while thinner cirrus is south-east of the storm. All of these clouds appear colder than the low cloud tops in the centre of the image. Over Canada, it is difficult to distinguish low clouds from land. Figures 6.5 and 6.6 show visible and infra-red images with resolutions near 2 km whereas figures 6.3 and 6.4 have visible images at 4 km and infra-red images at 8 km resolution. In the finer resolution infra-red images, clouds often show more texture than they do in the 5–8 km resolution of infra-red sensors on geostationary satellites. These images show cirrus clouds over Lake Superior and cumulus clouds to the north and east as seen by a USA Defense Meteorological Satellite Program (DMSP) spacecraft.

There are numerous reports on satellite image interpretation with excellent examples: the report of Anderson *et al.* (1966) has many examples of visible images, while those of Brimacombe (1981) and Colwell (1983) have more recent examples, including infra-red and geostationary images.

Within the last decade, infra-red images and frequent images from geostationary satellites have become generally available. These data have been used extensively for research on tropical storms, thunderstorms and fog dissipation, and for estimating rain from satellites. Visible and infra-red images can be used to estimate the stage of development of tropical storms and to follow their intensification and eventual weakening. From visible data, the shape and size of central dense overcast clouds, organization of outer bands and presence of an eye can be used to estimate wind speeds and central pressures. Enhanced infra-red imagery gives cloud-top temperatures and gradients towards the tropical storm centre, and these measurements also help to estimate storm intensity. Current estimates of the storm intensity are compared against estimates from previous days to check for unreasonable changes in intensity which would cast doubts on the current estimates.

Time sequences of geostationary satellite images have shown that severe convective storms tend to form in favoured locations. Long arc-shaped lines of cumulus congestus may form along the forward edge of thunderstorm clusters and mark the edge of a mesoscale high. Arcs which intersect with other arcs or air-mass boundaries are likely to have severe thunderstorms at the intersections. Areas which experience maximum heating due to clear skies during morning hours have also been shown to be favoured regions for convective growth after cloud lines move into these regions.

Regions of fog can be readily identified in visible images and their drift or dissipation followed from geostationary satellites. The fogs appear very uniform, with no texture, and are moderately bright. The fog persistence correlates well with the fog brightness in the visible images. At night fogs are much more difficult to detect in infra-red images; however, the onset of radiation fog can sometimes be seen in image sequences. Over cold land, at night, inversion conditions occur in which fogs and low stratus are often warmer (darker) than cloud-free land and have been described as 'black stratus'.

Satellite imagery has been used extensively for precipitation estimates. These estimates can be made by image interpretation, automation or with an interactive system. Visible, infra-red and microwave channels have been used. The image interpretation aspects for visible and infra-red are described here and automated aspects including microwave channels are described in Section 6.6.

Most of the Earth's surface lacks precipitation monitoring by rain gauges and weather radars. Satellite data are particularly helpful for inferring the horizontal extent of precipitating regions, particularly at low latitudes, or for identifying showery weather in mid-latitudes where most of the rain comes from convective systems which vary greatly in time and space. Geostationary satellite images reveal the growth and decay of these convective systems. Meteorologists have found that the clouds which are coldest in the infra-red and brightest in the visible produce the most rain. Clouds which are expanding in areal coverage produce more rain than those which are unchanging or decreasing in size. The merging of cumulonimbus clouds also increases the rainfall of the cloudy area. Finally, the most significant rainfall of a convective system occurs in the upwind portion, which can be identified from the satellite images. The satellite images, enhanced to show the coldest and brightest clouds, display the area of rainfall better than the intensity of rain. Rainfall estimates derived from convective weather have had mixed results in applications to higher latitudes where warm fronts, cold fronts and cyclonic storms may produce rain or snow from clouds with temperatures, reflectivities and growth rates significantly different from convective systems. Numerous authors have contributed to the growing applications of rainfall estimates from satellites. Many of these studies are summarized by Barrett and Martin (1981).

Image display techniques

All of the images described so far could be black and white displays on photocopy or a TV screen, with varying greyshades representing different cloud reflectivities or temperatures. Although clouds can be recognized quite well in these displays, it is frequently useful to enhance the images with colour. These enhanced images are often called 'false colour' since they do not correspond to the colours seen by the human eye. The colour enhancements are useful since the eye can recognize hundreds of colour gradations but only dozens of greyshades. Since satellite images may have 64 to 1024 greyshades, the colour enhancements can bring out useful details of cloud temperatures or reflectivities. Colour enhancement is also useful for merging images from different spectral bands to produce a single image for interpretation. Examples are given by Colwell (1983).

Another important way to improve the image is to use a stereographic display. The stereographic display gives the impression of three dimensions so that high clouds, thick or thin, stand out from low clouds. It is most often applied to visible images. The stereography requires overlapping images from two different views, as for example, from geostationary satellites which are separated by 60° or less longitude. The higher resolution of geostationary visible images and better clear/cloud contrasts of low clouds can be combined with cloud altitude estimates based on geometric rather than infra-red principles to produce a stereographic visible image with excellent cloud features. A more limited display can be generated from simultaneous visible and infra-red images using the infra-red to estimate cloud height. Stereographic observations and applications are described by Hasler (1981).

6.5. *Cloud motion and wind estimates*

The value of geostationary satellites as a source of improved cloud interpretation and wind observations was recognized soon after the launch in December 1966 of the first spin-scan camera aboard the US Application Technology Satellite. The accuracy of satellite-derived cloud-drift winds to represent the environmental wind field has been a continuing subject of research. A fundamental problem is that clouds are usually not a passive tracer of the wind field. Through the processes of development and decay, clouds interact with the environmental winds; consequently, tracking cloud elements does not consistently provide representative estimates of the wind field. The primary purpose of this section is to summarize information on (*a*) the nature of the problems encountered in inferring winds from cloud motion, (*b*) the methods used in measuring cloud motions, and (*c*) the spatial coverage and accuracy of the wind estimates. A detailed presentation of the topics included in this section is given by Hubert (1979).

Cloud and wind motions

Winds are derived from geostationary satellite observations by measuring the motion vector of cloud fields as displayed by a pair or a sequence of images. The two factors to be considered are, firstly, the nature of the cloud targets to be tracked with respect to image resolution, cloud target persistence and image frequency, and secondly, the relation between cloud motion and wind.

The resolution of the image at the satellite subpoint of the various geostationary satellites ranges from 1 to 3 km in the visible and from about 5 to 10 km in the infra-red ($10 \cdot 5 - 12 \cdot 5 \, \mu$m). The visible images can only be used during daylight hours, and this, combined with the inability to identify cloud features with only one pixel, leads to a situation of tracking patches of clouds up to 20 km across rather than the much smaller individual cloud. In addition, the automatic correlation techniques often used to determine cloud motion require patterns of clouds covering areas from 100 to 200 km on a side.

The routine operation of geostationary meteorological satellites produces a new image every 30 minutes. Small cumulus clouds and some clouds in the upper troposphere have lifetimes which are often shorter than 30 minutes. Consequently, in the lower troposphere the entities that are tracked at time intervals of 30 minutes are usually mesoscale patterns of clouds. Many of the upper clouds tend to be layered, and patches of these clouds may persist for hours and are good candidates for tracking. The considerable work using clouds for wind estimates has led to the conclusion that the relationship between observed cloud motion and wind is strongly influenced by the type of cloud and the resolution and frequency of the images.

Methods for cloud motion determination

Although efficient automatic cross-correlation techniques are available for pattern matching and subsequent cloud-drift determinations, it is still necessary for a skilled operator or meteorologist to be involved in selecting suitable cloud targets and in editing the final set of vectors. An additional essential part of the procedure involves the assignment of a height which is representative of the targets tracked. An effective manual method for selecting suitable cloud targets is to produce and use animated sequences of the images. The animation serves to eliminate the use of standing-wave clouds or gravity-wave clouds, since these will generally have quite different velocities than the local winds. A skilled meteorologist will usually be able to recognize those targets which are obviously poor representations of the wind field.

Generally, an infra-red channel of the geostationary satellite is used to estimate the height of the clouds being tracked. Thin cirrus clouds or scattered cloud fields in the middle and upper troposphere will usually lead to low height estimates unless corrections are made. The European Space Agency (ESA) has used a multichannel radiance analysis to help identify the presence and types of clouds. Further experience with the multi-channel analysis should identify the situations in which automated classification is appropriate; the remaining situations would still be handled using manual methods.

Once the candidate cloud target has been chosen, its displacement between successive pictures must be measured. Cross-correlation is a typical computer method of pattern matching. An array of pixels within a pre-defined reference window is selected from an image taken at one time, and this is correlated with the pixels of an image taken at a later time. The maximum correlation coefficient occurs for the lag position which corresponds to the best overall fit of the motion of the cloud pattern within the array. The lag position and the time interval between images yields a velocity vector which is assigned to a location at the centre of the array.

ESA, NESS and the Japanese Meteorological Satellite Centre each use slightly different correlation procedures to compute cloud displacements (Hubert, 1979). Regardless of the particular correlation procedure used, interpretation of the results is often a problem because (*a*) there is no obvious peak in the correlation function (the pattern was not sufficiently persistent), (*b*) the correlation values are all large with no identifiable maximum (probably caused by a large uniform overcast or nearly cloudless area within the reference window), or (*c*) there is more than one maximum (a fairly frequent occurrence which is caused by the pattern in the reference window being similar to more than one pattern in the larger search window).

Manual methods may also be used to track cloud patterns. These methods consist merely of identifying an initial position of a cloud and then designating the position of the same cloud on a subsequent image or sequence of images. Once the initial and end points have been specified, computers are used to calculate the cloud-motion vector. In general, this manual method is too slow for routine operations, but ESA, NESS and the Japanese Meteorological Satellite Centre all have interactive computer systems and displays which are used on occasions when the completely automatic system fails to give consistent or reasonable cloud-motion vectors.

Height assignment

The problem of the wind–cloud relationship has already been mentioned. Part of the problem is knowing a height to assign to the cloud-drift vector. The cloud top, which is usually the portion of the cloud being tracked, may not be moving at the velocity of the wind at the height of the cloud top. For many convective cloud patterns, it has been found that the velocity determined from cloud motions corresponds strongly to the velocity near the cloud base. For stratus, on the other hand, the wind vector determined from cloud motion may be appropriately assigned to the cloud top. Low-level clouds over oceans are assumed to be cumulus or stratocumulus, and at NESS a height of 900 mb is assigned to the winds determined from the low-level cloud motions. Hubert (1979) has shown that, in one test, the level-of-best-fit wind fell within ± 100 mb of 900 mb for about 65 % of the cases. The wind field used for the level of best fit was obtained from rawinsonde stations which were near the locations of cloud-motion determinations.

Targets at mid- and upper tropospheric levels tend to be comprised of layer clouds which more closely follow the winds at the cloud top. However, cirrus and some middle-level clouds may have low emissivities compared with the almost blackbody emissivities of convective clouds in the $10-12 \cdot 5 \mu m$ region. Consequently, cloud-top heights determined from the infra-red-sensed temperature of thin cirrus will be much warmer than the physical cloud temperature; using the infra-red-sensed temperature to infer the height of the cloud top will thus lead to an underestimate of the cloud-top height. Using a man – computer interactive system and co-located visible reflectances and infra-red temperatures, the emissivity of the cloud may be estimated. The principal disadvantage of this technique is that visible data are not available at night.

In some early research at NESS, it was observed that most extensive cloud patches contained smaller regions of denser clouds with high emissivities. These high-emissivity regions were tracked, and tracks of features in the surrounding thinner and warmer appearing regions were assigned the same height as the dense patches. ESA has developed an automated process for estimating cloud-top temperatures which uses the radiances measured simultaneously in both the window and water vapour channels. Since the extra channel provides additional information, improved estimates of cloud-top temperatures can be expected.

Errors in cloud tracking

In summary the sources of errors when obtaining winds from cloud motion are as follows:

1. Image resolution and temporal frequency.
2. Cloud pattern consistency over the time period between images.
3. Assignment of appropriate height for the measured cloud motion vector.

An indication of the reproducibility of wind estimates was obtained at the University of Wisconsin. It was found that different scientists using the same data set and interactive computer system could determine winds with an rms difference of 2 m s^{-1} for the cirrus cloud level and $1 \cdot 3 \text{ m s}^-$ for the cumulus level. However, when the height error is considered, the wind error becomes much larger. Hubert (1979) presents results which indicate that the standard deviation of the difference between rawinsonde and winds derived from cloud motions was about 5 m s^{-1} for upper-level clouds and about 4 m s^{-1} for lower-level clouds. He suggests tha 70–80 % of the cloud vectors represent the synoptic scale field of motion with about the same accuracy as rawinsondes, and consequently satellite and balloon observations should be combined in synoptic scale analyses Hubert (1979) also notes the following two characteristics: firstly, abou 10 % of cloud vectors deviate markedly from rawinsondes and the mos probable source of the error is in the satellite winds; and secondly, a significant number of these larger deviations are due to influences other than erroneous height assignments and are most likely to be due to tracking clouds which are not displaced by the wind.

A consequence of these characteristics is that more work needs to be carried out in order to improve the selection process of suitable clouds or cloud patterns for estimating winds.

An example of winds derived from satellite observations of clouds

Hubert (1979) presents examples of wind coverage from two geostationary satellites located at 75°W and 135°W. The coverage for wind estimates ranged between ±50° latitude and from 20°W westwards to 170°E longitude. Large portions of the region contained clouds which allow wind estimates. Generally, the coverage of winds from low clouds is larger than the coverage for high clouds and of course only about 10 % of the area is covered by data at both levels.

Figure 6.7 is an example of the satellite derived wind field over a limited area of the tropical Atlantic Ocean during a special observation period in 1974. The wind estimates were derived from special scan mode data in which images were obtained at 15-minute intervals. The 15-minute data gave much better results than data taken at 30-minute intervals. The arrows with circles in figure 6.7 are the surface winds reported by ships at 1200 GMT, which is 1 h earlier than the data used for the satellite derived winds. The agreement between the two types of data is excellent.

A well-defined convergence pattern is evident in the data shown in figure 6.7. The winds in the lower half of figure 6.7 are generally from the west or south-west whereas the winds in the upper half are, on average, from the north-east. The data illustrate the type of detail in the wind field which is often attainable from satellites provided that a dedicated effort is applied to the problem.

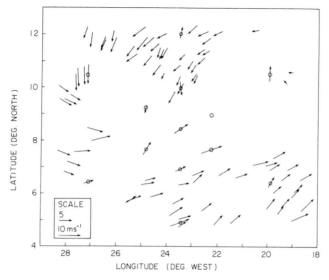

Figure 6.7. Low-level (about 950mb) winds for 1230 GMT, 5 September 1974 from 15-minute sequences of cloud images from the first Synchronous Meteorological Satellite. The arrows with open circles at their mid-point are surface winds reported by ships at 1200 GMT (from Martin, Chatters and Suchman, 1975, Univ. Wisconsin, NASA Contract NAS/5/2329).

6.6. Algorithms for objective cloud detection

This section describes how clouds are detected in automated processing of satellite imagery data. Automated imagery processing implies fully objective abstraction of cloud information from the image. Thus it is possible to distinguish between automated processing and the techniques described in Section 6.4. The earlier material discussed human interpretation and controlled objective processing, as in computer interactive systems, where the objective process is directly under the control of a human observer. The need for automated processing arises from the massive quantities of data provided by weather satellites. Only automated processing can handle this many data. A GOES satellite senses enough data in 30 minutes to generate nearly 700 different images on a TV screen if all of the visible data (1 km resolution) and infra-red data (8 km resolution) are used. A cloud analysis procedure requiring human interpretation of the images or extensive controls on interactive processing would obviously not keep pace with the flow of data from the satellite; however, many applications require cloud sensing over the whole globe using data from more than one satellite. Although automated cloud analysis has the advantage of speed and offers the obvious solution to requirements for global cloud analysis and climatologies, it is prone to errors due to optically thin clouds, poor contrasts between clear and cloudy regions, highly reflective or very cold backgrounds such as snow cover, etc. These limitations on automated cloud analysis are generating significant research efforts so that future improvements are expected.

In order to summarize algorithms for automated cloud analysis some terminology will be introduced from pattern classification and scene analysis. Duda and Hart (1973) describe the common elements of pattern classification systems as a transducer, which views the scene and provides raw data for computation; a feature extractor, which derives useful statistics or features from the raw data; and a classifier, which uses the features to decide what categories are in the scene. For cloud applications, the transducers are the imaging satellite sensors; the feature extractors consist of software for Earth-location, calibration and computation of image features; and the classifiers are the various algorithms to make clear/cloud decisions and estimate cloud properties.

Automated cloud analysis may use categorical variables such as clear/cloud, cloud types, cloud phase, precipitating/non-precipitating, or it may use continuous variables such as cloud altitude, coverage, optical thickness and rainfall rate. For the categorical variables, the goal of classification algorithms is basically one of partitioning the feature space into regions, one region for each category, with a minimum of wrong decisions. Figure 6.2 (p. 212), for example, shows how clear regions and cloud types tend to be separated in the two-dimensional feature space defined by infra-red temperatures and visible reflectivities. An automated classifier could be defined to classify an unknown sample by calculating its distance from all the category means and selecting the closest.

For the continuous variables, the algorithms must relate the appropriate region of the feature space to a numerical estimate. This is generally done using equations or look-up tables with the appropriate dimensions. The most common example is to estimate cloud altitude by converting its infra-red radiance to temperature using the Planck function and comparing it to a temperature–altitude relation for the appropriate climate.

Threshold techniques

Many algorithms for cloud detection use threshold techniques to define key features (see the discussion of cloud climatologies in Chapter 1). The portion of the image which exceeds the threshold defines the fraction of cloud cover in the image. A linear relationship is often assumed between the fraction of cloud and the difference between the observed radiance and a threshold radiance. Individual algorithms vary as to whether infra-red, visible or both images are used, how the thresholds are determined, and whether the threshold is applied to individual pixels or the total radiance in the image area. The visible thresholds are intended to represent cloud-free backgrounds so that brighter areas are clouds while the infra-red thresholds are intended to represent the apparent temperature of the background as the satellite sees it through the atmosphere. The areas in the infra-red images colder than the threshold are assumed to be clouds. Visible thresholds which are brighter and infra-red thresholds which are colder than the clear/cloud thresholds can be used as cloud features to distinguish precipitating from non-precipitating clouds.

Thresholds can be entered into cloud detection algorithms in at least four ways. Firstly, they can be constants so that the same infra-red and visible thresholds are applied to all samples of imagery data. Secondly, they can be derived from other weather or geographical data bases. Infra-red thresholds can be derived from surface reports of temperature, which are well-sampled over many land masses. Visible thresholds can be derived from the type of surface which is viewed. For most visible bands $(0 \cdot 5 - 1 \cdot 0 \, \mu m)$, oceans are darkest and forests, open land, deserts and snow cover are progressively brighter. Thirdly, thresholds can be derived from the image itself if the area includes clear areas. Clustering approaches, which will be described later, can help to establish the thresholds. finally, to ensure that an area is partly clear so that the thresholds represent clear/cloud boundaries and not cloud layer boundaries, the threshold can be determined from a series of images at different times and the extreme radiance (dark visible or warm infra-red) can be used as the threshold.

Fixed thresholds can be useful for restricted areas, such as oceans. The oceans are usually quite dark in the visible bands with the important exception of specular reflection (see the discussion of figure 2.5), known as sunglint, when the Sun and satellite angles are equal and in the same plane. The sunglints are usually brighter than clouds so these areas should not be processed in visible images. Otherwise, a reflectivity threshold of 0.10 or less can be used for visible bands. Sea surface temperatures vary slowly during the day and from day to day so that a fixed temperature threshold may be useful within restricted latitude ranges and times of year.

Coastlines and land areas have enough variability in background temperatures and reflectivities to make fixed thresholds very restrictive to automated processing so that other methods are needed for global cloud detection. The surface temperatures seen by satellites may vary by 30 K in a single day over desert areas with pronounced diurnal heating. Monthly mean surface temperatures vary from 300 K in tropical areas to 220 K at the coldest stations in the USSR or Antarctica. Deserts may have reflectivities exceeding $0 \cdot 40$ in visible bands. Open snow-covered regions such as ice caps may have reflectivities of $0 \cdot 80$ and appear brighter than most clouds in visible images.

The alternatives to fixed thresholds also have shortcomings. Temperature data are sparse over some land areas or may not represent adequately the cooler temperatures of mountains. Geographical or climatological types such as desert may have enough variation in reflective materials to make a single visible threshold inadequate for the type. Algorithms which look for minimum reflectivities or maximum temperatures in images may find only clouds in the image. Rapid changes in reflectivity, as when brown vegetation changes to green in the spring or snow falls for the first time in winter, may lead to unsatisfactory visible thresholds for periods of time. In general, the algorithms using thresholds have the computational advantage of simple cloud-feature extraction, but the logical problem of finding the best thresholds to use must be solved.

Statistical techniques

Various research groups have developed statistical techniques for cloud detection using image features more complicated than simple thresholds. Some of these techniques are called clustering algorithms. They attempt to find natural groupings of pixels in an image. If the image features are chosen well, then the image can be decomposed to separate clusters of points in the feature space and these clusters will each represent a clear area or cloud layer. The maximum number of clusters to be found may be limited but the algorithms are otherwise free to find as many clusters as the image suggests. This property is important because it distinguishes the clustering algorithms from classification algorithms, which use a previously determined set of clear and cloudy categories as a fixed set of choices. The clustering algorithms are sometimes described as unsupervised procedures that produce unlabelled categories. They are, however, very flexible and useful for coping with the great variety of cloud types and backgrounds encountered over the Earth. The classification algorithms are usually described as supervised procedures since they require a dependent or training set to associate image features with designated clear and cloud types.

An example of clustering using a one-dimensional histogram of infra-red temperatures is given in figure 6.8. Three clusters, labelled Modes 1, 2 and 3 were found in one 8×8 array of infra-red samples covering an area of about 40×40 km from an automated cloud analysis procedure known as 3-D nephanalysis, which is described more completely in Section 6.8. Greyshade 30 is 262 K while 62 is 212 K. The homogeneous groupings of warm and cold temperatures in Modes 1 and 3 represent a low and a high layer of clouds. The intermediate temperatures in Mode 2 may be mid-level cloud; however, at least some of the Mode 2 samples are likely to be high clouds which are optically thin or do not fill the

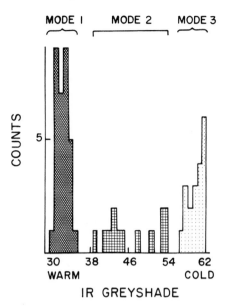

Figure 6.8. Example of a one-dimensional histogram of infra-red temperatures analysed for cloud layers using the 3-D nephanalysis algorithm.

sensor's field of view. This automated procedure allows cloud-top estimates for multiple layers. The coldest sample in a mode is generally used to represent the layer altitude. The fractional cloud cover is also estimated by counting the number of samples in each mode. This algorithm determines its own thresholds separating cloud layers and clear regions; however, the algorithm must compare the warmest mode with a global map of surface temperatures to check if it is a clear region or a cloud layer.

Figure 6.9 is an example from a clustering algorithm designed to use the visible, infra-red and water vapour channels of METEOSAT. It is designed to find natural groupings of image samples in the three-dimensional feature space defined by the METEOSAT radiances. A 200 × 200 array of pixels covering about 1000 × 1000 km was used over a disturbed area in the Intertropical Convergence Zone. Figure 6.9 shows the locations of five clusters found by the multi-spectral algorithm projected on the infra-red/visible plane. The water vapour channel helps to identify thin cirrus clouds. This algorithm not only benefits from the use of three spectral channels but also has the advantage of finding clusters which are not limited by straight lines or plane surfaces, as in threshold methods. Two-dimensional histograms constructed on the infra-red/visible plane often show oblong-shaped groupings of points with major axes not parallel to the axes of the diagram.

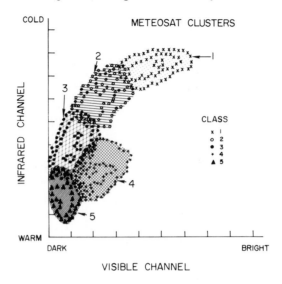

Figure 6.9. Example of a multispectral clustering algorithm designed for METEOSAT imagery channels (Desbois et al., 1982).

The classification algorithms use a previously determined set of cloud categories which are usually derived from the standard cloud typologies determined by the World Meteorological Organization. 'Clear' or 'no clouds' is used as a category along with the cloud categories. One set of categories which has been used in several studies is: (A) Clear, (B) Cumulus/Stratocumulus/Stratus, (C) Cumulonimbus, (D) Cirrus, (E) Cirrus and lower clouds, or a mixture of cloud layers. The categories which can be distinguished are limited as a result of the viewing geometry and relatively coarse resolution of data compared to an observer classifying clouds from a weather station. High cloud types are well-represented in the satellite cloud typologies while various types of middle and low clouds are not easily distinguished. Applications of automated classifiers have shown that accuracies improve when similar types like cumulus and stratocumulus are merged into a single category or even broader categories such as cumuliform or stratiform are used. Some unconventional categories such as ice or water cloud and precipitating or non-precipitating cloud have also been applied.

There is almost no limit to the number of features which can be extracted from the image samples and used as candidates for defining a feature space for the classifier. A few samples of features include: mean reflectivity or temperature over an array of image pixels; standard deviations, skewness and kurtosis over the distribution; gradients in various directions; the temperature or reflectivity at the 1 % or 99 % level of the distribution; as well as the number of pixels above or below a threshold. There are many other features which have been defined to measure the texture or apparent roughness of the image and these features are useful for distinguishing cumuliform clouds from stratiform clouds or clear regions. Features may be defined by a two-dimensional Fourier transform of the image. These features say something about the sizes and patterns of clouds and the highest wave numbers are useful for detecting small-scale clouds.

The various features which are considered useful for automated cloud classification define the feature space used by the classifier. The feature space is sometimes easy to visualize, as in figure 6.2 which uses only reflectivity and temperature to distinguish categories. Figure 6.2 shows mean values; an observer, guided by a scattergram of the individual cases for all the categories, could simply sketch the boundaries between the categories. A computer could classify new samples using the arbitrarily drawn boundaries or it could calculate the two-dimensional distance to each category mean and use the minimum distance as a decision rule for the best classification. On the other hand, when the feature space has many dimensions, then the observer finds it difficult to draw boundaries between categories. When the various features differ in units or numerical ranges, it is best not to trust a minimum distance classifier which does not attempt to scale or adjust the different dimensions of the feature space. For a complicated feature space, it is best to follow mathematical procedures (Duda and Hart, 1973) and use computers for making decisions. Bayesian probability theory is useful for making probability estimates of a category given a set of observed features (a feature vector). Discriminant functions can be defined as decision rules. For example, a classifier is said to assign a feature vector x to class i if

$$g_i(x) > g_j(x) \qquad (6.2)$$

for all $j \neq i$. Thus, the automated classifier computes a discriminant function $g_i(x)$ for each possible category and selects a most likely category corresponding to the largest discriminant. When the variables in the feature vector are assumed to have a multivariate normal distribution, discriminant functions are readily defined and rather diverse features can be combined with ease.

Although more than 100 features may be used, relatively few will produce most of the classification skill. Unfortunately, there is no consensus on the optimal combination of features for cloud classification. Simple features such as means or the coldest or brightest values within an area are among the most useful. Features to measure texture improve classification accuracies substantially when only one channel is available and are important for night-time analysis on infra-red data. Day-time analysis, using both visible and infra-red features, is more accurate and the availability of multispectral features reduces the utility of texture features. When both visible and infra-red data are used, classification accuracies exceeding 80 % have been achieved using automated classifiers on independent data.

Radiative transfer applications to algorithms

The cloud detection algorithms which have been described are based primarily on observation from satellites and make relatively few assumptions on radiative transfer in clouds. However, radiative transfer techniques can be used to develop new algorithms or improve the empirical algorithms. Although the visible $(0\cdot5-1\cdot0\,\mu m)$ and infra-red $(8-13\,\mu m)$ windows have very good data coverage in time, location and fields of view since 1974, the data coverage for the other channels in table 6.1 (p. 206) is much less satisfactory and radiation models have been used to simulate what the satellite would sense and what cloud properties could be retrieved if the data were available. In particular, many simulations and precipitation retrievals have been done for microwave radiometers of the Nimbus 5, 6 and 7 satellites. In the near infra-red windows at $1\cdot6$, $2\cdot1$ and $3\cdot7\,\mu m$, radiative calculations have been very useful for simulating changes in radiance due to cloud phase and varying particle sizes. During daytime, the $3\cdot7\,\mu m$ radiance includes reflected sunlight as well as thermal emission, and a radiation model is needed to sort out these properties. Radiation models for clouds in sounding channels are described in Chapter 5.

The same radiative transfer calculations which are used to simulate the cloud radiances can be used in cloud detection alogrithms in several ways. Firstly, the calculated radiances and corresponding cloud properties can be used in a look-up table so that measured radiances can be related to cloud properties. Secondly, various cloud properties such as horizontal coverage and altitude could be found in part by a statistical algorithm and in part by a radiation model so that the local radiation budget is consistent with the model. Thirdly, the radiative transfer techniques are easy to adjust for variations in viewing geometry and atmospheric attenuation, and these adjustments can be used in cloud detection algorithms.

The applications of radiative transfer techniques to cloud detection algorithms are diverse. The radiation budget of a pixel may be estimated by combining visible and infra-red measurements. Assuming a single cloud layer, the sums of the fractional contributions from clear and cloudy regions within the instrument field of view may be equated to the observed radiances in two simultaneous equations yielding cloud altitude and fractional cloud cover. The derived cloud parameters are consistent with both visible and infra-red measurements, but cloud reflectivity and emissivity must be assumed. It is also possible to assume that the cloud fills the field of view and to derive the cloud emissivity from the reflectivity. This approach will detect some cirrus clouds and assign reasonable altitudes to them. Cloud detection algorithms have been defined by solving a complete radiative transfer equation to produce tables of model radiances which are functions of viewing geometry, atmospheric attenuation, cloud optical thickness, emissivity and altitude. Measured radiances are compared to the tables of theoretical radiances. These techniques generally require fast computers with large storage. They also provide cloud estimates with consistent radiative properties, which are desirable for applications such as cloud effects on heating and cooling the Earth's climate.

Algorithms for detecting rain

Most algorithms to derive the mass of liquid water or rainfall rate from microwave channels use information provided by radiative transfer techniques. These algorithms operate primarily over the oceans since the oceans provide a convenient, nearly uniform background for the satellite-borne radiometer. Calculations are more readily made for the microwave channels than for visible or infra-red channels since ice clouds, with their irregular particle shapes, can usually be ignored and since the microwave channels are monochromatic. The emissivity of ocean backgrounds is low (about $0 \cdot 4$) so that clouds with substantial liquid water content or rain-drops appear warm due to thermal emission. At the relatively long wavelengths of the microwave channels, calculations are simplified since energy is directly proportional to temperature following the Rayleigh–Jeans approximation. Calculations of microwave temperatures require a model relating rainfall rate and drop sizes. The Marshall–Palmer model is frequently used. The calculations are sensitive to the altitude of the melting layer of snowflakes since it defines the depth of the rain layer. Supercooled water droplets may exist above the melting layer and contribute to the sensed temperature. The microwave algorithms have been verified by comparison with weather radar data, rainfall measurements and ground-based radiometers looking up. Useful estimates of rainfall rate are made in the approximate range of $2–25\,\mathrm{mm\,h^{-1}}$. Many rainclouds do not fill the field of view of the microwave radiometers, which may be as large as $50\,\mathrm{km}$, and they limit the accuracy of the algorithms.

Some of the estimates of rainfall based on infra-red and visible channels, which are described in Section 6.4, can be converted to algorithms suitable for automated rainfall analysis. In contrast to microwave imagery techniques to estimate rain, they are based on thresholds or statistics since the infra-red and visible radiation sensed by satellites comes primarily from the smaller ice and water particles in the upper parts of clouds rather than the large precipitation particles in the lower parts of clouds. However, the coldest cloud temperatures and highest reflectivities along with the growth of clouds derived from a time sequence of images are very useful features for estimating the areal extent of precipitating clouds and the amount of rain.

The infra-red and visible algorithms appear to work better than microwave algorithms for convective clouds at low latitudes when geostationary satellite data are available. The temporal and spatial resolution of the geostationary images are useful for monitoring the rapid changes and smaller sizes of tropical rainclouds. The infra-red and visible algorithms have the limitations of being tuned to particular regions or climates and are usually not transportable. For example, algorithms tested on tropical clouds may work poorly at higher latitudes where precipitating clouds have different temperatures and reflectivities. At night, when only infra-red data can be used, precipitation estimates are useful but less reliable.

Limitations to objective cloud detection

Objective cloud detection is not a mature technology and there is room for improvement in many areas. Comparisons of automated cloud analysis with human interpretation of the same images or cloud reports from weather stations can reveal shortcomings in all of the algorithms which have been discussed. The following comments relate to the infra-red and visible channels common to geostationary and polar-orbiting satellites. The comments are based on the experience in processing polar-orbiting data for the 3-D nephanalysis and geostationary data from METEOSAT in automated cloud analyses. Infra-red data often fail to detect low cloud tops over the oceans or land since cloud temperatures are often close to the surface temperatures. The problem is greatest at high latitudes or in the presence of temperature inversions. When infra-red data are processed at night and day (without reference to visible data) the night analyses tend to be poorer than the day analyses due to the colder temperatures at the Earth's surface. Cirrus clouds are often confused with opaque clouds at lower altitudes. In fact, none of the algorithms using any channels have yet demonstrated excellent results in detecting the thinnest cirrus clouds and assigning them the correct altitude. Some sounding channels discussed in Chapter 5 and the $6 \cdot 3 \mu m$ water vapour band available on METEOSAT can help to detect and locate cirrus clouds, but they remain a challenge to automated cloud analysis. Algorithms for visible data can confuse bright surfaces such as snow cover or deserts with clouds. Modifications to algorithms to avoid erroneous clouds over bright surfaces may restrict cloud detection over these surfaces. When the satellite images are improperly located, coastlines may be confused with clouds in automated processing of visible or infra-red data. These problems are most likely to occur when the land is highly reflective or when the coast has a strong thermal gradient. African dust storms have been mistaken for clouds. Noise in the raw satellite data can produce erroneous overcast or clear areas with peculiar shapes. Other limitations discussed in this book are viewing geometry problems (Section 6.7), sub-field-of-view clouds (Chapter 7), and clouds over snow or ice (Chapter 8). These limitations on cloud detection algorithms are now the subject of research studies in many countries.

6.7. Comparison of satellite and surface cloud estimates

It has been underlined throughout this book that information on clouds is essential for the preparation of weather forecasts, for climatological studies and for successful surface environment sensing from space. Before meteorological satellites all information on clouds was obtained by ground-based observers or from aircraft. With the current geostationary and polar-orbiting satellites, clouds may now be observed globally, and over large portions of the globe the observations may be obtained every 30 minutes. However, a continuing problem is to relate the ground-based observations of clouds to the data collected more recently by satellites. In one case the observer is looking upwards and assesses the sky cover; in the case of satellites, the sensors look downwards and provide an assessment of ground cover. This section deals with the differences in cloud information obtained from sky cover and ground cover.

Ground-based observations of clouds have a relatively long history, and information on the type, stratification, amount, opacity, direction of movement and cloud-base height is often available; but usually the ground-based observations of clouds are subjective, particularly with respect to cloud-amount estimates (for a review see the discussion in Chapter 1 and particularly table 1.2, p. 14). The type of cloud is often a key to estimating the height of the cloud base although some instruments, such as a ceilometer, a balloon or radar may provide objective estimates of the cloud base. Ground-based observations of clouds have high spatial resolution in the vicinity of the observer, but the density of the observations is inadequate over most of the Earth. The high spatial resolution helps in the identification of multi-layered clouds provided that the lower cloud or clouds are not opaque. Finally, ground-based cloud observations taken at night generally provide less detail than those taken during the daylight hours.

As a result of sensor requirements and data processing limitations, individual clouds usually cannot be resolved from satellites (see Chapter 7). In contrast to surface cloud observations, however, satellites may provide global coverage. The classification of clouds observed by satellites is not developed to the same degree as ground-based classification. Usually the cloud analysis is limited to the assignment of a cloud-top height based on an infra-red temperature or an estimate of the percentage of cloud amount over an area. The errors and difficulties of cloud height procedures have already been described. Cloud amount is a key parameter for meteorological and climatological analyses: estimates of cloud amount from the surface are usually made by observers and errors in the estimates depend on the experience of the observer, the location of the observing station and the method of observation. Malberg (1973) describes several factors which may contribute to an inaccurate assessment of cloud amount by a surface observer. In particular, because of the perspective of the observer, gaps between individual clouds which are several kilometers from the observer may go undetected and lead to an overestimate of the cloud amount (see figure 1.8 and the associated discussion in Chapter 1). Also, the areal representativeness of observations made at one point depends on the height of the clouds and on the frequency of low clouds. Malberg estimates that the radius of the region available for observation is limited to only 5·7 km when an overcast cloud at a height of 1 km is present; the corresponding radius increases to 68 km when the lowest cloud is at a height of 12 km.

The relatively poor resolution, difficulty of distinguishing the surface from low clouds, and the effect of thin clouds contribute to errors in estimating cloud amount from satellites. A problem for both surface and satellite observations of clouds is the angle or line-of-sight between the observer or sensor and the cloud. Figure 6.10 illustrates the magnitude of this problem. The left-hand graph (taken from Lund and Shanklin, 1973) shows the probability of a cloud-free line of sight from the surface as a function of elevation angle and of the observed total sky cover (an angle of 90° is looking vertically upwards). Note that even with clear skies as reported by the observer, there is a finite probability that there will not be a cloud-free line-of-sight at elevation angles less than 50°. On the other hand, for overcast conditions as reported by the observer, there are occasions when there is a cloud-free line of sight regardless of the elevation angle. As expected, the probability of a cloud-free line-of-sight decreases as one looks at elevation angles which move towards the horizon.

The right-hand graph in figure 6.10 demonstrates that the observed average cloud amount depends on the viewing angle from the satellite.

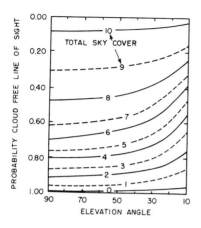
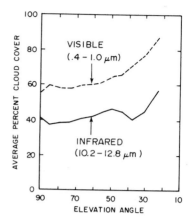

Figure 6.10. Probability of a cloud-free line of sight from the surface as a function of elevation angle and total cloud cover (left side) and the average of cloud cover estimates for satellite data (right side). An elevation angle of 90° is perpendicular to the Earth's surface (Lund and Shanklin, 1973; and the US Air Force Global Weather Central, Technical Services Division).

The data were obtained from the DMSP and include observations from 10 orbits covering portions of the Atlantic Ocean. The observations were mostly of low clouds but the results are considered to be representative of clouds over the oceans. An elevation angle of 90° is looking vertically downwards. The elevation angles represent the elevation of the satellite at the location of the cloud so that the viewing geometry is the same for both graphs of figure 6.10 except that the satellite is looking down while the observers are looking up. The basic data had a resolution of about 5 km. The visible sensor detects significantly more cloudiness than the IR sensor because the lowest clouds have infra-red temperatures which are close to those of the ocean surface and consequently are incorrectly associated with the surface. The observed average cloudiness increases as the viewing angle moves closer to the horizon; the increase is similar to that for the decrease in the probability of a cloud-free line of sight with smaller elevation angles for a surface observer. The effect demonstrated for both surface and satellite data in figure 6.10 is due to the increased probability of seeing clouds as the slant path through the atmosphere becomes longer. The effect is quite pronounced for the polar-orbiting satellites which are at an altitude of about 830 km and view a total swathwidth of about 3000 km. The geostationary satellites, however, are not immune to the problem since their elevation angle is about 22° for Earth locations which are at a latitude of 60° relative to the satellite subpoint.

Henderson-Sellers *et al.* (1981) have discussed the problems of acquiring cloud data for global climatologies. They used the 3-D nephanalysis and compared the cloud climatologies with those from surface observations. The 3-D nephanalysis is summarized into 'boxes' which are about 40 km on a side, and this is approximately the size of the area in which clouds can be adequately classified by surface observers. When the surface observations are used to represent larger areas, then the information may deviate from the analysis provided using satellite data.

Malberg (1973) found that surface observations of cloudiness over a three-year period from a single station assumed to be representative for 2·5° geographical sections in the northern Atlantic and Europe gave mean cloudiness amounts which were from 0·8 to 1·6 tenths greater than the mean cloudiness derived from satellite data for the same period. On the other hand, comparisons of cloud cover from all-sky cameras and from satellite data during the GATE experiment in 1974 found general agreement between the two types of data. However, when there were disagreements between the two types of analyses, the total cloudiness obtained from the satellite data was consistently larger by 5–10%. Henderson-Sellers *et al.* (1981) described the problems in comparing satellite and surface observations of clouds. They presented histograms of cloudiness derived from the 3-D nephanalysis and surface analysis and suggested that '. . . while there is some correlation between the two data sets, the level of agreement is not as high as might be hoped.' Since satellite data will form the basis of any future cloud climatology, it is essential to develop improved methods for interpreting and relating the data from both satellites and surface observations.

6.8. Climatologies of clouds detected from satellites

Chapter 1 summarized satellite cloud climatologies from 1960 to the present (see also Barrett, 1974). The earliest cloud climatologies made use of operational nephanalyses prepared by image analysts. They interpreted visible pictures using the elements of image interpretation described in Section 6.4. The boundaries of cloudy regions were drawn by hand and the fraction of cloud cover estimated in ranges such as 0–20%, 20–50%, 50–80%, >80%, or solid (100%). These nephanalyses were produced from US and Soviet weather satellites. The assembly of nephanalyses into climatologies began in 1962, although the early satellite data were sparse, with no coverage at high latitudes. Some locations had fewer than 10 observations per month due to the orbits and sampling characteristics of the early weather satellites. Image brightness varied considerably due to the satellite instruments and ground stations so that human interpretation was necessary. Some of the early climatologies covered low latitudes over the globe. The longest climatology produced from the manual nephanalyses was done at the University of Hawaii for the years 1965–1973 over the Pacific Ocean. The fractions of cloud cover for $2\cdot5°$ latitude–longitude squares were estimated each day and summarized by months. Areas of maximum cloudiness, such as the high latitudes of the North Pacific, the persisting stratus and stratocumulus of the eastern Pacific, and the deep convective cloudiness near the equator are well-represented in this climatology and their intra- and inter-annual variations can be followed.

By the late 1960s, satellite and ground systems were improved so that cloud climatology could become more automated. The visible images which were available then could be routinely mapped from the image format to conventional maps. Orbits could be composited into global maps of brightness from 12 or 13 consecutive orbits of data. Polar-stereographic projections have been used most frequently since they are also used in numerical weather prediction. A joint project of the US Air Force and National Environmental Satellite Service used these maps to derive a global atlas of relative cloud cover for the years 1967–1970. The fraction of cloud cover was estimated for a 512×512 matrix for each hemisphere, which has a spacing of about 40 km on the Earth's surface. Histograms of visible reflectivities within 40×40 km areas were converted to estimates of total fractional cloud cover by an empirical formula. The weights relating the brightness histogram to the total cloud amount varied from summer to winter to account for smaller cloud sizes in the summer months. In terms of automated cloud detection, the brightness histogram was the feature extracted from the raw satellite data and arbitrary thresholds and weighting factors defined the clear/cloud decision rule. This climatology had several advantages over previous climatologies, the chief advantage being the higher horizontal resolution of cloud cover estimates over most of the globe. On the other hand, it shared the limitation of providing no information on the diurnal changes of clouds since only the daytime passes of satellites were used. For example, convective clouds tend to appear and grow during daylight hours so that a climatology based on noon or afternoon passes of a polar-orbiting satellite could estimate more clouds for a region than a day and night average would find. Moreover, highly reflective areas such as the Sahara desert appeared as highly cloudy regions in the 1967–1970 climatology.

The US Air Force Global Weather Central automated cloud analysis model known as 3-D nephanalysis† has made cloud height estimates using infra-red data from polar-orbiting satellites. It is the only satellite

---

†The only description of this automated cloud analysis algorithm is contained in a *US Air Force Global Weather Central Technical Memorandum* entitled 'The AFGWC Automated Cloud Analysis Model', by F. K. Fye, 1978. This document is available from the National Technical Information Centre, 5255 Port Royal Road, Springfield, Va. 22161, by requesting document AD/AO57176.

climatology providing cloud altitude information at this time. When available, data from two polar-orbiting satellites are processed so that satellite cloud estimates are updated four times per day, a considerable improvement over earlier climatologies. Surface observations, vertical ascent data (radiosonde) and polar reports are also incorporated. The infra-red algorithm is a clustering algorithm operating on histograms such as figure 6.8 (p. 227). A surface temperature field separates clear zones from cloud layers and estimates cloud cover. Temperature altitude profiles convert the cloud temperatures to heights for each layer. The 3-D nephanalysis processes visible data for orbits near local noon, except over snow cover. The number of visible pixels with reflectivities greater than a threshold decides the fractional cloud cover; however, the location-dependent threshold is determined by complicated minimum brightness logic using past visible data. When infra-red and visible data are processed over the same area for the same observation time, a decision rule simply chooses the greater fractional coverage from the two estimates since the visible processor might not detect cirrus clouds or the infra-red processor might miss low clouds.

Cloud altitudes and coverage estimates are given on polar-stereographic grids with a spacing ranging from 25 km near the equator to 50 km near the poles. Cloud estimates are updated every three hours, if new satellite or surface data are available. Surface weather reports are entered into the 3-D nephanalysis through a separate processor and are merged with the satellite data. For climatological applications where a uniform set of estimates is preferred, the mixture of surface and satellite estimates is a significant problem. On the other hand, surface reports tend to be more accurate estimates of total cloud coverage than satellite estimates in cold or snow-covered regions and are usually more timely than the polar-orbiting satellite data. They have remained in the 3-D nephanalysis despite satellite improvements. Since most of the Earth's surface has no weather stations, the 3-D nephanalysis climatology is primarily a satellite climatology. The 3-D nephanalysis has processed both Northern and Southern Hemisphere data since 1973.

The 3-D nephanalysis provides operational cloud analyses which are saved for climatological purposes; therefore, some limitations for climatological applications are present in the data base. Aside from the mixture of sky cover (surface reports) and Earth cover (satellite data) in the cloud cover estimates, the Global Weather Central has changed the cloud detection algorithms frequently to correct limitations in the cloud analyses. The revised algorithms are applied to new data but the climatological record is not revised. The requirements for real-time global processing have dictated computationally simple cloud detection algorithms with limitations described in Section 6.6. Despite these problems recently archived information from the 3-D nephanalysis has been used successfully in global climatological studies (see Section 3.3, p. 103).

Visible and infra-red satellite data have been converted to rainfall estimates in various parts of the world to supplement raingauge and rainfall data (Barrett and Martin, 1981). Rainfall estimates over the tropical Atlantic have been made in support of a World Meteorological Organization project. The preparation of a global oceanic rainfall atlas based on Nimbus-5 microwave data (19 GHz) from December 1972 to February 1975 is also described by Barrett and Martin (1981). The rainfall data were derived over 1° latitude–longitude squares in oceanic areas and were then averaged over a 4° latitude by 5° longitude grid to yield monthly estimates of average rainfall rate.

All of the satellite cloud climatologies have limitations, so that research leading to improved cloud detection and climatologies is active. For example, different cloud climatologies over the Pacific Ocean during 1967–1970 disagree to an extent that estimates of cloud effects on the Earth's radiation budget are uncertain. Limited 'cloud truth' information frustrates evaluation of cloud climatologies. Cloud thickness or cloud cover is not measured with the accuracy that radiosondes can achieve for atmospheric temperature or humidity. The massive quantities of satellite data involved for climatological applications have favoured the use of coarse-resolution imagery data and simple algorithms for cloud detection.

As is described in Chapter 1, the World Climate Programme of the World Meteorological Organization is sponsoring the International Satellite Cloud Climatology Project (ISCCP) to use the international array of five geostationary and two or more polar-orbiting satellites and provide improved cloud estimates for climate research. The most important goal of the cloud climatology will be a globally and temporally uniform methodology to estimate various indices of cloud cover. Five years of climatology are planned, beginning in 1983. Geostationary satellite data will be used as much as possible to measure diurnal variations in cloud cover. Polar-orbiting data will be used at high latitudes and to normalize cloud detection by the various satellite sensors. Satellite data will be collected and pre-processed by participating nations. A global processing centre will produce the cloud climatology. The cloud information will include cloud fractional coverage, heights, equivalent temperatures and reflectivities. Cloud estimates will be provided every three hours for areas of about 250×250 km latitude and longitude. Algorithms for cloud detection are being tested and evaluated so that the most useful algorithm can be applied to the satellite infra-red and visible data.

The original sets of satellite data will also be provided after some horizontal averaging. Most existing satellite climatologies are difficult or impossible to compare with the original satellite data since the original data are no longer available or are in their original formats instead of the map used by the climatology. The ISCCP climatology will provide a cloud data base so that climate modellers can begin to understand and predict the long-term changes in cloud cover which may warm or cool the Earth.

References

Anderson, R. K., Ferguson, E. W. and Oliver, V. J., 1966, *The Use of Satellite Pictures in Weather Analysis and Forecasting*, W.M.O. Technical Note No. 75 (Geneva: World Meteorological Organization), 184 pp.

Barrett, E. C., 1974, *Climatology from Satellites* (London: Methuen), 418 pp.

Barrett, E. C. and Martin, D. W., 1981, *The Use of Satellite Data in Rainfall Monitoring* (London: Academic Press), 340 pp.

Brimacombe, C. A., 1981, *Atlas of Meteosat Imagery* (Noordwijk, the Netherlands: ESA Scientific and Technical Publications), 495 pp.

Colwell, R. N. (editor), 1983, *Manual of Remote Sensing,* Second Edition (Falls Church, Va: American Society of Photogrammetry), 2 Vols., 2400 pp.

Desbois, M., Seze, G. and Szejwach, G., 1982, Automatic classification of clouds on METEOSAT imagery: Application to high-level clouds. *Journal of Applied Meteorology* **21**, 401–12.

Duda, R. O. and Hart, P. E., 1973, *Pattern Classification and Scene Analysis* (New York: John Wiley & Sons), 482 pp.

Hasler, A. F., 1981, Stereographic observations from geosynchronous satellites: an important new tool for the atmospheric sciences. *Bulletin of the American Meteorological Society* **62**, 194–212.

Henderson-Sellers, A., Hughes, N. A. and Wilson, M., 1981, Cloud cover archiving on a global scale: a discussion of principles. *Bulletin of the American Meteorological Society* **62**, 1300–1307.

Hubert, L. F., 1979, Wind derivation from geostationary satellites. In *Quantitative Meteorological Data from Satellites*, W.M.O. Technical Note No. 166. (Geneva: World Meteorological Organization), Chapter 2, pp. 33–59.

Lund, I. A. and Shanklin, M. D., 1973, Universal methods for estimating probabilities of cloud-free lines-of-sight through the atmosphere. *Journal of Applied Meteorology* **12**, 28–35.

Malberg, H., 1973, Comparison of mean cloud cover obtained by satellite photographs and ground-based observations over Europe and the Atlantic. *Monthly Weather Review* **101**, 893–97.

7

Problems encountered in remote sensing of land and ocean surface features

Michael J. Duggin
State University of New York
Syracuse, NY, USA
and
Roger W. Saunders
European Space Agency
Darmstadt, FR Germany

7.1. Introduction

Satellite data are being used increasingly to map, monitor and quantify features on the Earth's land and ocean surfaces. In this chapter, the physical principles controlling the data recorded by the satellite sensors will be examined and the analytical procedures used to study surface features will be outlined. It will be shown that unresolved cloud and haze can strongly affect the accuracy of land feature mapping and retrieved sea surface temperatures. Methods of screening cloud-affected data will be discussed, as will present directions in research to remove the systematic part of this error source and to estimate the effects of random error on target discrimination and quantification.

7.2. Land surface sensing: effects of atmospheric distortion

The radiance incident on the sensing device depends on the reflectance properties of the ground in the visible and near (reflective) infra-red regions. In the thermal infra-red region, the radiance received by the sensor depends on the temperature and emissivity of the ground features studied. Most sensors use several discrete bandpasses. The nominal (half-

Table 7.1. Nominal half-power bandpass values for the Landsat MSS, the Thematic Mapper (TM) and the visible and near infra-red bands of the NOAA AVHRR.

Sensor	Half-power bandpass values			
MSS	$0 \cdot 5 - 0 \cdot 6 \mu m$;	$0 \cdot 6 - 0 \cdot 7 \mu m$;	$0 \cdot 7 - 0 \cdot 8 \mu m$;	$0 \cdot 8 - 1 \cdot 1 \mu m$
TM	$0 \cdot 45 - 0 \cdot 52 \mu m$; $1 \cdot 55 - 1 \cdot 75 \mu m$;	$0 \cdot 52 - 0 \cdot 60 \mu m$; $2 \cdot 08 - 2 \cdot 35 \mu m$;	$0 \cdot 63 - 0 \cdot 69 \mu m$; $10 \cdot 40 - 12 \cdot 50 \mu m$	$0 \cdot 76 - 0 \cdot 90 \mu m$
AVHRR	$0 \cdot 570 - 0 \cdot 686 \mu m$;	$0 \cdot 713 - 0 \cdot 986 \mu m$		

power) bandpasses of the Landsat multi-spectral scanner and thematic mapper, together with those of the NOAA-7 advanced very high resolution radiometer (AVHRR), are described in detail in Chapter 2 and are listed in table 7.1. The wavelength bands of sensing devices designed to study features on the Earth's surface have been selected to record radiance in those parts of the spectrum where the features to be identified differ in their reflectance or emissivity properties from other, surrounding targets. This point is illustrated in figure 7.1(a) for the visible and reflec tive infra-red regions. Here, it is seen that bands in the red and in the reflective infra-red regions are effective in discriminating healthy from stressed vegetation, due to the optical properties of the individual leaves and due to their geometrical disposition and density in the vegetative canopy. Bands at approximately $1 \cdot 5$ and $2 \cdot 5 \mu m$ (not shown in figure 7.1(a)) are useful in detecting the moisture content (turgidity) of leaves

Many aesthetically attractive books have been based on 'pictures' of the land surface of the Earth, e.g., *Mission to Earth* (referenced in Chapter 1, p. 44). Many early investigations of land surface features depended on image analysis techniques. These operations became quantified once computer rather than human interpretation was implemented. Still more recently, surveys of land surface information have utilized directly the digital radiance data retrieved from the satellites. All quantitative analyses of surface features may be affected by sub-pixel, and hence unidentifiable, clouds and aerosols. In this section some of the atmos pheric distortions which are all too easily overlooked are considered. The theory described here is applicable to most land-surface retrieval programmes currently in operation.

The discrimination of different targets would require the use of dif ferent parts of the spectrum, in order to effect satisfactory discrimination with (say) a 95 % probability of being correct. For identified wavelength regions, e.g., the 'visible', some targets, such as water, show a mono tonically decreasing reflectance (and therefore target radiance) with increasing wavelength, while others show a monotonically increasing reflectance: a typical example would be soil. A knowledge of the reflect ance properties of those targets to be discriminated as a function of

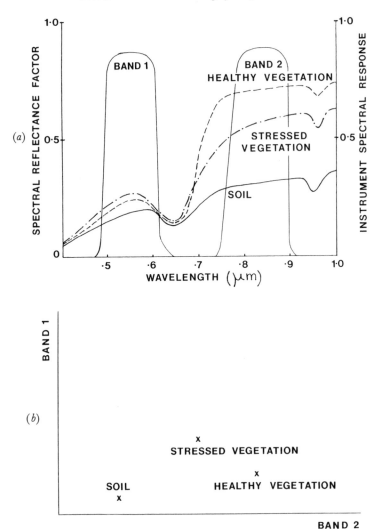

Figure 7.1. (a)Reflectance spectra of different ground classes. Typical sensor optical channels are illustrated. Discrimination can be seen to be possible.
(b)Illustration of the method of discrimination of ground cover classes using clustering of measurement vectors.

wavelength is necessary in order to be able to select those bandpasses needed for optimal target discrimination.

As described in Chapter 2, reflectance properties of the target also depend on the geometrical relationship between the Sun, the target and the sensor. Radiance, recorded from the ground target in several optical channels simultaneously, constitutes a 'measurement vector' for the

resolved element at ground level. Each element in the vector consists o
a measurement in one optical channel: one dimension in 'measuremen
space'. This column matrix may be manipulated to provide a 'featur
vector' which facilitates identification of resolved ground features a
different from each other. To illustrate this principle, a two-dimensiona
measurement vector is illustrated in figure 7.1(*b*) for the data measurec
in Bands 1 and 2 in figure 7.1(*a*), for targets filled with soil, health
vegetation and stressed vegetation. The addition of data points from a
large number of other, similarly resolved areas would produce clusters i
which the three original points are contained. If these clusters ar
insufficiently distinguishable (if they overlap) then mathematica
manipulation, changing the measurement vectors into feature vectors
might make the task of distinguishing the clusters from each othe
simpler. The methods of production and use of this figure are similar t
those described for figure 6.2 (p. 212).

It is necessary for computer analysis to know the nature of grounc
areas (typically features to be mapped) which are represented by larg
numbers (over 30, for statistical reasons) of pixels. Such known areas ar
called 'training areas'. Thus, each class of ground cover for which ther
is a 'training area' will have an associated cluster of sampled values (fron
each pixel used in the training process) in feature space. In analysing th
remotely sensed image, each pixel in the whole image is sequentiall
tested, to see where its feature vector places it in 'feature space'. That is
using some algorithm, a decision is made as to which cluster (anc
therefore which class) each pixel belongs. Generally, a map is producec
after each pixel in the image has been so classified: each class (grounc
type = cluster in feature space) is usually represented by a colour. I
arithmetically or geometrically based decision algorithms, the decision i
made as to which clearly divided portion of n-dimensional feature spac
each pixel belongs. Where statistical algorithms are employed, the deci
sion is made as to which portion of n-dimensional feature space each pixe
most probably belongs. Having produced a map, it is of course necessar
to test its accuracy, and so 'test areas' of approximately the same numbe
and distribution as training areas are used. The frequency with which tes
areas, of known nature, are correctly or incorrectly mapped is tallied anc
the accuracy with which each class has been mapped may be so assessed
This procedure is referred to as 'supervised classification'. This is becaus
the computer is 'trained' (supervised) to search for specific signature
which characterize ground types to be mapped. In cases where little i
known about the area surveyed, then the data are grouped into cluster
in feature space and a map is drawn of each 'spectral class': each pixe
is plotted according to which cluster (formed from all the data, not fron
training sets) it belongs. However, this operation assumes that eacl
spectral class is indeed related to a ground feature. This is different fron

using training areas on known ground features to define clusters (of pixel radiance measurement) in measurement space and to increase separability of those clusters by transforming from measurement space into feature space.

Sensing devices record radiance from ground targets under conditions which might introduce errors. That is, Sun–target–sensor geometry varies with scan angle, spacecraft track and Sun zenith angle, as shown in figure 2.6 (p. 58). Both ground reflectance and atmospheric backscatter are anisotropic and can strongly affect the level and spectral (i.e., between-band) distribution of radiance incident on the sensor. Indeed, in the case of the NOAA AVHRR, the sensor scan angle limits are ±54° and the Sun angle can vary by 23° across the image due to Earth curvature effects. The spectral reflectance factor (actually, the spectral hemispherical-conical reflectance factor) refers to the proportion of radiance reflected from a real target relative to that which would be reflected from a perfectly diffuse, white, standard reflector. The value of the spectral reflectance factor depends strongly on the view zenith and view azimuth angles. It also depends strongly on the polar distribution of the incident radiation field; that is, the spectral reflectance factor depends

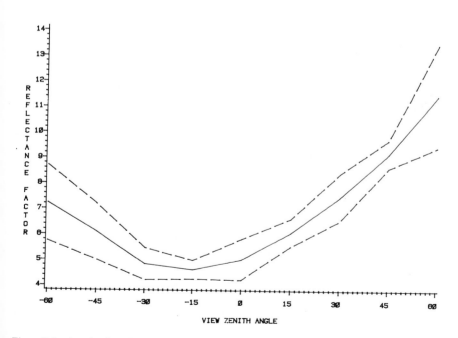

Figure 7.2. Angular dependence of the hemispherical-conical reflectance factor of wheat for the NOAA-6 AVHRR Band 1 (0·570–0·686 μm): mean and 95% confidence limits.

on the solar zenith and azimuth angles and also cloud and haze distribution. Under conditions where remote sensing data would be collected, the irradiation at the target consists mainly of illumination from the Sun; generally less than 20 % of the total irradiance consists of irradiance from other portions of the sky. An example of the angular dependence of the spectral reflectance factor of wheat is shown in figure 7.2.

The term hemispherical-conical reflectance factor, briefly explained, refers to the situation in which radiation from the whole hemisphere of the sky is incident on the target (irradiance) while radiation in a narrow cone (of dimensions determined by the area of the detector and by the distance from the detector to the target) falls on the sensor. Mention is made in the literature of the biconical reflectance factor and of the bidirectional reflectance factor. These terms refer, respectively, to the reflectance factor of a surface illuminated by a finite source, which is viewed by a finite detector, and to a surface which is illuminated by and viewed by an infinitely small source and detector.

There are not a great deal of experimental data on the goniometric† anisotropy of the spectral hemispherical-conical reflectance factor, but there are mathematical models by which the Sun–target–sensor angle dependence of the spectral reflectance factor may be calculated. Work is currently proceeding to calibrate these models, so that calculations may be extended to angular geometries descriptive of large scan angles and extreme Sun elevations and azimuths.

The anisotropy of the atmospheric backscatter is also strongly dependent on Sun–target–sensor geometry, as may be seen from the figures in table 7.2 calculated from a theoretical model of the non-turbulent atmosphere. Although much work has been done to study atmospheric scattering and absorption effects, while estimates of systematic goniometric variations in backscattered radiation may be made, the random effects are not calculable.

As described in Chapter 2, the spectral response of all sensors is wavelength dependent (see figure 2.12, p. 69). Since target spectral radiance can vary with Sun–target–sensor geometry, then so can the interaction between the spectral response of the sensor and the spectral radiance from the target. The spectral radiance which is incident on the target (i.e., irradiance) is also very strongly wavelength dependent. The radiance reflected from a target will be proportional to the irradiance on the target, to the reflectance of the target and to the transmission of the atmosphere between the target and the sensor. The backscattered radiance from the atmosphere which is recorded by the sensor will be additive to that reflected from the target. The radiance recorded by a sensor is

---

†The term goniometric refers to the Sun–target–sensor angle dependence.

Table 7.2. Path radiance (mW cm^{-2} sr^{-1} μm^{-1}).

Atmosphere	Scan angle	$\lambda = 0\cdot650\,\mu$m		$\lambda = 0\cdot900\,\mu$m	
		$\phi = 0°$	$\phi = 180°$	$\phi = 0°$	$\phi = 180°$
Clear	5°	0·894	0·977	0·330	0·344
	15°	0·845	1·099	0·325	0·368
	25°	0·841	1·276	0·337	0·415
Turbid	5°	2·915	3·047	2·590	2·624
	15°	2·938	3·379	2·599	2·734
	25°	3·187	4·102	2·762	3·058

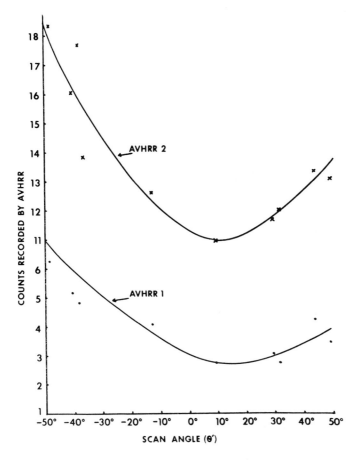

Figure 7.3. View-angle dependence of the recorded radiance from forest in the first two (i.e. visible and near infra-red) bands of the NOAA AVHRR: the same target viewed from different orbits, close-spaced in time.

dependent on Sun – target – sensor geometry due to the angular
dependence of the variables. An example of the angular dependence of
radiance data recorded for different view angles on sequential images
taken over a period of one week over the same forested area in Michigan
is shown in figure 7.3. This variation is due mainly to angular anisotropy
of the reflectance properties at ground level. Figure 7.4 shows the variation (with scan or view angle) of recorded radiance across an apparently
cloud-free image obtained over crop and forest land. However, the
presence of cloud and of cloud shadow can produce random fluctuations
superimposed on systematic variations of the type shown in figure 7.5.

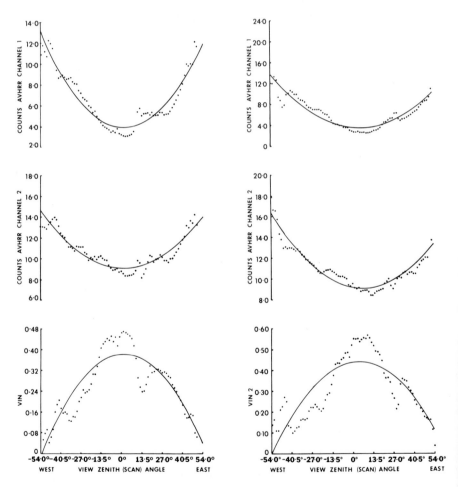

*Figure 7.4. Variation with scan angle of recorded radiance in NOAA AVHRR Bands 1 and 2 from
mixed cropland (left) and forest (right) on an apparently cloud-free image. The vegetation index VIN
(AVHRR 2/AVHRR 1) is also shown.*

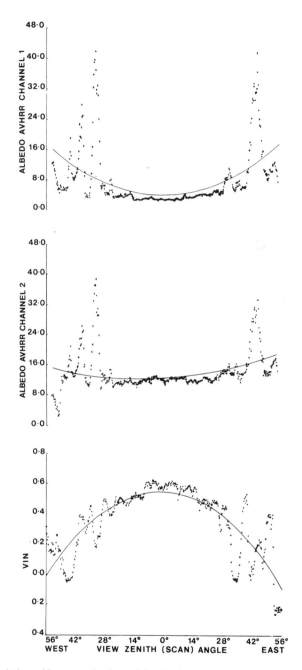

Figure 7.5. Variation with scan angle of recorded radiance in NOAA AVHRR Bands 1 and 2 over the same target as shown in figure 7.4 with the presence of slight scattered cloud and haze in the image. The effect of random radiance variation due to cloud on the systematic scan-angle dependence is seen.

While the presence of cloud can increase the radiance levels in the individual bandpasses in the visible and in the reflective infra-red regions, usually a decrease in temperature is recorded in the thermal infra-red region. Indeed, temperature is one of the methods of measuring the presence of cloud and it is anticipated that by studying simultaneously the visible, near infra-red and thermal infra-red regions, for each pixel, progress will be made in the detection of sub-pixel-sized (i.e., unresolved) cloud.

Unresolved cloud can cause difficulties in mapping and quantifying ground features. Since cloud detection algorithms are successful only for levels of cloud contained within a pixel which exceeds a certain value, undetected cloud and haze can distort the level and the spectral distribution of radiance from targets; that is unresolved cloud can alter the radiance at wavelength λ from a given pixel, because unresolved cloud can alter the apparent reflectance factor of the ground area imaged within a pixel.

The greater the percentage of the pixel covered by cloud, the higher the overall radiance levels. However, in the case of vegetation, due to the far lower reflectance of the plant canopy in the visible part of the spectrum than in the reflective infra-red, the presence of cloud can alter the ratio of the apparent near infra-red radiance to the visible radiance reflected from the area included within a pixel. Since vegetation is frequently monitored on a repetitive basis using satellite scanner data, methods have been developed by which combinations of radiance recorded in different scanner channels are combined into a 'vegetation index'. Such an index is not only more sensitive to the nature and vigour of vegetation than the radiance in any individual bandpass, but because it is one-dimensional information, it is also simpler (and cheaper) to analyse, especially for repeated coverage of large areas, than multichannel, multidimensional data analysed using supervised classification. A popular vegetation index is the ratio of the radiance recorded by the NOAA advanced very high resolution radiometer (AVHRR) in the near infra-red region to that in the visible part of the spectrum (see table 7.1, p. 242). This vegetation index will clearly be affected by the presence of unresolved cloud and haze. Another popular vegetation index, which is similarly affected, is

$$VIN = \frac{\text{near IR radiance} - \text{red radiance}}{\text{near IR radiance} + \text{red radiance}}$$

$$= \frac{\text{AVHRR 2 radiance} - \text{AVHRR 1 radiance}}{\text{AVHRR 2 radiance} + \text{AVHRR 1 radiance}} \qquad (7.1)$$

One algorithm which is currently used to screen the NOAA AVHRR data for cloud is (AVHRR Band 2 − AVHRR Band 1): if this expression falls below zero for any pixel, then that pixel is said to consist of cloud. However, calculations have shown that for some agricultural targets the pixel can consist of up to about 30 % cloud before detection occurs and the pixel is screened out. Cloud levels of much less than 30 % of the pixel area can cause serious distortion of vegetation indices: a simulated example for AVHRR 1 is shown in figure 7.6 for wheat (an erectophile canopy). Here, the atmosphere is considered to be 'clear' (horizontal visibility > 50 km) and yet there is a strong dependence of the vegetation index on both sub-pixel level cloud content and a smaller dependence on scan angle. The Sun elevation was assumed to be 45° in these calculations, and it was assumed that the Sun azimuth angle was the same as the larger view azimuth angle. The dependence of the radiance values in the visible part of the spectrum (AVHRR Band 1) on both the scan angle and on the percentage of unresolved (sub-pixel sized) cloud is shown in figure 7.6, while that for the reflective infra-red part of the spectrum (AVHRR Band 2) is shown in figure 7.7. As one might expect, the radiance levels show overall increases with the amount of unresolved (sub-pixel level) cloud. However, for both bandpasses and at all cloud levels, the radiance levels in each of the bands show asymmetry about the nadir position. This is due to the angle-dependent anisotropy of the target radiance, as mentioned above. The effects of unresolved cloud and of scan angle on the vegetative index (AVHRR 2/AVHRR 1) are shown in figure 7.8(*a*).

In the case of a 'turbid' atmosphere (horizontal visibility < 10 km), the scan-angle dependence and the sub-pixel sized (unresolved) cloud both affect the vegetation index less severely than in the case of the clear atmosphere. This is shown in figure 7.8(*b*) for the same wheat target. The scan-angle dependence of both the level and spectral distribution of the recorded radiance are less pronounced for the turbid atmosphere than for the clear atmosphere, because the level of radiance scattered into the sensor from (diffusely reflecting) cloud forms a higher proportion of the signal, which is thus less affected by surface reflectance angular anisotropy.

Current work is directed towards empirical studies on the effects of cloud (detected using existing algorithms) on data collected simultaneously in the optical reflective (visible and near infra-red) region and in the mid-infra-red† and thermal infra-red regions. These studies, coupled with interactive image analysis, may lead to the formation of improved cloud-detection (screening) algorithms. However, such image-specific studies must necessarily be parallelled by simulation studies of the scan-angle dependence and unresolved cloud-content dependence of recorded radiance. This dual approach is directed towards a generalization of

---

†Here the term 'mid-infra-red' refers to the 3–5 μm wavelength region.

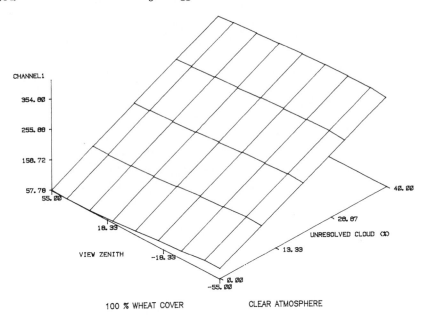

Figure 7.6. The simulated dependence of relative recorded radiance in AVHRR 1 from a wheat canopy on scan angle and on unresolved cloud level (% of pixel occupied by cloud). A clear atmosphere (horizontal visibility > 50 km) was assumed.

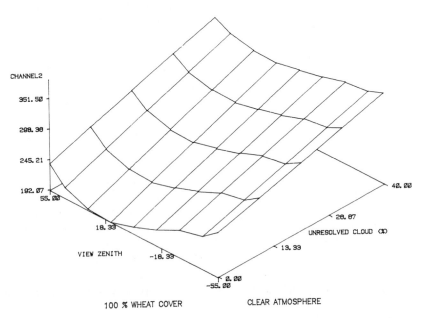

Figure 7.7. The simulated dependence of relative recorded radiance in AVHRR 2 from a wheat canopy on scan angle and on unresolved cloud level (% of pixel occupied by cloud). A clear atmosphere (horizontal visibility > 50 km) was assumed.

(a) VIN

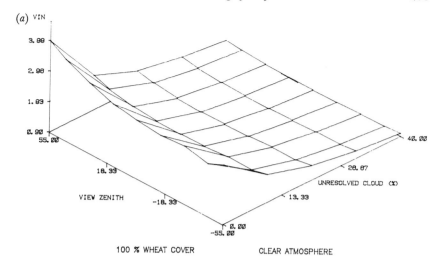

100 % WHEAT COVER   CLEAR ATMOSPHERE

(b) VIN

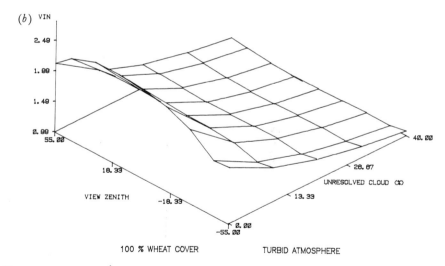

100 % WHEAT COVER   TURBID ATMOSPHERE

Figure 7.8. The simulated dependence of the vegetation index (AVHRR 2/AVHRR 1) for wheat on scan angle and on the level of unresolved cloud for (a) a clear atmosphere (horizontal visibility > 50 km); and (b) a turbid atmosphere (horizontal visibility < 10 km).

image-specific conclusions which are obtained from a finite number of empirical analyses.

Further, while such studies for typical, economically important targets are underway, it must be borne in mind that it is the contrast between the target to be discriminated (and possibly quantified) and its surroundings which is of prime importance. Therefore, it is the dependence of the contrast ratio between the target and its surroundings on scan angle and on unresolved cloud level which is of critical importance.

The discrimination of two targets on the basis of recorded radiance will not be limited by a systematic dependence of recorded radiance on certain variables, providing empirical correction algorithms exist to normalize the data. However, discrimination will be limited by random variations in target radiance caused by several factors. Factors controlling the separability of two clusters in feature space may be understood by considering factors determining the separability of two targets in one-dimensional measurement space.

The contrast ratio is defined in several ways. One popular definition (e.g., Slater, 1980) is the differential contrast. This is, at ground level

$$C_G = (L_T - L_B)/L_B \qquad (7.2)$$

where L_T is the target radiance; and L_B is the background radiance.

In fact, the differential contrast seen by the observer, C_0 is related to the differential contrast at ground level C_G by the expression

$$Y = C_0/C_G$$

$$= \left(1 + \frac{L_p}{L_B\tau}\right)^{-1} \qquad (7.3)$$

where Y is called the contrast transmittance; τ is the atmospheric transmittance and L_p is the atmospheric path radiance.

The differential contrast seen by the observer is

$$C_0 = \frac{(L_T - L_B)\tau}{L_B\tau + L_p} \qquad (7.4)$$

where the radiances are calculated for a selected bandpass. This restriction to a particular bandpass is applicable when data are collected using multichannel sensors.

Calculations of C_0 for wheat (at the boot stage, i.e., $3 \cdot 5$ on the modified Feeks growth scale) contrasted against a mixture of 30% moist soil + 70% wheat at the same growth stage were made for AVHRR Band 1, for two atmospheric turbidities: clear (horizontal visibility > 50 km) and turbid (horizontal visibility < 10 km). These calculations were made for various levels of unresolved cloud and are shown in figure 7.9. It is assumed here that the level of sub-pixel-sized cloud and haze is constant across the scene. It is clear from the figures that there is a systematic

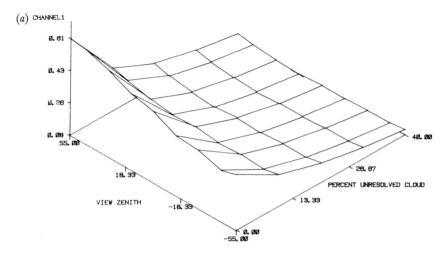

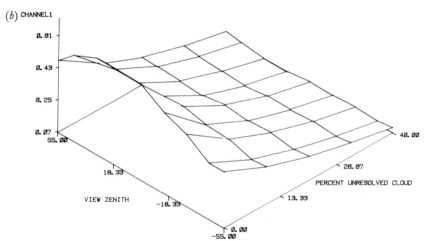

Figure 7.9. (a)*The simulated dependence on scan angle and on average unresolved cloud level of the differential contrast in radiance recorded in AVHRR Band 1 between wheat and 70% wheat + 30% moist soil. A clear atmosphere was assumed (horizontal visibility > 50 km).*
(b) The simulated dependence on scan angle and on average unresolved cloud level of the differential contrast in radiance recorded in AVHRR Band 1 between wheat and 70% wheat + 30% moist soil. A turbid atmosphere was assumed (horizontal visibility < 10 km).

dependence of the contrast in the radiance recorded in AVHRR 1 on scan angle and on the sub-pixel cloud level. Similar findings apply for calculations conducted for AVHRR Band 2. Thus, the contrast between target and background depends significantly on the mean level of unresolved cloud and haze.

While the above calculations relate to the effect of systematic variations in unresolved cloud, there are also random variations which remain uncompensated. Such effects may be taken into consideration, however, in determining the mean radiance difference which is necessary for the discrimination of two targets with a pre-determined level of confidence.

As a simple demonstration of the importance of unresolved cloud in determining variance in the radiance from a given target which is incident on the detector in a given bandpass, the radiance L_r incident upon the sensor in a selected bandpass r can be expressed as

$$L_r = E_r \, R_r \, \tau_r + (L_\mathrm{p})r \qquad (7.5)$$

where R_r is the hemispherical-conical reflectance of ground target A and cloud in bandpass r; τ_r is the atmospheric transmission in bandpass r; E_r is the irradiance in bandpass r; and $(L_p)_r$ is the path radiance in bandpass r.

These parameters all depend on the angular parameters shown in figure 2.6 (p. 58). The variance, σ_{LL}, in L_r is given by

$$\sigma_{LL} = [\sigma(L_r)]^2 \qquad (7.6)$$

where $\sigma(L_r)$ is the sample standard deviation which may be approximated following, for example, Davies (1961).

The variance in the radiance from two targets to be differentiated may be used to calculate the necessary difference in the mean radiance from groups of pixels in each of two targets, in order to perform a discrimination between the targets with 95 % probability of being correct. It should be noted that if any pixel can contain up to 30 % of unscreened cloud and if this percentage can fluctuate between 0 % and 30 %, then the path radiance term can be up to approximately 30 % of the total recorded radiance. This is because the undetected cloud could be present in variable quantity across the entire scene and could then backscatter into the sensor.

The necessary difference in the mean radiance values of groups of pixels in two targets to be differentiated depends on the variance in target radiance from each target. Clearly, the level and variation of unscreened (undetected) cloud and haze will affect target discriminability. This situation will be exacerbated by any variations in the accuracy with which systematic effects are removed from the recorded radiance data using empirical data normalization algorithms.

The implications are serious: severe signature distortion can result from the level and from fluctuations in the level of undetected cloud which is present in a scene. Furthermore, the difference between the mean pixel radiance values in a given bandpass which is needed to discriminate two targets will depend greatly on the level of and variation in undetected cloud. This points to a need for further studies leading to improved cloud screening algorithms.

In this section theoretical calculations have been used to demonstrate the considerable problems which have so far been avoided by those members of the remote sensing community interested in sensing land surface features. The vast majority of land surface surveys so far undertaken have been strongly supervised. Once automated routines are applied to large quantities of data, the types of problem identified in this section will become manifest. Higher precision is generally required for land surface investigations than for surveys of the oceans or the cryosphere, and thus it is data pertaining to land surfaces which will be most readily contaminated by sub-pixel clouds and the unidentified atmospheric aberrations.

7.3. Sensing oceanic features from space

The world's oceans are an extremely important source and sink of energy, and strongly influence our climate. In order to discover the detailed interaction between the oceans and the atmosphere, comprehensive global observations have to be made over a period of many years. The only effective way of achieving this is to monitor continuously the ocean and atmosphere from space.

Before the advent of meteorological satellites, observations of the sea surface were made from ships, buoys, experimental platforms or aircraft. These measurements are made routinely by weather ships, for instance, or for short periods of time many inter-related measurements have been made in a specified area as part of an organized research programme (e.g., GATE). The disadvantage with *in situ* measurements is that they are only representative of one particular point at one particular time. Also, the expense of operating many ships or buoys deployed over a wide area to get good global coverage becomes prohibitive.

In contrast, one satellite in a Sun-synchronous orbit can sample almost the entire globe twice a day with a scanning radiometer, permitting good spatial coverage to be obtained. The problem with satellite measurements is retrieving the sea surface parameters accurately enough to be of interest to climatologists. Uncertainties can be introduced from many sources, such as variable atmospheric absorption and scattering, clouds, sunglint, sea state and variable viewing geometry. Understanding and reducing these uncertainties is currently an important area of research in order to obtain accurate global sea surface parameters which are of use to the oceanographic and meteorological community. In the following sections some of the problems involved in making sea surface measurements (particularly sea surface temperature) from space will be discussed and various methods which are currently being used to overcome the sources of uncertainty will be illustrated. There have been a number of recent publications to which the interested reader is referred, e.g., Cracknell (1981; see Chapter 1, p. 44) and Gower (1981), which describe some aspects of sea surface remote sensing in more detail than given here.

7.4. The methods used to measure cloud-free sea surface temperature

One of the most important sea surface parameters to measure is sea surface temperature (SST). Global SST values with an absolute accuracy of $0 \cdot 5$ K would provide a useful data set as input for climate models and for validation of these models. If the SST values are made available in real time then the possibility exists for them to be used as input for forecasting models. Anomalies in SST may significantly influence the atmospheric circulation. If both variables are monitored at the same time, accurately, it should be possible to investigate how changes in SST affect the general circulation. There have been a number of recent papers on 'teleconnections' where SST anomalies have been shown to be correlated with periods of unusual weather. For instance, the severe winter of 1962/63 in western Europe was associated with unusually high (i.e., warm) SST values in the tropical Atlantic. During the drought over western Europe in 1975/76 there were unusually low (i.e., cold) SSTs in the Pacific Ocean. Oceanic circulation models will also benefit from accurate global SST data together with high-resolution studies of important phenomena such as the interaction of the Gulf Stream with the surrounding cooler water. The need for absolute accuracy over smaller scales (100 km) is not so important, but high relative accuracies are still required to study the formation and dissipation of tidal fronts between well mixed and stratified regions of the ocean. Many other interesting phenomena can be studied using the infra-red satellite data.

The sea surface temperature of terrestrial oceans typically varies between 272 K and 303 K. A blackbody emitting at these temperatures has a peak in the Planck function at a wavelength of about 10 μm. This emitted infra-red radiation can only be detected from space if the intervening atmospheric absorption at these wavelengths is sufficiently small to allow the emitted surface radiation to escape (see figure 6.1, p. 204). The primary atmospheric constituents which absorb radiation at these infra-red wavelengths are water vapour, carbon dioxide, ozone, nitrous oxide and methane. Also, additional absorption due to water vapour dimers and polymers, which varies slowly with wavelength and aerosol scattering, can also attenuate infra-red radiation significantly (as discussed in Chapter 2). A number of regions of high atmospheric transmission are apparent in figure 6.1. Conveniently there is an atmospheric window near the peak of the blackbody radiation from the sea surface at $11\,\mu m$. Most of the SST measurements from space have been made in this window region. Emitted radiation from the Earth detected above the atmosphere through such a window can come from various possible sources, as shown in equation 7.7 for a cloud-free radiance

$$E_T(\nu, \theta) = E_s(\nu, \theta) + E_u(\nu, \theta) + E_D(\nu, \theta) + E_o(\nu, \theta, Z) + E_A(\nu, \theta) \qquad (7.7)$$

where E_T is the detected radiation at the top of the atmosphere; E_s is the radiation emitted by the sea surface and attenuated by the intervening atmosphere; E_u is the upward radiation emitted by the atmosphere; E_D is the downward radiation emitted by the atmosphere and reflected up by the sea surface; E_o is the solar radiation attenuated by the atmosphere and reflected up from the sea surface; E_A is the radiation scattered into the beam by atmospheric aerosols; ν is the frequency; θ is the local zenith angle of the satellite; and Z is the solar zenith angle. In order to understand the various contributions to the upwelling radiance at the top of the atmosphere each of the terms on the right-hand side of equation 7.7 is considered in turn.

The radiation from the sea surface E_s (which by definition is normally the dominant term in a window region) can be expressed as

$$E_s(\nu, \theta) = \epsilon_s(\nu, \theta)B(\nu, T_s)\tau_0^H(\nu, \theta, w) \qquad (7.8)$$

where ϵ_s is the emissivity of the sea surface which is close to 1 (i.e. $0\cdot99$ at $11\,\mu m$) but depends on satellite zenith angle θ (when $\theta > 60°$ the change in the emissivity becomes significant); $B(\nu, T_s)$ is the Planck function for frequency ν, temperature T_s; and τ_0^H is the atmospheric transmittance from the surface ($z = 0$) to the top of the atmosphere ($z = H$). This transmittance depends on many variables as various atmospheric species can contribute to the absorption. At $11\,\mu m$ the principal absorber is water vapour so the transmission τ_0^H is a strong function of vertical water vapour amount w. The vertical atmospheric pressure and temperature profiles also influence the magnitude of line broadening and hence absorption. To detect a useful amount of radiation from the sea surface, $E_s(\nu, \theta)$ at the top of the atmosphere should be more than 50% of the total detected radiance. The transmittance at $11\,\mu m$ is higher than 50% at mid-latitudes and polar regions for a vertical path to space but in the tropics it can be lower than this due to the higher water amounts present. In addition, transmittance decreases with increasing zenith angle of observation. The window at $3\cdot7\,\mu m$ is less affected by water vapour absorption and as a result has a much higher transmittance in tropical regions. The relative merits of using the $3 \cdot 7$ and $11 \mu m$ windows for SST measurements have been mentioned in Chapter 5 and are discussed later in this chapter.

The second term in equation 7.7 is the upward atmospheric emission, which can be expressed as

$$E_u(\nu, \theta) = \int_0^H B(\nu, \ T(h)) \frac{\partial \tau(h)}{\partial h} \, dh \qquad (7.9)$$

where $B(\nu, \ T(h))$ is the radiation emitted by a layer of atmosphere at height h; and $\partial \tau / \partial h$ is the weighting function for an atmospheric layer at that height. Integrating this expression up through the atmosphere gives the total upwelling radiance from the atmosphere. For a mid-latitude atmosphere the contribution of this term to the total signal received at the top of the atmosphere at $11 \mu m$ can be approximately 20%.

The remaining terms usually have only a small effect but should not be neglected, especially at shorter wavelengths. The term which represents the reflected downward atmospheric emission is given by

$$E_D(\nu, \theta) = [1 - \epsilon_s(\nu, \theta)] \tau_0^H(\nu, \theta, w) \int_H^0 B(\nu, \ T(h)) \frac{\partial \tau(h, \theta)}{\partial h} \, dh \qquad (7.10)$$

assuming isotropic scattering by the sea surface with a reflectivity of $[1 - \epsilon_s(\nu, \theta)]$. At thermal infra-red wavelengths ($\approx 11 \mu m$) the reflectivity is small ($\sim 1 \%$) but at $3 \cdot 7 \mu m$ increases to 3% and for microwaves it increases to 50%. The reflectivity is more important for the reflected solar component given by

$$E_0(\nu, \theta, Z) = S(\nu) \cos(Z) \, d\tau_H^0(\nu, Z) \tau_0^H(\nu, \theta) \varrho(Z, \nu, \theta, \phi) \qquad (7.11)$$

where $S(\nu)$ is the mean annual solar insolation at the top of the atmosphere at a frequency ν; Z is the solar zenith angle; d is the Earth–Sun distance correction factor which is a function of the time of year; and ϱ is the anisotropic scattering function (i.e., the hemispherical-conical reflectance factor) for the sea surface. Radiation scattered from the sea surface is strongly peaked for observation angles which allow direct specular reflection to occur (i.e., in sunglint). At $3 \cdot 7 \mu m$ and microwave frequencies the solar reflection term in equation 7.11 can be appreciable in areas where sunglint is occurring. The final term in equation 7.7, $E_A(\nu, \theta)$, represents the radiation scattered into the detector by atmospheric aerosols and is small at $11 \mu m$ and microwave frequencies but becomes significant at shorter wavelengths (i.e., $3 \cdot 7 \mu m$). Problems arise when a large quantity of aerosols is suddenly injected into the atmosphere, as happens during some volcanic eruptions or large dust storms. This can seriously affect radiances measured at $3 \cdot 7 \mu m$ introducing errors in SST which are derived using this channel. This problem occurred recently when the eruption of El Chichón in Mexico on 4 April 1982 caused considerable additional aerosol scattering at both visible and $3 \cdot 7 \mu m$ wavelengths in a latitude band containing this volcano. The sea surface temperature products derived from NOAA-7 (AVHRR) were significantly degraded for some time after this eruption.

The various contributions to the total measured upwelling radiance E_T are described above. In order to infer the SST (T_s in equation 7.8) the contributions to the measured radiance from the other terms have to be measured or at least estimated. The measured brightness temperature T_B of the upwelling radiance will be less than the true SST by an amount dependent on these various atmospheric contributions. For a radiometer operating in a single frequency band the effect of the atmosphere $\tau(\nu, \theta)$ (i.e., the various terms in equation 7.7) has to be modelled for that frequency knowing the vertical temperature and humidity profiles.

Many studies have been carried out using single-channel ($11 \mu m$) data to derive SST. The atmospheric correction is calculated either from coincident satellite soundings, radiosonde ascents or profiles obtained in a numerical forecast model. Alternatively, a mean model atmospheric profile can be assumed. The values of SST obtained are only accurate to a few degrees (± 2 K) because of uncertainties in the actual atmospheric corrections at the time of observation. For example, water amount, which strongly influences the transmittance, is highly variable. Coincident radiosonde ascents occur infrequently and so unless satellite soundings (see Chapter 5) are available there will always be an uncertainty in the atmospheric transmittance for many of the points over the globe where SST measurements are made.

Recently, multichannel radiometers have been launched into space (e.g., AVHRR on NOAA-7) which measure the upwelling radiance at a number of visible and infra-red window wavelengths. The atmospheric transmittance varies appreciably with wavelength, as shown in figure 6.1 (p. 204). Therefore the relative contributions from each of the terms in equation 7.7 will be different for each channel. By measuring the radiance at many different wavelengths the magnitude of the atmospheric absorption and emission terms can be inferred without prior knowledge of the atmospheric humidity and temperature profiles.

Consider the radiances in terms of brightness temperature $T_{3.7}$ and T_{11}. If the difference between the true surface temperature and the brightness temperature at the top of the atmosphere is plotted as a function of vertical water vapour amount it increases with water amount as shown in figure 7.10. The effect of increasing water amount is greater at $11 \mu m$ than at $3 \cdot 7 \mu m$. To take advantage of this multichannel technique

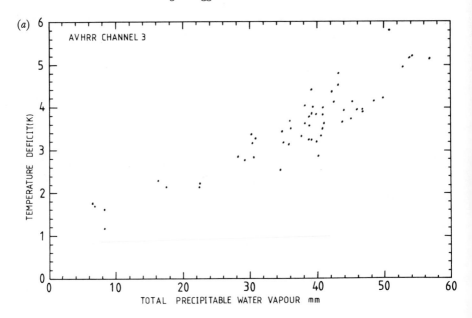

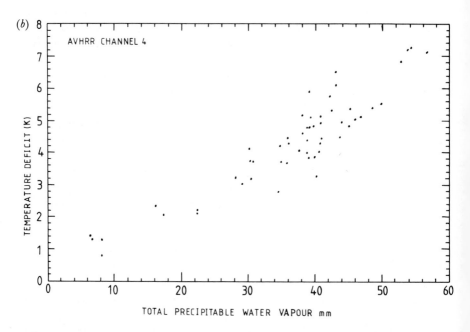

Figure 7.10. The effect of increasing the water vapour amount on the difference between surface temperature and measured brightness temperature for (a) the AVHRR Channel 3 (3·7 μm) and (b) Channel 4 (11 μm). Each point is for an atmospheric profile measured by a radiosonde close to or over the sea surface.

a regression relationship is computed which relates SST (T_s) to the measured brightness temperature in each of the available channels

$$T_s = C_0(a) + C_1(a)T_{\nu 1} + C_2(a)T_{\nu 2} + C_3(a)T_{\nu 3} \qquad (7.12)$$

where $T_{\nu i}$ is the brightness temperature at a frequency ν_i; and $C_i(a)$ are the regression coefficients which are a function of air mass a. The values of the coefficients are obtained either empirically, by comparing ship or buoy SST measurements with coincident satellite brightness temperatures, or by simulations for a range of atmospheric conditions. The latter method is useful for understanding the effects that different atmospheric parameters have on the coefficients. A range of representative profiles (of temperature and humidity) of the atmosphere can be taken, and for each one the brightness temperature at the top of the atmosphere is computed for a given SST, taking into account contributions from each term in equation 7.7. This involves having a good knowledge of the atmospheric absorption and emission characteristics at the frequency of interest. The coefficients are then determined by minimizing the difference between the input SST and that computed from the brightness temperatures at the top of the atmosphere for all the profiles considered. If the model is an accurate representation of the atmosphere's interaction with infra-red radiation, over the range of conditions considered, a combination of measured brightness temperatures $T_{\nu i}$ will allow the regression coefficients $C_i(a)$ to be computed using equation 7.12. There will always be a small uncertainty in the retrieved surface temperature using this technique ($\sim 0 \cdot 2$ K) because of atmospheric variability about the mean of the representative profile, effects of non-linearity and instrumental noise.

The three wavelengths currently being used on AVHRR are centred at $3 \cdot 7$, 11 and 12 μm. All these channels are used to derive T_s at night, but the magnitude of the reflected solar contribution E_0 at $3 \cdot 7$ μm can be large in regions of sunglint so only the 11 and 12 μm channels are used during the day. These multichannel methods for measuring SST from space have reduced the uncertainties from 1 or 2 K for single channel measurements to less than 1 K. Difficulties in comparing a point measurement from a ship with a satellite value averaged over a large area of sea surface (50×50 km) make more accurate comparisons difficult. It should be remembered that the satellite measures the radiative temperature of the top sub-millimetre 'skin' of the sea surface whereas the ship measures a bulk surface temperature representative of the top metre of ocean. It has been shown that these temperatures differ by up to $0 \cdot 2$ K under normal conditions. The sea surface skin is normally cooler than the bulk water temperature and this difference has been measured and modelled (Grassl, 1976).

A further improvement to multichannel techniques is currently being developed. The concept is for a radiometer to view an area of sea surface directly below it (nadir view) and then, a few minutes later, along the orbit, it views the same area at an oblique angle (oblique view) of about 55°. Assuming a plane-parallel atmosphere which can be considered to be constant over a time period of a few minutes, then the difference between the nadir view and oblique view radiances will be due primarily to atmospheric absorption and emission. The basic principle can be shown by considering equation 7.7 and neglecting the contributions to upwelling atmospheric radiance. For the nadir view ($\theta = 0$)

$$E_{nad} = \epsilon(\nu)B(\nu, T_s) \exp\left(-\tau_0^{\prime H}\right) \tag{7.13}$$

and for the oblique view ($\theta = 60°$ to simplify equations)

$$E_{obq} = \epsilon(\nu)B(\nu, T_s) \exp\left(-2\tau_0^{\prime H}\right) \tag{7.14}$$

assuming the sea surface emissivity ϵ is the same for both views. The transmittance is replaced by the vertical optical depth $\tau_{h_1}^{\prime h_2}$ through the atmosphere between heights h_1 and h_2. The only unknown apart from T_s is $\tau_0^{\prime H}$ and this can be eliminated by combining equations 7.13 and 7.14 to give an expression for T_s in terms of E_{nad} and E_{obq}. For an accurate calculation of T_s all the terms in equation 7.7 have to be included and so a regression relationship has to be derived in a similar way to the multichannel regression. The advantage of this double viewing angle method is that it measures the atmospheric correction. Multichannel techniques break down when abnormal amounts of aerosols or absorbing gases (i.e., water vapour) are present in the intervening atmosphere.

Another method for measuring SST from space is to detect the emitted surface radiance at much longer wavelengths (>1 mm) where there is another window region. An important advantage in the microwave region is that most types of cloud (non-rain-bearing) do not scatter radiation at these wavelengths so that the problems of cloud contamination discussed in the following section are not always applicable to microwave measurements. However, there are disadvantages with this method. The sensitivity $\partial B/\partial T$ of the Planck function to temperature change is much less for microwaves (Rayleigh–Jeans region) than at infra-red wavelengths. Therefore a high temperature resolution is harder to achieve at longer wavelengths. The emissivity of the sea surface in the microwave region ($\sim 0 \cdot 5$) is variable and depends on the sea state, introducing another uncertainty. Also, sunglint can degrade the signal at these frequencies because of the high transmission of the atmosphere and high reflectivity of the sea surface. The longer wavelengths also mean that to obtain a high spatial resolution a large antenna has to be used. The scanning multichannel microwave radiometer (SMMR) on SEASAT had a maximum resolution (at 8 mm wavelength) of 28 × 18 km. It was found that for distances less than 600 km from land the retrievals were contaminated by surface emission from land in the side lobes of the antenna pattern. Therefore, for coastal SST studies only the infra-red measurement method can be used. Away from land and avoiding rough seas and rain-bearing clouds, the mean standard deviation between ship and microwave satellite estimates can be reduced to 1·2 K (Hofer *et al.*, 1981). Microwave measurements of SST will be useful over areas of persistent cloudiness (e.g., South Atlantic) where infra-red measurements can only be made at irregular intervals. Microwaves can also be used to retrieve many other geophysical parameters, such as surface windspeed and atmospheric water vapour amount (see Section 7.7).

7.5. The detection and removal of cloud-contaminated pixels for ocean areas

The methods described in the previous section all assume a cloud-free view of the sea surface, so that only atmospheric absorption and scattering modify the sea surface radiance. However, at any one time approximately 50 % of the ocean surface is covered with cloud, necessitating correction of the infra-red radiance for cloud effects (see Chapter 5). One approach is to use only those pixels which are cloud-free. The problem is to determine whether a certain pixel is contaminated with cloud or not. If it is contaminated its radiance value will not be representative of the cloud-free situation considered in the previous section and so the regression relationship will not be applicable. This is very similar to the disturbing effect of undetected cloud cover over land areas described in Section 7.2. The situation is complicated by the fact that various cloud types can have different effects at different frequencies.

If atmospheric absorption/emission and scattering are neglected, the effect which cloud tops of high optical thickness have on the measured radiance of one pixel can be shown by considering the following relation

$$E_{\text{pix}} = \int_{\nu_1}^{\nu_2} \left[\alpha \epsilon_{\text{cl}} B(\nu, T_{\text{cl}}) + (1 - \alpha) \epsilon_s B(\nu, T_s) \right] \psi(\nu) \, d\nu \qquad (7.15)$$

where $B(\nu, T)$ is the Planck function; α is the fractional cloud amount within the pixel; ϵ is the emissivity; and $\psi(\nu)$ the normalized filter profile. The measured radiance is a combination of the warm cloud-free sea surface radiance and cold cloud-top radiance. Figure 7.11 shows the effect which different cloud amounts and four cloud-top heights have on the measured brightness temperature at 3·7 and 11 μm using this relation. For high cold cloud tops (230 K) the effect of increasing cloud cover at

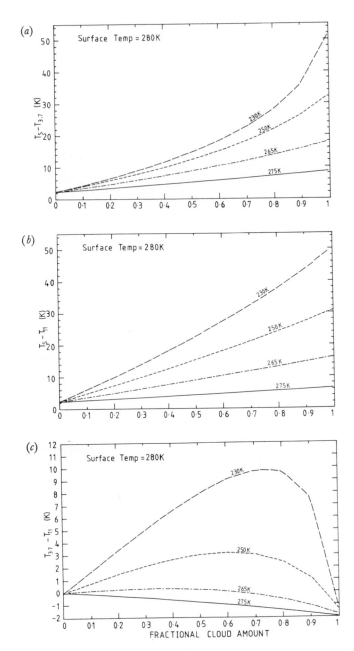

Figure 7.11. The effect of increasing the cloud amount on the radiances at (a) 3·7 μm and (b) 11 μm, including atmospheric absorption for a mid-latitude atmosphere. Four different cloud-top temperatures were used. The difference between the brightness temperature measured in the two channels is shown in (c).

Figure 7.12. TIROS-N AVHRR Channel 3 (3·7 µm) image for 12 July 1979. The higher radiances appear dark (data courtesy of the University of Dundee, processed on IPIPS, University College, London).

$3 \cdot 7\,\mu$m is much less than at $11\,\mu$m. This is due to the greater sensitivity of the Planck function to temperature at $3 \cdot 7\,\mu$m. When looking at the brightness temperatures measured by the AVHRR at night over high cloud, the $3 \cdot 7\,\mu$m brightness temperature can be up to 10 K higher than the corresponding $11\,\mu$m brightness temperature. For low warm cloud tops the effect at $3 \cdot 7\,\mu$m is similar to that at $11\,\mu$m, but high cloud amounts give a warmer $11\,\mu$m brightness temperature in contrast to the high cold cloud-top case. Again, AVHRR measurements at night confirm this by showing the $11\,\mu$m brightness temperature to be $1-2$ K warmer than the $3 \cdot 7\,\mu$m brightness temperature over areas of optically thick low cloud. These results also show that only a small amount of cloud (1%) will affect the brightness temperature by $0 \cdot 1$ K so that any cloud clearing methods must be sensitive to these small amounts of cloud.

During the day reflected solar radiation can also contribute to the total radiance. At $3 \cdot 7\,\mu$m the incident solar radiance is a significant proportion of the total signal. Low clouds have an albedo of $\sim 0 \cdot 70$ at $3 \cdot 7\,\mu$m, so a considerable proportion of the total radiance is reflected solar radiation. This results in low clouds appearing brighter than cloud-free sea surface or warm land, as shown in figure 7.12 where bright areas appear dark. High cold clouds have a lower albedo and low emitted flux so they appear white (low radiance). Cloud-free sea surface has a radiance in between with a high emitted flux but low albedo, except in areas where specular reflection can occur (in sunglint). The appearance of the $11\,\mu$m infra-red image (figure 7.13) is more familiar with low radiances appearing white and high radiances appearing black (to make the clouds appear white, which closely matches the familiar visible appearance of the scene). At these longer wavelengths the solar contribution is less and the albedo of the cloud and sea surface is also lower. Therefore only emitted radiances contribute to the signal and the clouds will all appear colder (whiter than the sea surface). These infra-red images are widely used for forecasting purposes because active cloud patterns can easily be discerned both by day and night.

There have been a variety of different methods developed for detecting and removing cloud-contaminated pixels in infra-red satellite images. Some of these methods can only be used over the ocean whereas others can equally well be applied over the land. Clouds over snow and ice are particularly difficult to detect and methods using multichannel radiometers have to be used. The methods described here concentrate on the problems of removing cloud contamination from infra-red pixels over the sea surface to allow an accurate ($\pm 0 \cdot 2$ K) sea surface temperature to be obtained.

The most commonly employed method to detect cloud-contaminated pixels is to use a visible channel which measures scattered and reflected solar radiation. The pixels of both infra-red and visible channels should

Celtic Sea Front   Flamborough Head Front

Figure 7.13. TIROS-N AVHRR Channel 4 (11 μm) image for 12 July 1979. The higher radiances appear dark. The positions of well marked SST fronts are indicated. Note the anomalously warm area of sea surface to the north of the Netherlands (data courtesy of the University of Dundee, processed on IPIPS, University College, London; after Saunders et al., 1982).

be colocated and of the same size. All clouds except thin cirrus have a relatively high reflectivity ($>0\cdot50$; e.g., table 6.1) when compared with the sea surface ($<0\cdot05$ outside sunglint regions) so that even small amounts within the field of view are readily detected. To illustrate this a 50×50 pixel array of $1\cdot1$ km NOAA-7 AVHRR data was selected which contained both low cloud, high cloud and cloud-free sea surface. Histograms of radiances or brightness temperatures of this array are shown in figure 7.14. The first point to note about the 11 μm infra-red (Channel 4) radiance histograms is that they exhibit a marked cloud-free radiance peak (this can be compared with figure 6.8, p. 227). The cloud-free peak in the visible histogram of reflected and scattered solar radiances from the cloud-free sea surface shows that only the lowest radiances contribute to this cloud-free peak. This allows a threshold technique to be applied to the visible channel data so that only infra-red pixels with corresponding reflected radiances within the visible cloud-free peak are assigned to be cloud-free. As the peak is usually a well-defined minimum radiance value it is possible to devise an automated procedure which

M. J. Duggin & R. W. Saunders

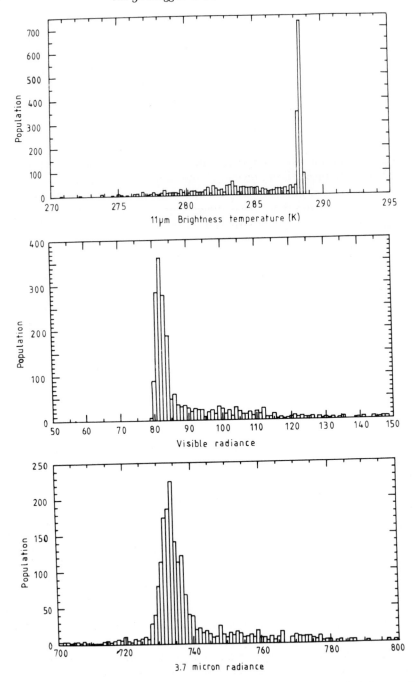

Figure 7.14. Histrograms of radiances in different AVHRR channels for a 50 × 50 pixel array which contained both low and high cloud and cloud-free sea surface.

detects if a peak is present and if so to define a threshold value to determine whether pixels are part of the cloud-free peak, or not. The
minimum radiance value of the cloud-free peak is a function of solar
zenith angle and of how far the satellite observation angles are from the
point where specular reflection can occur. Hence a simple fixed visible
radiance threshold cannot be applied except over very localized regions
(see discussion in Chapter 6). This cloud detection method has the major
advantage of being able to discriminate unambiguously between low
clouds and cloud-free sea surface, because of their markedly different
albedos at visible wavelengths. There are many disadvantages to this
visible threshold method however. One major disadvantage is that it cannot be used at night, causing an inconsistency in the data processing
scheme. During the day, areas viewed close to the specular point where
sunglint occurs is another situation where the visible threshold method
cannot be used. The magnitude of the sunglint is a function of the sea
state. Calm seas will only have a small range of observation angles when
sunglint can occur so only a small area is affected. In contrast, choppy
seas will present a much larger range of angles for which direct specular
reflection can occur so that the area of sunglint will be larger. In fact local
variations in sea state can lead to marked differences in the reflected
radiances. These conditions seriously distort the minimum radiance
peak, making automated cloud detection very difficult. The magnitude of
the sunglint is a good measure of the sea surface roughness and hence
surface wind speed.

Sunglint can occur for both geostationary and Sun-synchronous
satellite orbits. For geostationary satellites sunglint is present along the
latitude band which is half of the solar declination angle. For instance, in
the Northern Hemisphere's winter, sunglint will occur in a latitude band
south of the equator. However, for full disc images sunglint is only present in a small fraction of the data. For Sun-synchronous satellites such
as the NOAA series, sunglint can be present in a much larger area of the
image if areas well away from the nadir point are viewed. The difference
between local noon and local time of satellite overpass is one of the major
factors which governs how far from the sub-satellite point the sunglint
occurs. The closer the daytime pass is to local noon the closer the specular
point is to the sub-satellite track. For instruments scanning perpen-
dicularly to the satellite track, sunglint will enter the field of view when
the radiometer is pointing far enough away from the nadir (towards the
Sun) to allow direct specular reflection. The position of the specular point
in the image will vary with latitude. For NOAA-7 AVHRR images, up
to 40% of the scan line can be affected by sunglint. To avoid sunglint a
Sun-synchronous orbit with a local overpass a few hours from local noon
should be chosen. If the sea surface is then only viewed close to the sub-
satellite point or away from the Sun then the specular point will not be

viewed. Unless the sea surface is perfectly calm, however, the sunglint can be spread up to 30° from the specular point making the limitations on viewing geometry even more restrictive.

The visible threshold also becomes difficult to apply for large solar zenith angles (when the Sun is at grazing incidence) because clouds will cast shadows across the sea surface, especially if they have significant vertical structure (e.g., cumulonimbus). These shadows over the sea appear darker than the cloud-free sea surface so that the visible histogram minimum radiance peak will be split into minimum cloud-free radiances and areas of shadow. Other nearby clouds may also have shadows cast on them, reducing their scattered radiance as viewed from space. Therefore it is wise to restrict the visible threshold technique to solar zenith angles of less than 75°. This problem is commonly encountered at mid-latitudes during the winter months.

Finally, another problem with cloud detection at visible wavelengths is due to variable concentrations of aerosols in the atmosphere, which increase the radiation scattered back to the satellite. Over the cloud-free ocean the radiation scattered from the sea surface is small so that the atmospheric aerosol scattering accounts for a significant proportion of the measured radiance. The problem is that the concentrations of these aerosols can be highly variable, making the cloud-free radiance peak variable. For instance, the area of the Atlantic Ocean to the west of Senegal is often subject to high turbidity due to high aerosol concentrations caused by Saharan dust being lifted up into the atmosphere and then blown westwards over the ocean. METEOSAT images have allowed these large areas of dust to be tracked for days as they propagate westwards. Another source of atmospheric aerosols is volcanic activity, which can suddenly inject vast quantities of aerosols into the upper troposphere or stratosphere in a narrow latitude band. These aerosols then appear to spread out latitudinally. The eruption of El Chichón, which resulted in large quantities of aerosols being released, made cloud detection more difficult at visible wavelengths, as the backscattered cloud-free radiance was much higher.

These limitations of the visible threshold method, especially the constraint of only being able to operate in daylight, have led to a number of alternative or additional methods which do not rely on reflected radiances. One method which has been developed from the 'spatial coherence method' described by Coakley and Bretherton (1982) relies on the assumption that over small horizontal distances the sea surface emitted radiance values from each pixel will be virtually constant. The emitted radiance values of cloud tops are assumed to be variable from pixel to pixel. The method consists of defining a small 'local' array (i.e., 3 × 3 pixels) and calculating local means and standard deviations over each array of brightness temperatures or radiances. Over cloud-free sea surface (with no horizontal temperature gradients) the local standard deviation in brightness temperature was found to be small, whereas for cloud-contaminated pixels it was normally much higher. All local arrays with standard deviations higher than a predefined threshold value are assigned as cloud contaminated. The pixels which survived the local standard deviation test can then be used in calculating the mean brightness temperature of a larger array (i.e., 50 × 50 pixels) or can be subjected to another test for cloud contamination. For 3 × 3 pixel local arrays of 1·1 km AVHRR data a local standard deviation of 0·1 K was found to be a good threshold value for 11 μm brightness temperatures.

The cut-off standard deviation will depend on the noise-equivalent temperature of the radiometer and on the digitization interval for the radiometer counts. The size of individual pixels (projected onto the surface) and the number of pixels in the local array will also influence the value of the cut-off. Finally, higher cut-offs should be used over areas of the ocean which are known to have strong horizontal gradients in SST.

The method is illustrated in figure 7.15, where the 11 μm pixel values in a 50 × 50 pixel array (55 × 55 km) are plotted as a histogram. The array contained cloud-free sea surface at a uniform temperature with broken low stratus and high cirrus cloud over part of the array. The uppermost histogram consists of all the 2500 pixels within that array. The lower histograms are made up from pixels which survive the local standard deviation test for various cut-off values. The local array size was 3 × 3 pixels in this case. As the local standard deviation cut-off was decreased, more and more pixel values were removed from the histogram but, as figure 7.15 shows, the colder cloud-contaminated pixels are removed in preference to the cloud-free pixels. The cloud-free radiance peak of the histogram remains virtually unchanged down to cut-offs of 0·1 K. As the digitization interval is 0·1 K the method is not applicable for cut-off values less than this. For the optimum cut-off of 0·1 K the histogram consists almost entirely of cloud-free brightness temperatures. A small fraction of cloudy pixels at lower temperatures have survived the test because in a few cases the cloud-top temperature was uniform over the 3 × 3 array. Fortunately uniform cloud layers normally have a variable cloud-top temperature which is colder than the sea surface so that in most cases the cloudy pixels can be recognized and removed. A disadvantage of this local uniformity method is that over areas where an appreciable horizontal gradient in SST is present (e.g., ocean fronts, edge of the Gulf Stream, etc.) the pixels are rejected when in fact they are cloud-free. However, strong SST gradients make up only a small fraction of cloud-free SST pixels.

The visible threshold method and local uniformity method complement each other to some extent. The former can detect uniform low cloud and allow SST values to be inferred over regions where horizontal gradients

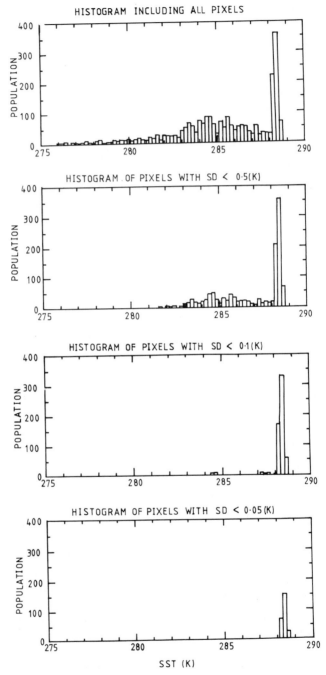

Figure 7.15. The effect on the histogram of 11 μm brightness temperatures of decreasing the local standard deviation cut-off value. The local array size was 3 × 3 pixels.

in SST exist. The latter is better for detecting cirrus cloud and can be used at night and in areas of sunglint. Ideally both methods should be applied, first the local uniformity method and then the visible threshold method. All pixels which do not pass the cloud screening test are zeroed and not used in the subsequent calculation of mean SST.

Another method which has been applied to infra-red images over the sea surface is the 'histogram technique'. This method relies on the following assumptions. Firstly, the SST must be uniform ($\sim 0 \cdot 2$ K) over the array of pixels from which the mean SST is calculated. This constraint means that the array size must be less than 50×50 km on the Earth's surface because SST is known to vary latitudinally by more than $0 \cdot 1$ K for distances greater than 50 km. Secondly, the radiometer and atmospheric noise are assumed to have a Gaussian distribution and the digitization interval must be significantly less than the changes in radiance due to noise. Finally, all cloud tops are assumed to be cooler than the sea surface.

If these conditions are fulfilled, then by considering the histogram of infra-red brightness temperatures shown in figure 7.14, the warm side of the cloud-free peak should have a Gaussian distribution. A Gaussian curve can be fitted to the warm side of the peak (Crosby and Glasser, 1978) in order to find the value of the brightness temperature at the peak, which will be the mean brightness temperature of the cloud-free pixels which make up the array. If significant horizontal gradients in SST are present within the array, as invariably happens where fronts are occurring, the curve will be fitted to the population of pixels with the highest brightness temperature. The histogram technique therefore gives the maximum cloud-free brightness temperature over the array, which in many cases will be close to the mean brightness temperature. An advantage of this technique is that it can be used at night and is not affected by sunglint. When used in conjunction with the local uniformity method an accurate night-time cloud-free radiance value should be obtained. It should be noted that the histogram method can be most useful over areas where only small gaps in the cloud occur because over these small areas the SST is likely to be uniform. Over larger areas horizontal gradients in SST can lead to significant errors. A simplified, but less accurate version of the above method is to assume the cloud-free brightness temperature to be the maximum value which has a reasonable population of pixel values measured in the array. This could be described as a dynamic threshold method applied to the infra-red channel.

Two-dimensional histograms are often employed to determine the cloud-free sea surface radiance. Figure 7.16(*a*) shows a two-dimensional histogram constructed by combining the one-dimensional histograms of the visible radiance and 11 μm brightness temperature, shown along the axes. The histograms are constructed from all the pixels in a 50×50

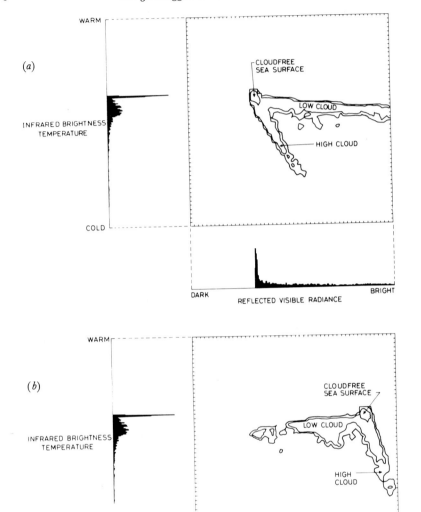

Figure 7.16. Bidimensional histograms for 50 × 50 pixel arrays of (a) visible reflected radiances and 11 μm brightness temperatures; and (b) 3·7 μm radiances and 11 μm brightness temperatures. The contours correspond to populations of 2, 5 and 50 respectively, so the cloud-free peak is well pronounced. The corresponding one-dimensional histograms are also shown.

array. The cloud-free peak in the two-dimensional histogram is well defined. Regions of low and high cloud are well separated into different areas of the two-dimensional histogram, although these areas are more spread out than the cloud-free peak. This is due to the many different cloud heights present and pixels which are only partially filled with cloud. A cluster analysis can be performed on this two-dimensional histogram to extract the cloud-free radiance. The disadvantage of this method is that like the visible threshold method it can only be used during the day. However, other wavelength combinations can also be used such as $3 \cdot 7 \mu m$ radiances and $11 \mu m$ brightness temperatures. Figure 7.16(b) shows this combination during the daytime when the low cloud appears on the opposite side of the cloud-free peak to the high cloud. During the day both terrestrial-emitted and solar-scattered and -reflected radiances contribute to the $3 \cdot 7 \mu m$ signal so that whereas cold cirrus cloud has a low radiance (low emission and lower reflectivity), low stratus cloud has a high radiance (higher emission and high reflectivity). The sea surface away from sunglint has a low reflectivity but a high emitted component so that the overall radiance at $3 \cdot 7 \mu m$ is more than for high cloud but less than for low cloud. At night the solar-reflected/scattered component disappears so that the $3 \cdot 7 \mu m$ radiance is from emitted radiances only. The appearance of the night-time $3 \cdot 7 \mu m$ images is similar to the $11 \mu m$ images. In contrast, the daytime $3 \cdot 7 \mu m$ image shown in figure 7.12 has a confusing appearance, with low clouds apparently warmer and high clouds colder than the sea surface.

It is also possible to use the solar-reflected component of the $3 \cdot 7 \mu m$ radiance as a discriminator between low clouds and sea surface. However, difficulties do arise since two thresholds (an upper and a lower) have to be determined to define the cloud-free peak instead of one, as shown by the histogram in figure 7.14. Further, a more serious limitation is that if high cloud is also present the possibility exists of a pixel being partially filled with low and high cloud which can have the same radiance as the cloud-free sea surface. Therefore the difference between a cloud-free radiance and a low and high cloud mixture cannot be distinguished for the $3 \cdot 7 \mu m$ channel alone during the day. Of course comparisons with $11 \mu m$ radiances can help resolve this problem. Finally, in common with the visible channel, sunglint areas cannot be used for cloud detection and these areas can be larger than those affected at visible wavelengths.

At night the $3 \cdot 7 \mu m$ brightness temperatures may be more useful than the $11 \mu m$ brightness temperatures for detecting low cloud. The results in figure 7.11 show some differences between the two brightness temperatures for low cloud due to the different emissivity of the cloud at different wavelengths. Monitoring the difference in brightness temperature at the two wavelengths may be a good indicator of the presence of low cloud. Similarly, differences between the 11 μm and 12 μm channels on AVHRR, for instance, may also help to identify low cloud at night.

Finally, a method which can be used on a series of images of the same area, over a reasonably short time period (a few days) is to take the maximum 11 μm radiance from a time series for each pixel. Assuming the SST remains constant over this time period then as long as the pixel is cloud-free at least once over this period it will contribute a cloud-free radiance to the spatial average. This method differs from the Gaussian fit technique which applies to many pixels over a uniform area at one time. Here consideration is made of one pixel for a long time period (a few days). For most of these observations the pixel will be covered with cloud but, hopefully, for at least one observation it will be cloud-free. The method, therefore, is an attempt to increase the number of cloud-free pixels over a given area at the expense of only being able to derive a mean SST averaged over the time period chosen for the series. It is best suited to geostationary orbiter data where many observations can be made in one day (48 for METEOSAT) and the sea surface is always viewed from the same point in space.

All of the various methods described above have been applied to satellite radiance data with varying degrees of success. Each method has its own merits and disadvantages. The optimum procedure (for effectively removing clouds) is a combination of some of the different methods described. Operationally, of course, this may be prohibitive in computer time so that an assessment has to be made of which methods are the most important. A useful technique to reduce computer time is to pass the infra-red data initially through a crude cloud test (i.e., if the brightness temperature is more than 10 K less than the climatological surface temperature then reject the data as cloudy) which will in fact remove a substantial number of pixels (~ 50 %) before detailed processing begins.

7.6. *The interaction between clouds and sea surface temperature*

The bulk sea surface temperature remains virtually constant throughout one day, in contrast to the surface temperature over the cloud-free land which has a marked diurnal variation. In coastal regions this can result in low clouds or fog over the land being readily dispersed in the morning due to surface heating, whereas over the sea the cloud will remain. A good example of this is shown in figure 2.14(a) (p. 74) where low cloud/fog can be seen over the English Channel but not over the adjacent land, heated by insolation. The corresponding infra-red image, figure 2.14(b), which has been enhanced to show up small-scale variations in SST, shows that this cloud-top temperature is close to the sea surface

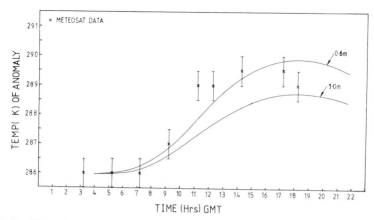

Figure 7.17. Observed (from METEOSAT) and calculated temperature values (for mixing depths of 0·6 and 1·0m) at the centre of the temperature anomaly as a function of time (after Saunders et al., 1982).

temperature. It is also interesting to note that there are many marked variations in SST not associated with cloud. These illustrate how erroneous the assumption of a constant SST can be, even over small areas. The difference in temperature between the lighter (colder) and darker (warmer) areas of the sea surface is about 2 K.

Another interesting feature, evident in figure 7.13 (p. 269), is the warmer (+ 3 K) patch of sea surface to the north of the Netherlands. This cloud-free area was under a ridge of high pressure at the time and the surface winds in this area were light. The lack of surface mixing allowed a marked diurnal thermocline to be set up where the top 60 cm was heated by solar insolation without mixing to the layers below. The diurnal heating cycle of this area was observed by METEOSAT, as shown in figure 7.17. The solid lines represent the results of a simple radiative model for different surface mixing layers. The diurnal variation of a few degrees (K) in SST suggests observations at 1500 LST may give an unrepresentative daily mean SST for areas which are cloud-free and have a low surface wind speed.

The global distribution of cloudiness over the oceans is in general highly variable. However, there are certain areas which are persistently cloudy, for instance, the South Atlantic between S.W. Africa and Brazil. Visible METEOSAT images, such as figure 2.3 (p. 51), shows this area to be persistently covered with low cloud. The Inter-Tropical Convergence Zone (ITCZ) is less marked and can be highly variable over the ocean. Over the adjacent land masses where surface heating in the late afternoon initiates intense convective storms it can be more stable. At temperate latitudes, the passage of large-scale frontal systems over the ocean dominates the cloudiness observed from space.

The North Atlantic to the west of Ireland is an area which is subject to a continual passage of warm, cold and occluded fronts in various stages of formation or decay. Other areas which are subject to persistent cloudiness are regions where warm currents such as the Gulf Stream meet cold currents such as the Labrador Current. Where the two currents converge, the warm moist air associated with the Gulf Stream mixes with the cold air from the Labrador Current and the water vapour condenses producing low cloud or fog.

7.7. Other oceanographic parameters measured from space

Ocean surface parameters from a radar altimeter

Observations of significant wave height (SWH) can be made from space using a radar altimeter which can also measure the mean height of the radar above the sea surface and the surface wind speed. A radar altimeter basically consists of a microwave transmitter which sends out a series of pulses at a wavelength of ~ 2 cm towards the sea surface. The returned signal, after being reflected back by the sea surface, has been modelled so that the shape and amplitude of the returned waveform and the time taken between transmission and return give an indication of the sea surface state. The majority of clouds are transparent at these wavelengths allowing good global coverage to be obtained.

For significant wave height, the shape of the leading edge of the pulse is the important parameter. Returns from wave crests will arrive before returns from wave troughs and the greater the SWH the bigger this time difference will be. This causes the leading edge of the returned pulse to have a slope which will be inversely related to the SWH in the radar's 'footprint'. Comparisons with buoy measured wave heights have shown (Fedor and Brown, 1982) that with a suitable algorithm global maps of SWH averaged over suitable time periods can be produced. The SEASAT radar altimeter produced maps of SWH demonstrating the potential of satellite data for continuously monitoring ocean surface parameters globally.

The radar altimeter, as its name infers, is primarily designed to measure the height of the radar above the sea surface by measuring the time interval between transmission and return of the microwave pulses. To infer the height of the sea surface relative to the global gravimetric geoid from these altimeter measurements, the orbit of the spacecraft and the geoid have to be accurately known to greater than the accuracy of the altimeter measurements. Unfortunately, at present the orbital position and geoid accuracies are much less than this (2–3 m). However, it is still possible to study the mesoscale features due to ocean circulation where the height of the sea surface varies over small distances (100 km). Figure 7.18

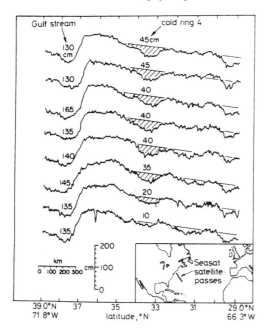

Figure 7.18. Passes of the SEASAT radar altimeter across the Gulf Stream at three-day intervals from 17 September 1978 to 8 October 1978 (from Cheney and Marsh, 1981). Note the cold eddy which passed out of the satellite's track. The vertical scale (in cm) and horizontal scale (in km) are also shown.

shows an example for the SEASAT radar altimeter over the western Atlantic across the Gulf Stream. The depression of the sea surface due to the Gulf Stream is clearly marked. Also, a transient cold eddy to the east of the Gulf Stream, a feature occasionally observed in infra-red images, shows up as a depression of up to 45 cm. As it gradually moved westwards, out of the sub-satellite track, the depression disappeared. This sea surface topography is a new way of studying ocean circulation patterns.

Finally, surface wind speed is also estimated from the radar altimeter return. For low surface winds the power of the reflected pulse will be high because the small-scale (10 m) sea surface roughness will be small. However, when the surface wind increases, the surface roughness will also increase and cause some of the incident microwaves to be scattered out of the beam. The power of the returned pulse is therefore inversely related to surface wind speed.

Measurement of surface wind speed and direction

Another instrument used for remotely sensing sea state parameters is an active microwave scatterometer system. This transmits two orthogonal

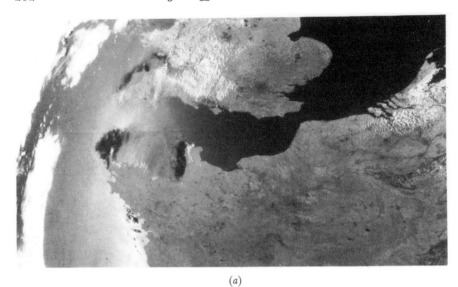

(a)

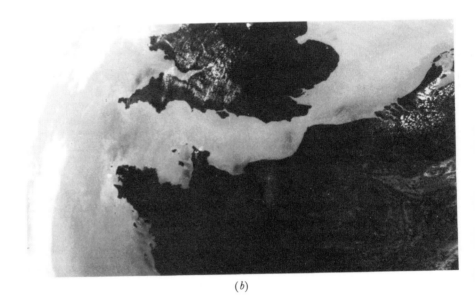

(b)

Figure 7.19. A TIROS-N visible (a) and infra-red (b) set of images for 1436 GMT, 22 July 1980, showing information about the sea surface in both channels (data courtesy of the University of Dundee, processed on IPIPS, University College, London; after Saunders et al., 1982).

beams at a frequency of 4·6 GHz incident to the sea surface, making an X-shaped pattern. In the same way as for the radar altimeter, the measurement of surface wind is based on the interaction with microwaves by short (1 cm) sea surface capillary waves. The strength of the back-scattered microwaves is proportional to the amplitude of the surface capillary waves and hence wind speed. The scattering is anisotropic so that with observations for different azimuths the wind direction can also be found, although there can be up to four possible solutions. These solutions all have similar wind speeds but different directions. The correct wind direction can be found by comparisons with surface observations. The pattern with which the microwaves illuminate the sea surface can be divided up into individual resolution cells. For instance, the SEASAT scatterometer system had a resolution of 50 km and was able to retrieve wind speeds to an accuracy of $2\,\mathrm{m\,s^{-1}}$ and wind direction to $\pm 20°$. As with the radar altimeter, the scatterometer can operate through all cloud types which are not rain-bearing.

Another way of inferring surface wind speed over the sea surface is to use the visible reflected solar radiance in the vicinity of the specular reflection point. For areas of sea surface close to the point where sunglint can occur if the surface roughness is high, the waves will present many possible inclined surfaces, some of which will allow direct specular reflection. The TIROS-N visible image in figure 7.19(*a*) illustrates this where the left-hand part of the image is subject to sunglint. However, there are well-defined areas within this sunglint which are dark. These are regions where the sea is much calmer and so direct specular reflection is not possible.

To demonstrate the effect surface wind speed can have on measured SST, the coincident 11 μm infra-red image is also shown in figure 7.19(*b*). The calm seas appear to have slightly higher SST values than those areas which have a higher surface wind speed. This is consistent with the observations shown in figure 7.17 which show that a diurnal heating cycle can occur when winds are light, reducing mixing of the surface layers.

Measurement of ocean turbidity

The measurement of suspended sediments, chlorophyll or pollutant concentrations from space has been shown to be possible by detecting reflected and scattered radiation in several narrow bands at visible wavelengths. The coastal-zone colour scanner (CZCS) is an example of an instrument designed to measure ocean turbidity. It has five narrow-band ($\sim 0\cdot02\,\mu$m) channels between $0\cdot43$ and $0\cdot8\,\mu$m with a spatial resolution of 825 m. There are many different algorithms currently in use that retrieve water turbidity from the reflected intensities in each channel.

One of the main problems, as with the infra-red measurements, is attenuation of the reflected surface radiance by the atmosphere. At short wavelengths ($0 \cdot 4 \mu m$) only 20 % of the reflected radiance is from the sea surface due to atmospheric aerosols scattering the radiation. The atmospheric effect can be removed by combining the radiances from two channels into one equation. The longer wavelength channel provides the atmospheric attenuation information for correcting the shorter wavelengths. Of course sunglint can also be a problem, so areas are only viewed well away from the specular point. Having obtained a true surface radiance for more than one channel, these radiances are then combined (e.g., ratioed) to provide information on ocean turbidity. Preliminary results (e.g., Gower, 1981) are encouraging, using CZCS data where marked variations in ocean turbidity have been observed. More advanced instruments have been proposed, such as the ocean colour monitor (OCM) which would have better spectral resolution and more channels to improve the accuracy of the atmospheric correction and turbidity determinations.

7.8. *Developments in remote sensing of the ocean surface*

Multichannel techniques have recently permitted SST to be measured more accurately from space than was possible with the single channel method used previously. It is interesting to consider whether the most suitable wavelength channels are currently being used for SST measurement. Although the $11 \mu m$ window performs well over mid-latitude and polar regions where the atmospheric water vapour amounts are low, over the tropics the $3 \cdot 7 \mu m$ window can be more suitable. One possible suggestion has been to have a number of narrow wavelength channels in the $3 \cdot 7 \mu m$ wavelength atmospheric window (Chahine, 1981). Uncertainties due to atmospheric absorption, sunglint and sea surface emissivity could be removed by comparing the signals in each channel. (In Chapter 5 a similar procedure is used to account for sunglint using two HIRS2 channels at $3 \cdot 7$ and $4 \cdot 0 \mu m$). Near infra-red channels ($\sim 1 \cdot 6 \mu m$) have also been shown to be useful for discriminating between low clouds and snow surfaces.

Additional channels in other atmospheric windows may also increase the SST accuracy. A combined microwave–infra-red instrument would be useful, the infra-red radiances providing the accurate high resolution values over cloud-free areas and the microwave radiances filling in the gaps. Microwaves at frequencies close to a water-vapour absorption line can also be used to determine the vertical water-vapour amount between the satellite and sea surface. This can be used to increase the accuracy of the multichannel regression relationship. SEASAT demonstrated the potential of active and passive microwave instruments for remote sounding of the sea surface, despite its relatively short lifetime. The optimization of these instruments will be carried out in the near future together with more onboard data processing, so that, if desired, sea surface parameters can be transmitted directly from the spacecraft.

There are a number of missions planned for the next few years to make oceanographic measurements from space. The European Space Agency (ESA) is proposing to launch an ocean remote sensing satellite, ERS-1, in 1988 into a polar orbit. It will carry active microwave instruments for SST studies. A synthetic aperture radar (SAR) is one of the main components of the active microwave instruments. It will provide high-resolution images of the surface waves through nearly all types of cloud. The SAR on SEASAT demonstrated the all-weather capability of this instrument and showed up interesting phenomena in the surface wave patterns. A wave scatterometer will measure the directional spectra of surface waves. A wind scatterometer will measure the surface wind field in the same way as described in the previous section. There will also be a radar altimeter for significant wave height measurements and the overall height of the sea surface with respect to a predefined geoid. Another experimental package, called PRARE (Precision Range And Range-rate determination Experiment), will help to determine accurately the orbital parameters for the altimeter. It is also hoped that the altimeter can be used to infer the properties of ice and snow in the polar regions (see the fuller discussion in Chapter 8).

The infra-red instrument on ERS-1 is the along-track scanning radiometer (ATSR) which has the three infra-red channels at $3 \cdot 7$, 11 and $12\,\mu$m, similar to those on the AVHRR of NOAA-7. The novel concept of this instrument is that it makes two near-coincident (within a few minutes) measurements of the same area of sea surface through different atmospheric path lengths. The difference in the brightness temperatures of the area viewed through different atmospheric paths will be a measure of the atmospheric absorption. Coupled with the multichannel techniques described in Section 7.4 to infer atmospheric absorption, a more accurate determination of SST should be possible with this instrument. It is hoped that the SST measurements from the ATSR will be accurate to $0 \cdot 33$ K when averaged over a 50 km square. It is planned to launch another satellite, ERS-2, a few years after ERS-1, which will have similar instruments on board.

Another oceanographic satellite, which is planned by the Japanese for a 1985/86 launch into a polar orbit, is MOS-1. It contains microwave sensors similar to those on ERS-1 and conventional scanning infra-red and visible channels. The USA is currently investigating the possibility of enhancing the radiometers on the NOAA orbiters by adding a number of narrow visible channels similar to those flown on the coastal-zone colour scanner on Nimbus-7. This will allow the turbidity of the ocean surface to be monitored at the same time as the SST from the infra-red channels. The geostationary meteorological satellites METEOSAT (ESA), GOES (USA), GMS (Japan) and INSAT (India) can also contribute to oceanographic studies. Short-lived or rapidly changing phenomena are best studied from geostationary orbit where measurements can be made once every 30 minutes.

Improvements in both hardware and data processing (both on board and on the ground) will allow more accurate sea surface parameters to be obtained in the future which will be of both commercial and scientific use. The collection and dissemination of parameters such as SST, significant wave height and surface wind direction in real time will undoubtedly benefit activities such as fishing, shipping and flood prediction and monitoring.

Successful retrieval of information relating to land and ocean areas from satellite data can only be achieved when atmospheric contamination can be operationally removed. Viewed from space, the Earth is a predominantly blue planet, its surface being relieved occasionally by patches of browns and greens. Over 50 % of the planet is cloud-covered. These dynamic and transient atmospheric features must be better understood before surface environmental remote sensing can be fully developed.

References

Cheney, R. E. and Marsh, J. G., 1981, Seasat altimeter observations of dynamic topography in the Gulf Stream region. *Journal of Geophysical Research* **86**, 473–83.

Chahine, M. T., 1981, Remote sensing of sea surface temperature in $3 \cdot 7\,\mu m$ CO_2 band. In *Oceanography from Space*, edited by J. F. R. Gower (London: Plenum Press), pp. 87–95.

Coakley, J. A. and Bretherton, F. P., 1982, Cloud cover from high resolution scanner data; detecting and allowing for partially filled fields of view. *Journal of Geophysical Research* **87**, 4917–32.

Crosby, D. S. and Glasser, K. S., 1978, Radiance estimates from truncated observations. *Journal of Applied Meteorology* **17**, 1712–15.

Davies, O. L. (ed.), 1961, *Statistical Methods in Research and Production*, 3rd Edition (London: Oliver and Boyd).

Fedor, L. S. and Brown, G. S., 1982, Wave height and windspeed mesurements from the Seasat radar altimeter. *Journal of Geophysical Research* **87**, 3254–60.

Gower, J. F. R. (ed.), 1981. *Oceanography from Space* (London: Plenum Press).

Grassl, H., 1976, The dependence of the measured cool skin of the ocean on wind stress and total heat flux. *Boundary Layer Meteorology* **10**, 465–74.

Hofer, R. E., Njoku, J. and Waters, J. W., 1981, Microwave radiometric measurements of sea surface temperature from the SEASAT satellite. *Science* **212**, 1385–87.

Reeves, R. G. (ed.), 1975, *Manual of Remote Sensing* (Falls Church, Va.: American Society of Photogrammetry), 2 vols.
Saunders, R. W., Ward, N. R., England, C. F. and Hunt, G. E., 1982, Satellite observations of sea surface temperature around the British Isles. *Bulletin of the American Meteorological Society* **63**, 267–72.
Slater, P. N., 1980, *Remote Sensing Optics and Optical Systems* (Reading, Mass.: Addison-Wesley), p. 204 *et seq.*

8

Cloud–cryosphere interactions

Andrew M. Carleton
Arizona State University
Tempe, AZ, USA

8.1. Introduction

The monitoring of global snow and ice coverage is assuming increasing importance with the recognition that cryosphere variations accompany large-scale atmospheric circulation anomalies on seasonal time-scales. The main thrust of research into cryosphere–climate interactions using remotely-sensed data has been largely to document snow and ice extent and its seasonal and annual variations. There has been little emphasis on providing concurrent satellite information on high-latitude cloud cover, either on the synoptic (individual cyclone) scale, or more particularly with respect to climatological (mean) cloud conditions. The need to investigate cloudiness variations near the snow and ice boundaries is particularly crucial, since various non-linear surface–atmosphere feedback processes important to global climate appear to be amplified there (Polar Group, 1980). These boundaries are highly sensitive to temperature and mark strong latitudinal discontinuities in surface albedo and energy balance that could be significant for inter-annual climatic variations. Variations in the planetary albedo, which is essentially a function of cloudiness, are also likely to be significant. However, the relative importance of cloud cover compared with snow/ice cover (reflectivity) variations at the synoptic and macro scales is not known.

This chapter focusses on the various satellite products, in particular high-resolution visible and infra-red imagery and passive microwave data, that have been used to investigate the interactions between cloud and the cryosphere in both hemispheres. This has traditionally been a very difficult task due to the similar radiances of snow and cloud in both the visible and thermal infra-red wavelengths and the lack of solar illumination for part of the year at high latitudes. The characteristics of the major snow and ice satellite monitoring systems and the theory and techniques used to derive data products pertinent to the study of climate dynamics are surveyed, particularly with respect to the newer passive microwave systems. Emphasis is laid on the difficulties involved in identifying concurrent cryosphere/cloud variations using different sensors and the resulting lack of studies in this area. A review is then made of studies which utilize satellite imagery to derive synoptic climatological information on cyclonic activity in the vicinity of cryosphere boundaries from pattern recognition of cloud vortex signatures. This approach, which has been extensively applied to the analysis of Antarctic sea-ice variations, represents a useful substitute for climatological (mean) cloud cover data. Recently, this method has also been applied to sea-ice and continental snowcover variations in the Northern Hemisphere using visible and infrared imagery of high spatial and temporal resolutions.

A key theme of this chapter involves identifying, at least qualitatively, the dominant synoptic-scale surface–atmosphere feedbacks. This is undertaken by using statistically-derived characteristic intensities of satellite-observed cloud vortex types applied to the observed variations in the extent of snow and ice. The observations provide some verification for the results generated in recent ice–atmosphere modelling experiments and in empirical analyses employing conventional data.

8.2. The importance of cloud–cryosphere observation at high latitudes from satellites

Determination of the role of clouds in climate and its variations and their detection by satellites is a primary objective of major climate monitoring programmes (Henderson-Sellers *et al.*, 1981, see Chapter 6). Similarly, the global extent of snow and ice constitutes an index of the climatic state, and cryosphere variations are believed to be important in the development of atmospheric circulation teleconnections (Polar Group, 1980). Surface–atmosphere interactions appear to be particularly crucial between latitudes 60 and 70°, which constitutes the zone of greatest seasonal range of surface albedo, especially in the Northern Hemisphere. Operational satellite surveillance of snow and ice extent has provided a major source of data for large-scale cryospheric and atmospheric studies, both empirical and modelling, although rapid advances in sensors and retrieval techniques mean that these data sets often lack homogeneity (Polar Group, 1980).

Until recently, the emphasis has been on satellite monitoring of snow and ice cover variations without the interference of clouds. However, from the standpoint of clouds in climate, much of the cloud actually occurs in the vicinity of the sea-ice margin, with this boundary representing the most dynamic part of the sea-ice cover. Also, the coupled cryo-sphere–atmosphere anomalies are best expressed at the regional level. Thus, the ice–ocean boundary has received considerable attention in satellite sensing and ice–climate programmes and is to be the major focus of the forthcoming MIZEX (Marginal Ice Zone Experiment) mesoscale programme in the Arctic. The recommendations of the working group for the ICEX (Ice and Climate Experiment), originally planned for the mid-1980s, underline the need to analyse the cloudiness factor within the context of the dynamics of snow and sea-ice cover. In particular, these involve examination of the relationships between sea ice, snowcover and clouds on both synoptic and climatological time scales, as well as deter-mination of the connection between cyclonic storm systems and the sea-ice margin. These relationships necessarily involve energy budget considerations (see, e.g., Polar Group, 1980).

In the Arctic, surface heterogeneities (e.g., open water areas) and cloud cover are important controlling factors for regional variations in the disposition of global radiation. Surface reflectivity is important in augmenting reflectance of global radiation from Arctic stratus cloud decks (Polar Group, 1980), i.e., planetary albedo increases where the surface is snow or ice covered. Also, a strong relationship has been found between ice albedo and cloud amount. Surface reflectance under cloudy conditions may be 4 – 6 % greater than for clear sky conditions, irrespective of the age of the ice or snowcover. In conjuction with atmospheric water vapour, clouds also affect the long-wave components of the heat budget. For example, the persistent and extensive cloud cover of the ice-free Norwegian Sea helps to reduce, by back-radiation, the pronounced sur-face net radiation deficits in this region in winter compared with less cloudy continental interior locations at the same latitude (Vowinckel and Orvig, 1970).

A major research focus (Polar Group, 1980) is the occurrence of exten-sive multi-layered stratus cloud decks over the Arctic pack ice in summer. These clouds have a mid-summer maximum of 90 % of the total cloud amount (Vowinckel and Orvig, 1970), and both modelling and observa-tional studies indicate their importance in the surface energy balance of the Arctic Basin. It is postulated that Arctic stratus results from the modification of warm and moist continental polar air masses moving across the pack ice, whereby condensation is initiated by long-wave emis-sion to space and turbulent transfer of heat downwards to the ice surface. These lower-level clouds are not easily separated in satellite visible and infra-red imagery from the underlying pack ice surface. In addition to their role in enhancing albedo, Arctic summer stratus and stratocumulus clouds affect the surface heat balance by increasing back-radiation while reducing direct solar radiation receipts. In winter, cloud cover over the pack ice is critical in lessening the intensity of the polar surface inversions, particularly over the Beaufort Sea and north-west of the Canadian Archipelago. The formation of a cloud deck near the inversion level is especially favourable in terms of surface energy gain, since enhanced long-wave emission from the atmosphere raises surface temperatures, leading to a decrease of the inversion maximum temperature. Over the elevated Antarctic ice sheet, cloudiness is low, particularly in winter, and surface and free-air temperatures are considerably lower than in the Arctic. However, extensive cloudiness characterizes the Antarctic sea-ice zone. The presence or absence of clouds and the dominantly occurring types and thicknesses are therefore crucial to surface and atmospheric energy budgets in polar regions. A few satellite studies (e.g., Streten, 1974) have used simultaneous very high resolution visible and infra-red images for cloud cover conditions in the vicinity of the summertime pack-ice boundary of the North American Arctic coasts. They confirm the existence of pronounced horizontal temperature gradients across zones of heterogeneous surface conditions, arising from variations in the amount

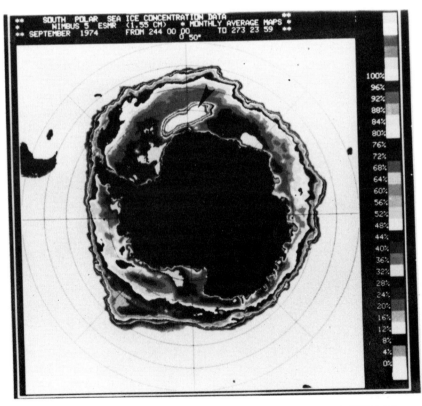

Figure 8.1. South Polar sea-ice concentration information derived from Nimbus-5 ESMR. The arrow indicates a polynya.

of absorbed solar radiation, and the close association with synoptic cloud features comprising the climatological Arctic front.

The presence of a sea-ice cover, which is itself a product of ocean – atmosphere interaction at high latitudes, dominates the radiative and turbulent exchanges of heat, moisture and momentum by limiting direct contact between the ocean and the polar atmosphere (Polar Group, 1980). Thus, the yearly cycle of sea-ice extent in the seasonal sea-ice zones, which is ultimately controlled by the annual march of solar radiation, has important effects on surface energy budget regimes. The occurrence of open-water leads and polynyas within the sea ice exerts an important surface control on the regional energy balance. These open water areas constitute a major source of heat and moisture to the atmosphere, especially in winter. This is achieved even with 1 % or less open water within the pack. Polynyas are particularly important as regions of positive heat exchange in the Barents Sea and Baffin Bay areas of the Arctic. They are detectable in high-resolution visible and infra-red imagery (Landsat, DMSP, TIROS-N), but only in the absence of cloud or through very tenuous clouds (e.g., figure 3.10, p. 104). For example, Landsat images have been used to study meteorological aspects of recurring springtime polynyas in coastal Alaska. Analysis of these important sea-ice features in the Antarctic has been made using satellite visible imagery composited over five-day periods in order to 'remove' the more transient cloud effects (see Section 8.3). Three-day averaged passive microwave data have also been used (e.g., Carsey, 1982). These data from the Nimbus-5 ESMR (Electrically Scanning Microwave Radiometer) sensor totally exclude cloud cover effects at high latitudes (Section 8.3). Figure 8.1 illustrates a persistently recurring polynya in the winter ice cover of the Weddell Sea, Antartica. Possible synoptic scale meteorological effects of the recurring Weddell Sea polynya from satellite analysis (Carleton 1981*a*) are outlined below.

Like sea ice, snowcover is an important factor in regional and large-scale energy budgets. A fresh snow layer is characterized by low heat conductivity, high reflectivity in the visible wavelengths ($0 \cdot 3 - \sim 0 \cdot 9 \, \mu m$), high thermal emissivity and low surface moisture. These energy exchanges are evidently accentuated by feedbacks in the vicinity of the snowline. As a result of this and the strong relationship between snowcover and tropospheric temperatures, it seems essential to monitor concurrent snow extent as well as cloudiness variations on a synoptic scale. Cloud – snow interactions play a major role in the surface energy budget of the Arctic pack ice in summer. The stratus cloud decks absorb solar radiation and reradiate heat to the snow surface which, together with multiple reflection effects, assist snowmelt and the formation of meltwater puddles. The resultant lowering of surface albedo then increases the absorption of short-wave radiation and accelerates the surface melting process.

8.3. Theory and techniques of cloud–cryosphere satellite monitoring

Over the last decade satellite visible and infra-red imagery has provided the major data source for studies of snow and sea-ice extent. However, since both clouds and ice masses possess similar reflectances and the main emphasis has been on monitoring the variations of the cryosphere rather than those related to cloud cover, techniques have been developed that attempt to exclude the 'interference' of clouds at high latitudes. These techniques include compositing (averaging) procedures for operational visible and infra-red cloud imagery, and passive and active microwave sensing. The need to obtain concurrent satellite-derived information on cloud and associated snow and ice conditions important in climate studies has led to the recent evolution of a range of new satellite sensors and data retrieval techniques. Some of these are still in the experimental stage.

Discrimination between the cryosphere and cloud cover using visible imagery in early satellite studies involved direct photo-interpretation. As a consequence of their similar reflectances, variations in the structural patterns exhibited by the two surfaces are crucial. For cloud cover at high latitudes, low solar illumination can help to identify these structural differences via shadow effects. Studies of this type have been relatively limited owing to the often subjective nature of the procedures, although higher resolution data from Landsat have permitted many studies of ice and snow surfaces under cloud-free or near cloud-free conditions. These include estimates of the ice-water fraction, detection of leads and different ice types, features of the near-shore fast-ice zone and areas of pack ice break-up, all of which are important in assessing the role of the cryosphere in climate via surface energy budget variations (see Gloersen and Salomonson, 1975). However, the poor temporal coverage together with cloud obscuration severely restrict operational use of these data.

A more reliable and objective method of identifying cryosphere extent and its variations in visible imagery involves computer processing of successive operational SR (Scanning Radiometer) and AVHRR (Advanced Very High Resolution Radiometer) visible images to produce multi-day Composite Minimum Brightness (CMB) displays. This technique relies on the relatively conservative nature of snow and ice fields between individual scenes and the fact that cloud tops have higher albedos than most surfaces. Using greyscale enhancement methods, only the lowest reflectances observed per geographical grid element during the period are retained in the final composite; thus transient cloudiness effects are suppressed and the sea-ice margin or snowline can be delineated. The most common compositing period is five days, although longer periods have been used for some climatological studies. Difficulties with this technique occur (a) when the sky may have been at least partly cloudy for the entire period, so that ice concentrations are artificially increased;

(*b*) when cloud-free conditions occurred during the whole period or at least during the satellite passes, giving apparently greater open-water areas if ice advection is occurring; and (*c*) when it is not possible to determine from which of the days in the period the brightness value at a particular point was taken (Wendler, 1973). Nevertheless, utilization of this type of data has permitted, for example, mean monthly albedo maps at the synoptic scale for Arctic sea-ice zones to be constructed and studies to be undertaken of polynyas in Antarctica, which would be frequently cloud obscured on individual images (Streten, 1974; Carsey, 1982). Annual variations in snow and ice extent can therefore be monitored from CMBs, except in winter.

A similar compositing technique, applied to thermal IR imagery, utilizes the highest radiative temperatures observed at each geographical grid element during the period and permits sea-ice extent to be monitored during the periods of polar darkness. This method is of less value for snowcover determination since there may often be little thermal contrast between the snow cover and adjacent bare land compared with the sea-ice–ocean boundary. These Composite Maximum Temperature (CMT) displays can be greyscale calibrated and even enhanced to distinguish between areas of below- and above-freezing temperatures. One contamination problem is that surface temperature inversions, particularly over the extensive Greenland and Antarctic ice caps, can result in signatures for cloudy areas being up to 20 K warmer (darker) than for non-cloudy areas. The analysis and mapping of polynyas from higher resolution thermal infra-red imagery has proved particularly valuable.

While both CMBs and CMTs effectively suppress cloudiness in the vicinity of cryosphere boundaries, this parameter may be of the greater interest. In this case, Composite Average Brightness (CAB) displays can be generated which reveal the persistence of cloud for the compositing period. However, they are of no use in providing information on the types of clouds in the region of snow/ice boundaries, which is an objective of proposed climate monitoring programmes (see Chapters 1 and 6).

The deployment of passive microwave radiometers on the later Nimbus satellites, beginning with Nimbus-5, has facilitated collection of various geophysical data. Measurements of sea-ice parameters, such as ice/water concentration, ice type and the location of the ice margin, have been available from the single-channel (19 GHz) Electrically-Scanning Microwave Radiometer (ESMR) on Nimbus-5 and the multi-frequency SMMR (Scanning Multichannel Microwave Radiometer) on Nimbus-7 and are denoted by variations in surface brightness temperatures. Passive microwave sensors permit all-weather and all-season analysis of sea-ice conditions in both hemispheres (e.g., figure 8.1), unaffected by variations in cloudiness. On the other hand, simultaneous cloud cover information is not available from this sensor.

The intensity of passive microwave radiation sensed by a satellite gives a brightness temperature T_B (from the Rayleigh–Jeans approximation of the thermal emission from a blackbody at temperatures typical of the Earth and atmosphere). The T_B of a given pixel represents contributions from both the surface and atmosphere, including water vapour and liquid water. The T_B (K) of sea ice is a function of the emissivity ϵ of the surface which, if isothermal, is related to its physical temperature T by the equation

$$\epsilon = T_B/T \qquad (8.1)$$

ϵ is dependent on the composition and physical structure of the surface, which *in situ* investigations have shown to include such variables as ice compaction, thickness, density, porosity and roughness characteristics as well as the temperature profile within the ice. Since the variations in surface emissivity for sea ice are largely a function of the age of the ice (salinity), passive microwave analysis can be used to differentiate ice types. ϵ also varies with the wavelength of radiation considered and the viewing angle of the scanner. Studies (e.g., Troy *et al.*, 1981) indicate that emissivity is a more fundamental ice parameter and exhibits less sample-to-sample variation than the brightness temperature, and that optimal differentiation of ice type is achieved between frequencies of about $22 \cdot 2$ and $31 \cdot 4 \, \text{GHz}$.

Emissivities vary greatly between ice and water, and at the Nimbus-5 ESMR frequency (19 GHz) they are approximately $0 \cdot 4$ for a water surface, $0 \cdot 84$ for second and multi-year sea ice and $0 \cdot 95$ for first-year and young ice. These give brightness temperatures in the range $130 - 160 \, \text{K}$, near $210 \, \text{K}$ and about $235 \, \text{K}$ respectively. A simplified radiative transfer formula for a water/ice mixture can be given by

$$T_B = (C_w \epsilon_w T_w + C_I \epsilon_I T_I) \exp(-\tau') + T_A \qquad (8.2)$$

where ϵ_w, T_w and C_w are the emissivity, surface physical temperature and relative concentration of water respectively; ϵ_I, T_I and C_I are corresponding values for sea ice; τ' is the total atmospheric optical thickness; and T_A is the atmospheric contribution comprising direct upwelling radiation, downwelling radiation reflected from the surface and radiation from space reflected by the surface (Comiso and Zwally, 1982). T_A must be separated into contributions over water and over ice, due to the different reflectances. In the dry polar atmosphere, attenuation of the T_B signal by atmospheric water vapour is negligible, and the cirrus and stratus clouds normally occurring there are quite transparent at microwave wavelengths. The determination of ice concentration from an observed T_B is complicated for mixtures of different ice types by the variations of emissivity and the lack of an independent measurement of the physical temperature T_I of the sea-ice surface in non-isothermal conditions. The appropriate equation for the ice concentration C_I from equation 8.2 (Comiso and Zwally, 1982) is given by

$$C_I = \frac{T_B - T_O}{\epsilon_I T_{eff} - T_O} \tag{8.3}$$

where $T_O = \epsilon_w T_w \exp(-\tau') + T_{Aw}$ is the observed brightness temperature of the water; and $T_{eff} = T_I \exp(-\tau') + T_{AI}/\epsilon_I$ is the effective surface temperature.

Information on ice parameters from microwave data include the delimitation of the ice–ocean boundary, estimates of ice concentration and the open-water fraction within the pack, areas of snowmelt, differentiation of ice types, and relative ice thicknesses. The latter is, however, only an extrapolation from the identification of first-year or multi-year ice. The sea-ice margin is identified by relatively abrupt T_B changes between the ocean ($T_B \approx 135\,K$) and the pack ice ($T_B > \sim 190\,K$) resulting from the strong emissivity differences between the two media. Ice concentrations across the ice–ocean margin (figure 8.1) reportedly increase from about 10 % to 70 % over distances of the order of 100–150 km (Carsey, 1982). The ice margin is broader in summer than in winter. Short-term synoptic-scale variations in the ice edge can be readily identified and are apparently the result of atmospheric and oceanic fluctuations.

Open-water areas and low ice concentrations within the sea ice (e.g., the polynya in figure 8.1) are crucial to the estimation of the turbulent heat fluxes (oasis effect), the summer retreat of the ice margin and consequently for large-scale climate modelling. Generally, lowered T_B in the microwave implies an increasing open-water fraction. Comparison of estimates of ice concentration from Nimbus-5 ESMR, for example, with visual data from Landsat for the predominantly first-year ice in the Antarctic show errors in estimating ice concentration of about 10–15 % (Comiso and Zwally, 1982). This is expected to improve to less than 5 % for multi-frequency sensors on Nimbus-7 (SMMR) and subsequent satellites. Analysis of ESMR data has shown that there are apparently extensive areas where ice concentrations are lower than previously thought, even in winter. This is important from an energy budget standpoint. However, care must be taken to differentiate between polynyas (or genuine open-water areas), as in the case of that recurring in the Weddell Sea in late winter/spring, and other areas of apparently low ice concentration which may be summer melt puddles on the ice surface. Other studies have shown the apparent influence of synoptic (mainly cyclonic) systems on ice concentration with wave-like variations in T_B propagating through the pack in summer. These are of the order of 100 km per week and are superimposed on the much slower variations related to ice type. Acquisition of multi-frequency microwave data, particularly where they are of a higher resolution, will allow clarification and more reliable assessment of these processes.

Satellite microwave sensing also indicates the highly heterogeneous nature of the sea-ice cover, which is an important variable in the energy budget climatology of polar regions. Differentiation of ice types into 'first-year,' consisting of new ice and refrozen leads, and 'multi-year' (old) ice is obtained from emissivity variations which relate primarily to surface salinity and pore size differences. These give the range of T_B variations of ice to be 200 K to 240 K. The optical depth of the emanating microwave radiation becomes greater with increasing age (decreasing salinity) of the ice, and with consequently greater internal scattering within the brine pockets. For young ice (one year or younger) the radiation emanates mainly from a thin top layer (1 – 3 cm), and for older (second- and multi-year) ice, microwave emission mostly occurs within the top 15 cm. Transitional types, mainly in the near-shore zone, have been determined in the Arctic from ground-truth and aerial reconnaissance observations. Ice-type differences are most evident in winter, in the absence of surface meltwater. In this respect, a surface snowcover is essentially transparent and has minimal effect on sea-ice T_B except in summer, when the presence of free water from melting lowers the T_B. In the coarser resolution Nimbus-5 ESMR data (approximately 30 km × 30 km) percentage estimates of first-year and multi-year ice (and open water) can be made where a mixture occurs and where individual floes are smaller than the radiometric 'footprint' of the sensor, provided that the assumption of approximate constancy of ϵ for a given ice type remains valid and that there are no strong gradients of surface temperature (Carsey, 1982). In the presence of only one ice type, a 1 % change in concentration is evident as a T_B change of about 1 K. The use of multichannel data from Nimbus-7 SMMR should permit more reliable assessment of ice concentration in the presence of different ice types to be made.

The problem of obtaining concurrent data on cloud – cryosphere interactions is particularly acute for continental snowcover. Variations in snowcover extent occur more rapidly as the result of an individual cyclonic storm than is generally the case for sea ice. Thus, there is a strong need to monitor snowcover variations at small temporal and spatial scales. Concurrent information on the distribution and frequency of cloud in the vicinity of the snowline can then be used to judge existing cloud analysis models.

The identification of snowcovered regions has also relied heavily on visible channel imagery, both from polar orbiting and geostationary satellites. NOAA/NESS Northern Hemisphere snowcover charts, for example, distinguish three categories based on surface reflectivity. Complications arise because the reflectance is influenced by Sun angle variations, whether or not a snowcovered region is forested, and the skill of the chart analysts. The separation of low clouds from snow is particularly difficult, but should become easier if the snow – cloud discriminator sensor tested on a DMSP satellite is adopted for future cryospheric programmes.

The deployment of a near infra-red snow – cloud discriminator on the DMSP satellite in July 1979 was an experimental step towards achieving this aim, and followed earlier snow – cloud experiments from Skylab and US Air Force aircraft flights. Automated separation of cloud from snow in sunlit scenes is possible owing to the low reflectivity of solar radiation by snow $(8-16\%)$ in the near infra-red $(1 \cdot 51 - 1 \cdot 63 \, \mu m)$ range, compared with water clouds (see the discussion of table 6.1, p. 206). Some contrast is also achieved between snow and cirrus (ice) clouds, since the latter reflect $16-30\%$ of the incident solar radiation. Increasing age of the snow results in lowered reflectivities at these wavelengths. The DMSP Special Sensor C (SSC), which is a snow – cloud discriminator system, optimizes the contrast between snow and cloud scenes in near-subsolar terminator regions. Problems with the experimental system include sensor saturation from liquid water clouds as well as from arid or semi-arid surfaces, and restricted spatial and temporal availability of the data. However, the data show good separation of snow-covered and cloud-covered scenes when applied to a minimum brightness format for the period from October to December 1979 and when compared with NOAA/NESS snowcover charts. An estimated reliability of 90% in cloud/no-cloud decisions would probably improve as the sensor resolution is increased to match the $4 \cdot 1$ km resolution of the DMSP visible band. SSC data may be particularly useful in determining the presence of snow in forested areas, where visible-band imagery may fail to detect the snow beneath a tree canopy. Operational application of snow – cloud discriminator data clearly holds great promise for cloud – climate studies (see also Chapter 6) as well as for a variety of other environmental disciplines and was initially planned for ICEX. This programme has since been dropped, but components may be incorporated into future research studies.

The following sections document the analysis of cloud – cryosphere interactions in both hemispheres. The emphasis with respect to the cloudiness component has been generally on synoptic scale cloud systems (vortices). Cloud vortex analysis has proved a useful substitute for observations of cloud amount, type and height, and important relationships between ice and snow extent variations and the atmospheric circulation (cyclonic activity) are demonstrated.

8.4. Satellite-derived ice-atmosphere interactions in the Southern Hemisphere

The annual cycle of growth and decay of Antarctic sea ice, which is of the order of $2 \cdot 5 \times 10^6$ to $20 \times 10^6 \, km^2$, is believed to be the most significant climatic variable on monthly and seasonal time-scales for the Southern Hemisphere. In the Northern Hemisphere, this role is evidently filled by seasonal snowcover variations (Polar Group, 1980). A change in the extent of the Antarctic pack ice of only $1°$ latitude amounts to a change of about 15 % of the total sea-ice cover of the Southern Hemisphere, and is therefore likely to be crucial to the large-scale energy balance. Many studies have established the relationship between the seasonal variations in Antarctic sea ice and various meteorological elements such as temperature, pressure and wind speed at higher southern latitudes (see Carleton, 1981b for a review). Similarly, interannual variations in these quantities have been linked with ice variations in certain key longitude sectors. Until recently, a major difficulty in the identification of Antarctic sea ice – atmospheric circulation interactions was the acute lack of suitable spatial and temporal surface and upper air data over the Southern Ocean, particularly at the higher latitudes. Thus, only a general assessment could be made regarding the role of the sea-ice cover in Southern Hemispheric climate. This lack of data was particularly crucial for real-time weather analysis and forecasting, until the advent of operational satellite cloud imagery. Recently, the deployment of an extensive network of data buoys over the Southern Ocean during the First GARP Global Experiment (FGGE) not only assisted weather forecasting but permitted details of the atmospheric circulation at higher southern latitudes to be resolved. These data are now the basis for dynamic climatological studies, particularly of the active winter season months.

The availability of once-daily visible-channel mosaics ($0 \cdot 5 - 0 \cdot 7 \mu m$) of cloud cover conditions led to their application to problems of the atmospheric circulation in the Southern Hemisphere (Carleton, 1981b). A major component of this research was the identification of a synoptic-scale index of the circulation obtained from analysis of cloud vortex systems. Even given the relatively coarse resolutions of the data or their degradation when transmitted via facsimile machine between meteorological offices, one of the most obvious features of this imagery is the extensive organized cloud systems associated with extratropical depressions in various stages of development.

The cloud vortex classification system developed by Troup and Streten (described in Carleton, 1981b) is a method of data analysis that has been employed in an attempt to quantify the characteristic signature patterns in terms of associated fields of surface pressure and upper air anomalies. Although statistical in nature, this information was applied operationally to chart analyses and numerical forecasting models of the Australian

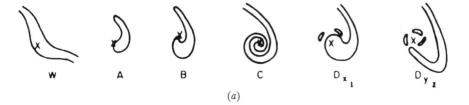

(a)

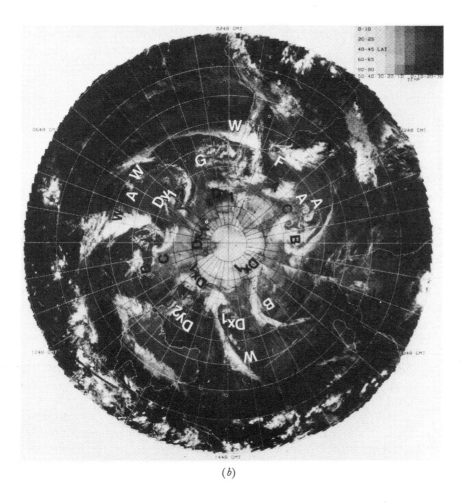

(b)

Figure 8.2. (a)Troup and Streten's classification scheme for Southern Hemisphere extratropical cloud vortices (described in Carleton, 1981a; see also figure 8.9).
(b) Composite infra-red image of the Southern Hemisphere illustrating the use of the Troup and Streten classification scheme.

Bureau of Meteorology. The classification of extratropical vortices is given in figure 8.2. Cyclogenesis (cyclone formation) may occur either as a wave development on a pre-existing frontal cloud band (W), or as an inverted comma-shaped (polar low) formation in isolation (A), frequently to the rear of a major frontal cyclonic vortex (see Section 6.4). Subsequent vortex development (B → C → D → decay) is deduced from the degree of organization of the vortical cloud and corresponds, respectively, to developing, maximum development (maturity) and dissipating stages of cyclonic evolution (figure 8.2). Maturity (C) occurs with the appearance of a distinct spiral of clear air in the cloud vortex (figure 8.2). The onset of dissipation occurs with the filling-in of the C-stage spiral, either as a cloud-covered (D_x) or a cloud-free (D_y) vortex. Further classification of the D-stage vortices was subsequently undertaken to account for the strong differences in orientation of the associated frontal cloud band in this stage, and this is incorporated into figure 8.2. Eventually, the old depressions lose their major frontal cloud band and decay (F/G vortices), usually at quite high latitudes.

While it is quite simple and based on earlier Northern Hemisphere satellite interpretation studies, this cloud vortex classification system has provided a framework for meteorological and climatological analysis of Southern Hemisphere cyclonic activity circulation patterns at middle and high latitudes. Carleton (1981*b*) used twice-daily NOAA infra-red imagery (figure 8.2(*b*)) for a five-year period (1973–77) to develop a synoptic climatology of cloud vortex types in the winter season (June to September). The original visible channel cloud vortex classification was found to be sufficiently versatile to apply to infra-red imagery, as vortices and frontal cloud bands appear geometrically similar in both image types. The basic features of the circulation identified for the other seasons were also found for winter, although some important regional differences were evident (Carleton, 1981*b*). In particular, a zone of high-latitude cyclogenesis was identified in the south-east Pacific (figure 8.3) and subsequently related to the relatively stable sea-ice margin in this region, in association with the Ocean Polar Front (OPF). These enhanced surface temperature gradients seem to explain the high frequencies of new cloud vortices in this area and also in the Ross Sea (figure 8.3).

The ability to distinguish successive stages of extratropical cyclonic development on the cloud imagery, based solely on pattern recognition of major recurring signature types, is the chief advantage of using these data compared with a synoptic chart series. This also helps to explain why previous workers had been unable to find convincing evidence of a relationship between Antarctic sea-ice extent and cyclone tracks, such as had been noted earlier for the North Atlantic. Most Southern Hemisphere extratropical cyclones form as wave disturbances (vortex type W) over middle latitudes in association with the major upper-air long waves in the

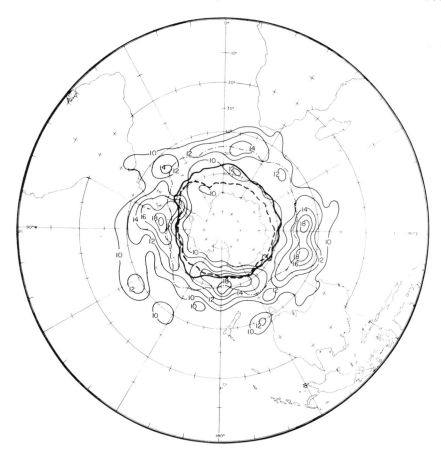

Figure 8.3. Mean monthly distribution of cyclogenesis (Troup and Streten W, A and B cloud vortices) for winters 1973–77, showing relationships with the Oceanic Polar Front (dot-dashed line) and the five-winter mean positions of the sea-ice margin for June (heavy dashed line) and September (heavy solid line). Isopleths give area-normalized cloud vortex frequencies in each 5° latitude by 10° longitude unit (from Carleton, 1981a, b).

vicinity of the OPF (figure 8.3). However, *in situ* cyclogenesis of the polar low (Type A) variety occurs generally at higher latitudes. It is the polar low comma cloud vortex that shows the close association with the ice boundary in the south-west and south-east Pacific (Carleton, 1981a) seen in figure 8.3.

The Antarctic sea ice is observed to undergo substantial interannual variation in terms of extent, rate of growth and the timing of maximum and minimum extent. Of particular interest is the relationship between these variations and the atmospheric circulation over middle as well as high latitudes. Non-satellite studies find that variations in the strength of the zonal westerly winds and the intensity of the major low pressure centres around Antarctica correspond to regional variations in ice extent. Schwerdtfeger and Kachelhoffer (see Carleton, 1981b) investigated climatological relationships between the frequencies of Southern Hemisphere cloud vortices identified on NOAA hemispheric visible mosaics and the seasonal locations of the Antarctic sea-ice margin in February/March (minimum extent) and September/October (maximum extent). They showed a statistically significant relationship between the expansion of the pack ice between autumn and spring and the accompanying equatorial shift in cyclonic activity. Strong increases in cyclone frequencies were also found for the Ross and Weddell Sea sectors, both of which show pronounced ice extent increases between autumn and spring. This link between the ice extent and cyclonic activity is believed to occur because the latitudinal temperature gradient at higher latitudes is stronger when the ice is nearest the equator (spring). These enhanced surface temperature gradients may amplify through much of the troposphere and be sufficiently unstable to generate more cyclones at that time. On the other hand, a greater number of lows probably helps advect the ice further towards the equator, at least in certain longitudes.

Substantial differences in Antarctic ice extent were observed for the winters of most contrasting hemispheric frequencies of cyclonic cloud vortices (1974, 1976) in the 1973–77 period (Carleton, 1981a). In the winter of least activity (1976) the zonally averaged ice extent was substantially less than in the winter of greatest activity (1974). Furthermore, there was virtually no increase in ice between August and September 1976 compared with these months in 1974. This anomaly probably represents a substantial change in the contribution of the pack-ice zone to the seasonal surface energy budget of the Southern Ocean and this may have been evident in the cyclonic frequency differences. In addition, longitudinal differences in cloud vortex frequency, particularly those in the later stages of cyclonic evolution (Types C and D) occurring at higher latitudes, were found to influence not only the regional extent of the ice but also the relationship between adjacent sectors through ice advection processes. Evidence of a synoptic feedback to the atmosphere by the ice was also found for the regional variations in cloud vortex frequencies in the Weddell Sea between the winters of 1974 and 1976, and particularly for cyclogenesis variations in the South Pacific. Satellite-observed frequencies of the W and A cloud vortices were greater in the 60–70°S zone in this region in 1976, when the winter ice edge was further south, than in this zone in 1974. However, greater frequencies of cyclogenesis occurred in the 50–60°S zone in 1974, consistent with the more extensive ice in that winter. This observation seems to bear out the importance of the sea ice for variations in surface temperature gradients and cloud vortex activity. An intriguing example of surface–atmosphere feedback

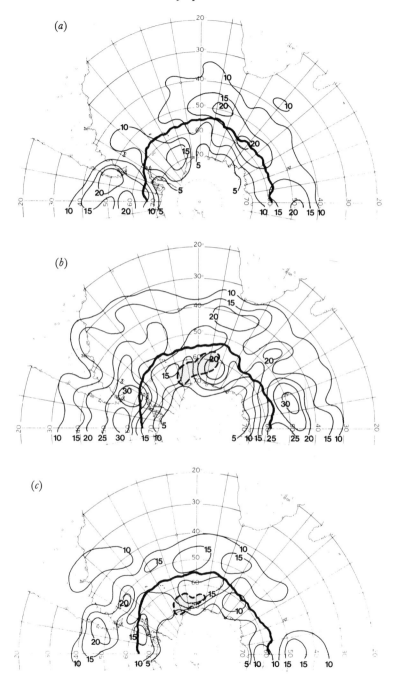

Figure 8.4. Mean monthly distributions of cyclogenetic cloud vortices for the South Atlantic sector, for winters (a) 1973, (b) 1974 and (c) 1975. Vortex frequencies are area-normalized to 45°S latitude. The monthly-averaged locations of the Antarctic sea-ice boundary in September (heavy solid line) and the Weddell polynya (heavy dashed line) are also shown for the relevant winters (from Carleton, 1981a).

is suggested by the patterns of cyclogenesis in the Weddell Sea for those winters when the open-water polynya was present (figure 8.4). Figure 8.4 shows that regional cyclogenesis was reduced in the winter of 1973, when the polynya was absent, compared to when the polynya was present (in 1974 and 1975). Apparently, the presence of such a feature can increase substantially the oceanic heat loss to the atmosphere during the winter and thereby provide more energy for the formation of new lows. Most of these vortices are of the Type A (polar low) variety.

In terms of the year-to-year (winter-time) variations of Antarctic sea ice and the occurrence of satellite-observed cloud vortices (depressions) over middle and high southern latitudes, some interesting relationships are observed (Carleton, 1983). These are summarized in figure 8.5, which shows the correlations r between the cloud vortex frequencies and the zonally averaged ice-edge latitude for each month of the five winters 1973–77. In general, between June and August of most years the number of lows increases as the ice increases (latitude decreases), but in September the frequencies of vortices decrease as the ice approaches its maximum extent. The table on the right of the diagram shows the relationship between both variables (ice, cloud vortices) for each full winter, and shows a trend from strongly negative (ice increases, lows decrease) to strongly positive (ice increases, lows increase) between 1973 and 1977. This also confirms a trend to generally less extensive winter ice observed during these years. Figure 8.5 also suggests a link between the higher-latitude location of the ice in 1977 and the greater cyclonic activity resulting from more ocean surface being exposed to the atmosphere in this winter. In the extreme ice-extent winter of 1973, however, the

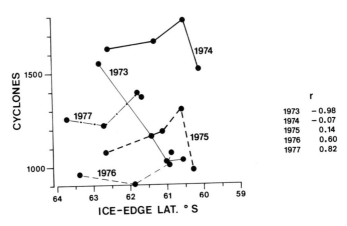

Figure 8.5. Plots of the monthly frequencies of cyclonic cloud vortices over the Southern Hemisphere (30–75 °S) in relation to latitude-averaged ice-edge position for the five winters 1973–77. Corresponding correlation coefficients r are also given (from Carleton, 1983).

opposite occurred (figure 8.5). Monthly cyclone frequencies decreased during the season in response to the reduced ocean – atmosphere heat fluxes at this time. Analysis of satellite-observed zonal cyclogenesis distributions in relation to the ice extent showed, in each year, decreases over middle latitudes and over the pack ice during winter, but an increase over ocean latitudes adjacent to the advancing ice edge (normally the 50 – 55°S zone). These results seems to confirm the winter mean relationship between ice extent and cyclonic activity.

Various studies have shown that interannual ice extent changes occur most strongly in the major Antarctic embayments and in East Antarctica. Similarly, the greatest regional variations in cloud vortex frequency also occur in these longitudes. Three key areas for such ice – atmosphere interaction are the Ross Sea (160°E – 140°W), the Weddell Sea (60°W – 0°) and East Antarctica between 90°E and 150°E. Carleton (1983) found, from statistical correlation analysis, that the least interannual variation of the three regions occurs in the East Antarctic. Figure 8.6(*a*) indicates that the difference between a heavy ice winter (1975) and a light ice winter (1977) in this region is relatively small in terms of the total numbers of dissipating (Type D) cloud vortices (greater in 1977). Conversely, maximum inter-annual variability in ice extent occurs for the Weddell Sea, but this appears to be related only secondarily to cyclonic activity. The importance of ocean circulation effects for the sea-ice regime of the Weddell region is confirmed. Stronger ice – cyclonic relationships are evident for the Ross Sea, and demonstrated for the winters of 1973 and 1976 (figure 8.6(*b*)). The maps show that in a winter of more extensive ice in this region (1973), the 'mean' low in the embayment is much stronger than in a winter of less extensive ice.

The relative significance of interannual variations in cloudiness and those associated with snow and ice extent are crucial to the planetary albedo. Earlier studies using multi-day averaged visible satellite imagery showed that the zones of maximum cloudiness delineate the hemispheric troughs over the Southern Ocean and, consequently, the major hemispheric storm tracks. More recent work has suggested that the Southern Ocean may be considerably more cloudy in winter compared with the Arctic Ocean. In order to assess, at least qualitatively, the possible effects on the planetary albedo of concomitant sea ice/cloud cover variations occurring on climatological time scales, the extreme ice extent winters of 1974 (heavy) and 1977 (light) were analysed. Satellite-observed depression centres in 5° latitude zones were obtained for the winters of 1973 – 77 as an indicator of relatively organized seasonal cloud cover and the cyclone frequencies for 1974 and 1977 were expressed as percentage departures from those five-year means. The latitudinal patterns are shown in figure 8.7. The 'cloudiness' differences between the two extreme ice years are particularly pronounced in the 50 – 55° S zone; over ocean

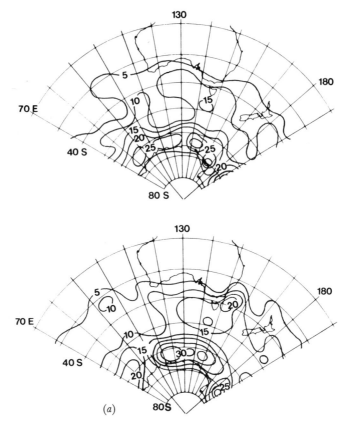

Figure 8.6. Spatial distributions of the frequencies of dissipating (Streten-type D vortices) (a) for the East Antarctic sector in winters 1975 (upper) and 1977 (lower); and (b) for the Ross Sea in winters 1973 (upper) and 1976 (lower). Isopleths are area-normalized to 45°S for each 5° latitude by 10° longitude unit (from Carleton, 1983).

latitudes just north of the sea-ice margin, and comprise the zone experiencing maximum seasonal variation in hemispheric albedo. Greater zonal ice extent in 1974 was associated with cyclonic activity (cloudiness) increases of 41%, whereas the lighter ice winter of 1977 had 10% less cloudiness in this zone compared with the five-year mean. Also in 1977 an increase of relative cloud amount in the 55–60°S zone occurred (towards the ice edge), but a decrease in this zone occurred in 1974. Relative cloudiness decreased in both years towards high latitudes. Figure 8.7 implies that 'cloudiness', related to organized synoptic cloud systems, may be greater over sub-Antarctic latitudes in a heavy ice winter than in a light ice winter, which may imply an increase in the planetary albedo in the climatically sensitive 50–60°S zone. However, this relationship is

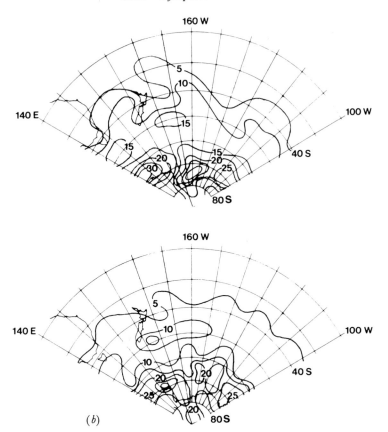

(b)

only based on very few years' satellite data and does not consider at all variations in unorganized and lower-level cloud as opposed to synoptic scale organized cloud systems of generally marked vertical thickness.

8.5. Relationships between snow and ice and synoptic cloud activity in the Northern Hemisphere

The question of cloud variability and feedback in the vicinity of cryosphere boundaries in the Northern Hemisphere is now receiving increased attention. Crucial to an understanding of cryosphere – cloud

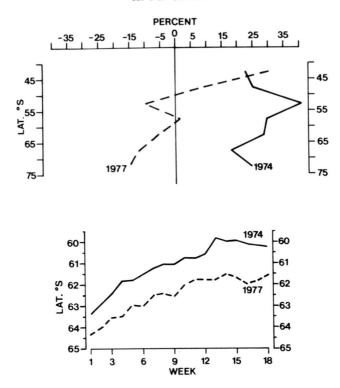

Figure 8.7. Latitude variations in 'organized' (vortical) cloud cover for the Southern Hemisphere in extreme ice winters (1974, 1977). Each value represents the percentage departure from the mean five-winter (1973–77) cloud vortex frequency for that particular 5° latitude band. The maximum variation in cloudiness occurs equatorward of the sea-ice margin in both years (50°–55°S). Also shown (lower) are the weekly latitude-averaged locations of the Antarctic sea-ice boundary for the two extreme winters. Week 1 corresponds to the ice chart closest in time to 1 June and week 18 is the chart at the end of September/beginning of October in each year.

interactions is the dominant direction and magnitude of feedback processes. Nimbus-5 microwave brightness temperature signatures of Arctic sea ice for a full annual cycle (1973/74) have been related to grid-point data on surface air temperature and atmospheric pressure. The direction of the ice–atmosphere forcing is found to vary seasonally and regionally, as well as between the permanent (multi-year) pack ice and the seasonal (mainly first-year) sea-ice zones. Similar studies using Nimbus-5 microwave data are now being conducted for the Antarctic sea ice (e.g., figure 8.1 and Comiso and Zwally, 1982).

In order to interpret reliably the feedback results from studies of time-averaged ice–atmosphere interaction, the role of individual synoptic events occurring on short (daily) time-scales needs to be assessed. An example, using two-day averaged ESMR digital surface brightness

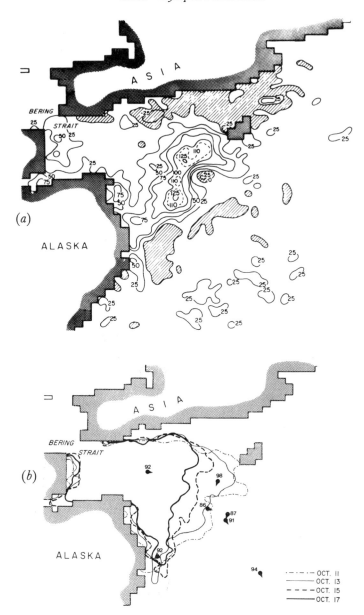

Figure 8.8 (a) Numbus-5 ESMR brightness temperature (T_B) difference map for the Chukchi and western Beaufort Seas, 17 October minus 11 October 1976; and (b) two-day averaged locations of the sea-ice – open water margin for the same period. The ice edge is taken to be the 210K isopleth in the ESMR data. Daily locations of the dominant low-pressure system, together with directions of movement and central pressures (900 + mb) are superimposed for the period 10 – 16 October 1976. T_B increases of the order of 110 – 125K are evident in (a) and are associated with the rapid 'advance' of the ice edge. The feature off the Arctic coast of Asia in the pack-ice zone is Wrangel Island.

temperature data for the Chukchi and East Siberian Seas (Western Arctic), is given for the first part of October 1976 (figure 8.8) — October is representative of the season of ice advance. The atmospheric circulation in the western Arctic is known to be important for the synoptic ice dynamics of this region and for the observed interannual variations in ice extent. Complementary DMSP infra-red cloud imagery for this sector was used to track the development of an intense and slow-moving cyclonic system for 10–16 October 1976. The extensive cloud vortex of this depression kept the Chukchi sea-ice cover obscured for the entire period in the infra-red data, but the sea-ice was determined from the all-weather ESMR sensor. The strong cold north to northwesterly airflow associated with the cyclone caused a rapid advance of the Chukchi ice edge of the order of 200 km from 11 to 17 October. This 'advance' doubtless represents the combined effects of wind-induced advection of the pack and new ice growth at the margin. Figure 8.8 shows the spatial patterns of the surface T_B changes occurring during the period, and the daily locations and intensities of the low pressure centre. T_B increases of 75 K are associated with the advance of the Chukchi ice margin, but locally these are as high as + 110 K and correspond to new-ice signatures. A feature of interest is the area of low ice concentration behind the advancing ice edge (figure 8.8(*a*)). The development and expansion of this polynya after 13 October is evidently a response to dynamic forcing, implying divergence of the pack under favourable wind and current patterns. The virtual stationary position of the polynya seems to emphasize the role of the ocean circulation in the vicinity of Wrangel Island, at the same time as the ice margin migrated rapidly southwards in response to forcing by the synoptic wind field. Clearly, the location and movement of the cloud system on successive days was the dominant control on the ice configuration and advection in this region. More problematical is the detection of surface feedback to the atmosphere, although the sharp deepening of the cyclone's central pressure (998 to 986 mb) on 12–13 October at the time of closest proximity to the ice margin (figure 8.8) is suggestive of enhanced tropospheric temperature gradients. The cyclone weakened on ensuing days as it traversed the pack ice (figure 8.8(*b*)).

On the climatological time-scale, a longstanding curiosity has been the apparently close regional relationship between Arctic snow and ice boundaries and synoptic (mainly cyclonic) activity. This is a feature reproduced in recent climate modelling experiments. The availability of sets of high resolution (3·7 km) DMSP visible (0·4–1·1 μm) and IR (8–13 μm) imagery has permitted this problem to be investigated using a satellite-derived climatology of cloud vortex signature types for the Northern Hemisphere (poleward of 20°N) for the mid-season months (January, April, July, October) of two years (1977, 1978/79). A new cloud vortex classification system applicable to high resolution imagery

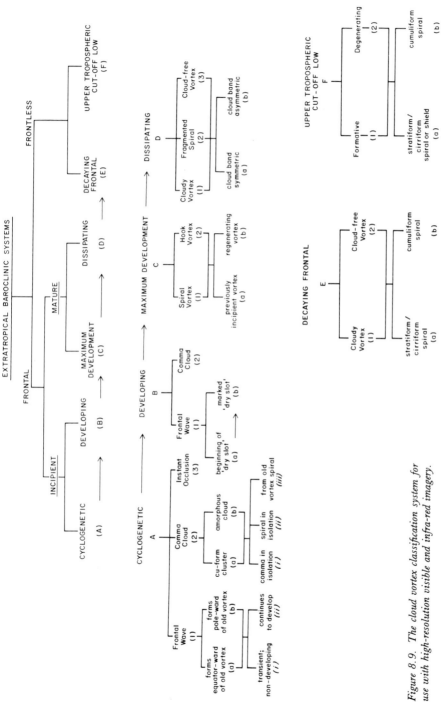

Figure 8.9. The cloud vortex classification system for use with high-resolution visible and infra-red imagery.

was developed for this purpose. The scheme for classifying frontal depressions is presented in figure 8.9. This system owes much of its basic framework to earlier cloud vortex classifications, such as that shown in figure 8.2 (p. 301) for the Southern Hemisphere, but the higher resolution of the available data allowed greater detail to be obtained at the secondary and tertiary levels. In particular, the polar low (A2) type is found to exhibit diverse signature configurations. Cloud vortices are grouped into the four primary development classes: cyclogenesis, A and B; maturity, C; dissipating, D; and decaying, E; for comparing with cryospheric variables. Three cryospheric boundary types are distinguished: sea-ice margin, continental snowline and snowcovered land/ocean margin; as well as two non-boundary classes (pack ice, snowcover). The boundary types are chosen because of the likelihood of strong horizontal temperature gradients being associated with them.

Table 8.1 gives mean daily frequencies of Northern Hemisphere satellite-observed depressions classified according to signature type (figure 8.9) for mid-season months. A mid-winter cyclonic maximum (two-year

Table 8.1. Mean daily frequencies of cloud vortex types for mid-season months, north of 20° N.

	January	April	July	October
Cyclogenetic				
A1	2·4	2·2	2·0	2·1
A2	2·1	0·6	0·2	1·0
A3	0·2	0·1	0·0	0·2
Developing				
B1	2·9	2·3	2·6	2·9
B2	0·8	0·3	0·1	0·6
Mature				
C1	1·6	0·9	1·1	1·0
C2	1·1	0·7	0·7	1·1
Dissipating				
Da	2·5	2·2	2·4	2·7
Db	0·5	0·7	0·8	0·7
Decaying				
E	2·4	1·6	2·2	1·6
Cut-off				
F	0·2	0·1	0·1	0·5
Tropical cyclonic				
TC	0·0	0·0	0·3	0·5
Hemispheric mean per day	16·7	11·7	12·5	14·9

total = 2079) is followed by a minimum in spring (1390), with cyclonic frequencies increasing through the summer (1565) and autumn (1826). The April minimum at higher latitudes involves the dominance of a seasonal polar anticyclone. Table 8.1 also shows that the ratio of non-frontal (A2) to frontal wave (A1) cyclogenesis for the hemisphere is at a maximum in January, decreasing through the spring to a summer minimum. A similar seasonal dependence of these cyclogenetic vortex types is also observed for the Southern Hemisphere.

Changes in the frequency of occurrence of most successive cloud vortex stages are found to be more strongly seasonal than latitudinal. This represents a strong difference from the Southern Hemisphere, where cyclogenesis is dominant over middle latitudes and cyclonic maturity and dissipation occur at progressively higher latitudes in all seasons, especially in winter. The Northern Hemisphere pattern seems to reflect the seasonal variation in thermal state of the underlying surface and is to be expected from consideration of changes in the dominant hemispheric heat sources and sinks during the year.

In January, cyclonic activity occurs over a broad range of ocean latitudes with a zone of maximum occurrence at 40–60°N. The strongest latitudinal zonation of vortex frequencies occurs in spring, near the time of maximum sea-ice extent, with marked activity over middle latitudes. In July, with the seasonal retreat of sea ice and snowcover, cyclone frequencies increase at higher latitudes (50–70°N) and over land. In autumn (October), total depression frequencies decrease between 50° and 70°N, particularly over land, but increase at lower–middle ocean latitudes. This pattern continues into winter.

In order to confirm this association between the seasonal (and latitudinal) synoptic circulation changes and the cryosphere boundary variables, a statistical study was undertaken for all mid-season months combined and for individual months (either snowcover or sea ice). The results are summarized in tables 8.2, 8.3 and 8.4. The χ^2 test used gives an idea of the degree to which the observed seasonal and latitudinal distribution of cloud vortex types associated with surface categories are

Table 8.2. χ^2 values and significance levels for seasonal cryosphere – cloud vortex associations.

	Pack ice	Sea-ice margin	Open ocean	Snowcovered land	Snowline	Snowfree land
χ^2	17·33	30·34	69·81	9·60	4·36	14·06 (21·26)[a]
Sig.(%)	5	0·1	0·1	Not sig.	Not sig.	5 (1·0)[a]

[a]Values in brackets represent included cyclone frequencies for July for the snowfree land surface group.

Table 8.3. χ^2 values and significance levels for sea ice–cloud vortex associations.

	January	April	July	October
χ^2	18·33	28·10	21·28	19·30
Sig.(%)	1·0	0·1	1·0	1·0

Table 8.4. χ^2 values and significance levels for snowcover–cloud vortex associations.

	January	April	October
χ^2	34·18	7·04	25·26
Sig.(%)	0·1	Not sig.	0·1

more likely to be due to chance or random factors than to indicate a real and significant relationship. Thus, 0·1 % chance is more significant (less random) than 5 % chance. There is a strong seasonal relationship (statistically significant at 0·1 %) between the sea-ice–ocean margin and cyclonic activity (table 8.2). Surface–atmosphere interactions for the ocean are also highly significant. For the pack-ice zone, seasonal vortex associations are slightly less significant (5 % level) and apparently correspond to the reduced thermal control by these surface types.

No statistical significance can be ascribed to the seasonal variations in cyclone frequencies over snowcover (table 8.2). This reflects a dominant subsidence of air over extensive snowfields. While the lack of a significant association at the snowline is perhaps surprising (table 8.2), the significance level increases to 5 % for snowfree land, possibly implying a seasonal displacement of cyclonic activity south of the snowline.

The latitudinal relationships between cyclonic cloud development and cryosphere groups (tables 8.3 and 8.4) show, in the case of the sea-ice variables, a χ^2 significance for the hemisphere that is highest in April (0·1 %). This corresponds with the time of maximum extent of Arctic ice and indicates a generally close association between cyclone tracks and the sea-ice margin in spring. For snowcover, the strongest relationships occur in January and October (significant at 0·1 %), but the association is not statistically significant in April. A strong regional correlation has been noted between the position of the snowline in Eurasia and an alternative synoptic index (the planetary upper frontal zone) in both transition seasons (Afanas'eva et al., 1979).

Since there is a generally close relationship between the cryosphere and cyclonic cloud vortices in the Northern Hemisphere, is it possible to determine, for climatological time scales, if the extent of snow or ice is helping to determine the cyclone distributions rather than the reverse? This feedback question is looked at in figure 8.10, which shows latitudinal

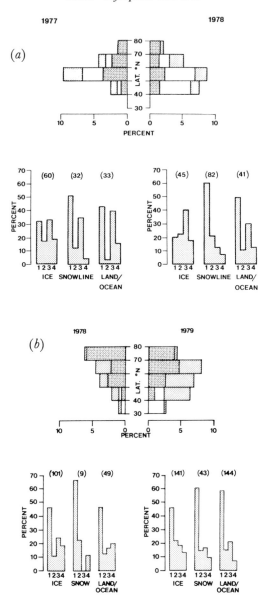

Figure 8.10. Relationships between cryospheric boundary variables and successive stages of cloud vortex development (a) for April 1977 and 1978; and (b) for January 1978 and 1979. Latitude frequencies of cryosphere variables (upper) are for the sea-ice margin (stippled), continental snowline (blank), and snowcovered land/ocean margin (hatched), and are presented as the percentage of all surface types (cryosphere and other) in 10° latitude zones. Cloud vortex stages (lower) are shown for each cryosphere boundary variable over all latitudes. Cyclone groups are: (1) cyclogenetic and developing vortices, (2) mature, (3) dissipating, (4) frontless. Numbers in brackets give actual numbers of cyclones associated with each cryosphere variable.

frequencies of occurrence of the three cryosphere boundary variables associated with cloud vortices for April 1977 and 1978, together with the respective cyclone frequencies. The latitudinal variations in surface types between the two years are statistically significant. Greater numbers of cyclones were associated with the snowline over middle latitudes in April 1978, especially at 40–50°N, compared with April 1977. Also, the ratio of cyclogenesis to cyclolysis (dissipating) vortex types was greater in 1978. Thus, a stronger surface to atmosphere feedback is implied from the enhanced cyclogenesis in 1978, arising from the more southerly location of the snowline. This is expected in view of the fact that, in general, cyclogenesis dominates at lower–middle latitudes of the Northern Hemisphere, especially along the eastern coasts of North America and Asia. The greater extent of snowcover appears to enhance the satellite-observed cyclogenesis patterns, even on a hemispheric scale.

Similar surface-to-atmosphere associations for both the snowcovered land/ocean margin and the sea-ice boundary, have been found for the two January months (1978, 1979; figure 8.10). The differences between years were also significant at the 0·1 % level. Enhanced temperature gradients along the seaboard areas, arising from more extensive snowcover extent in January 1979, are reflected in greatly increased cyclogenesis. Such a satellite-based result confirms, on a hemispheric scale, that previously

Figure 8.11. DMSP infra-red image for 12 January 1979 showing two cyclones positioned close to the sea-ice edge in the Greenland and Norwegian Seas.

suggested for eastern North America during winters of anomalously large snowcover extent. Since daily fluctuations in sea-ice extent are generally much less than those occurring for continental snowcover, the result for this boundary variable in January 1979 compared with 1978 (figure 8.10) is consistent with the assertion that greater sea-ice extent in a given month may enhance cyclogenesis for the hemisphere as a whole. Figure 8.11 shows this effect synoptically. Two cyclones are travelling east adjacent to the sea-ice edge of the Norwegian and Greenland Seas, having formed at the ice-ocean boundary 24 hours earlier. Similarly, the frequencies of cloud vortices in the later (dissipation) stages of development are observed to decrease in January 1979. Analysis of further years' satellite data is clearly warranted to confirm these synoptic feedback relationships. The patterns of cryosphere – atmosphere interaction shown here may be important in the development or augmentation of large-scale climatic anomalies on monthly and seasonal time scales, and have implications for cloud – climate modelling experiments involving the higher latitudes.

Variations in the frequencies and spatial distributions of higher-latitude cyclonic cloud vortices are evidently important for the generation of sea-ice anomalies on monthly and seasonal time-scales. However, a more quantitative approach is to consider the effect of intensity of cyclone types occurring in different latitude bands in different seasons. Such a satellite-based method has been applied recently to the Northern Hemisphere and facilitates comparisons with the results obtained from numerical climate modelling and previous observational studies of ice – atmosphere interactions.

Statistical models of surface pressure and upper-air height patterns associated with Northern Hemisphere cloud vortex types, based on the classification given in figure 8.9, have been derived from computer processing of extensive amounts of synoptic data (land surface and ships) for four mid-season months for 1977 and 1978. The cloud vortex models are presented as composite anomalies from the climatological monthly means of pressure and height which are seasonally, latitudinally and longitudinally dependent, according to pressure level. Spatial patterns of pressure ($\overline{\Delta p}$) and 500mb height ($\overline{\Delta \Phi}$) anomalies, grouped according to vortex signature type and for $10°$ latitude zones, have been derived. The technique is similar to that used in earlier work for Southern Hemisphere cloud vortices. The models show a general intensification and expansion of cloud vortex systems with development and also reveal a strong seasonal dependence, with most types being deepest in winter and shallowest in summer.

The mean surface pressure ($\overline{\Delta p}$) and upper air, 500mb ($\overline{\Delta \Phi}$) anomalies characteristic of each vortex type were assigned to the monthly cloud vortex census obtained from analysis of twice-daily DMSP imagery. Anomalies were then accumulated ($\Sigma \overline{\Delta p}$, $\Sigma \overline{\Delta \Phi}$) for the month

(a)

(b)

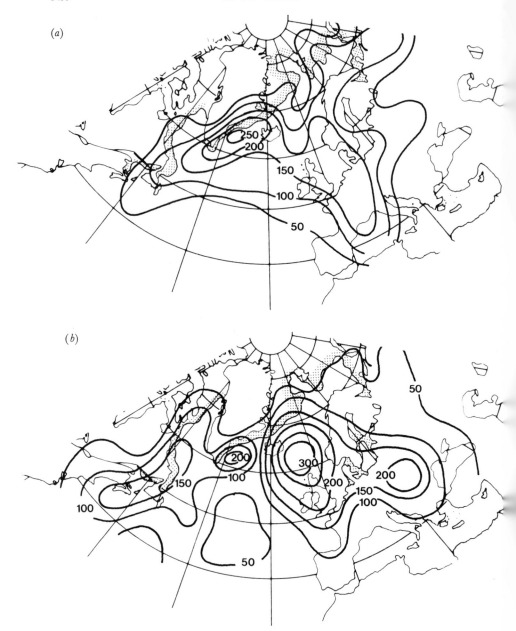

Figure 8.12. Monthly cumulative negative surface pressure anomalies based on statistically derived cloud vortex intensities, for the eastern Arctic in (a) January 1978 and (b) January 1979. The location of the sea-ice margin at the end of each month is also shown. Isopleths are in mb and refer to values in each 10° latitude/longitude unit.

in each 10° latitude × 10° longitude unit. In this way, monthly statistical distributions of surface pressure and 500mb height anomalies could be mapped according to the occurrence of specified cloud vortex signature types. They represent an initial attempt to compare systematically the observed interactions between regional sea-ice conditions and the different stages of cyclonic development, where 'mean' vortex size and intensity are implicit, with the coupled circulation/ice-extent anomalies generated in numerical climate modelling experiments.

Comparisons of the $\Sigma\overline{\Delta p}$ and $\Sigma\overline{\Delta\Phi}$ patterns for the two years showed the seasonal migration of the major pressure/height anomaly centres and their variations in intensity with regional changes in extent of the sea ice. This is particularly apparent for the semi-permanent Icelandic and Aleutian lows and is also a feature simulated in modelling experiments.

The question of the dominant direction of feedback between the ice and the cyclonic activity obviously becomes important in the interpretation of such patterns, and it is generally not possible to determine this adequately from current numerical models. However, this feedback can be evaluated here from the regional geostrophic wind patterns inferred from the distributions of negative anomalies. For example, the deep anomaly centre at the sea-ice margin south-west of Greenland in January 1978 (figure 8.12(*a*)) resulted in a dominant cold north-easterly airflow and more severe regional ice conditions in Davis Strait, in contrast to the milder southerly flow associated with the anomaly near Newfoundland in January 1979 (figure 8.12(*b*)). Similarly, the intense negative anomalies over north-west Europe in January 1979 (figure 8.12(*b*)) were associated with the extensive ice in the Baltic Sea in that month compared with the January of the previous year.

Strong differences in satellite-derived cumulative anomaly patterns were also observed between other corresponding months of the two years studied and these can be related to the variations in sea-ice extent. Figure 8.13(*a*) shows the difference pattern of surface pressure anomalies between April 1978 and 1977 in the North Pacific. Negative surface and upper-air pressure/ height anomalies in the Bering Sea were greater in the April of more extensive ice (1977) than in the lighter ice April of 1978. This is consistent with numerical modelling results for this region. However, less ice occurred in the Okhotsk Sea in April 1977 compared with 1978, which is associated with deeper regional anomaly patterns at the ice edge. Regionally dependent patterns are also evident between the two January months (1978, 1979) in figure 8.13(*b*). The less extensive ice in the Bering Sea in January 1979 was associated with much deeper anomalies to the south (cf. April), although ice was more extensive to the west in the Okhotsk Sea at this time. Similarly, the pressure anomaly at the Okhotsk ice edge was deeper in the winter of less extensive ice (1978). These

A. M. Carleton

(a)

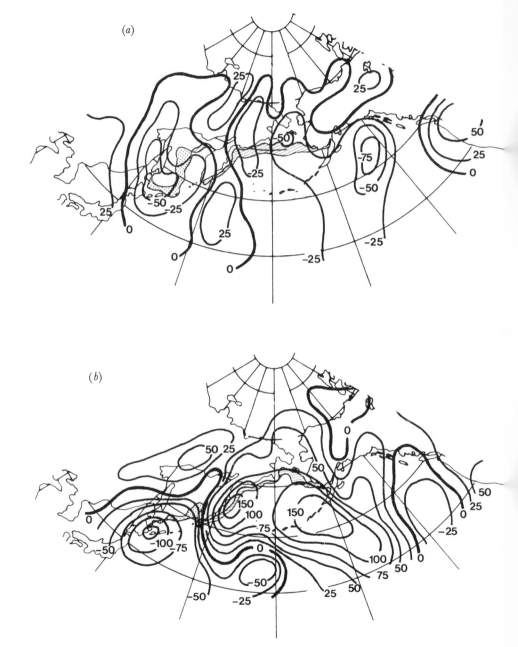

(b)

Figure 8.13. Difference patterns of monthly cumulative negative surface pressure anomalies for the western Arctic: (a) April 1978 minus April 1977; and (b) January 1979 minus January 1978. Isopleths are drawn at 25mb intervals. Differences in sea-ice extent are also shown stippled for greater ice in the latter year, and hatched for greater ice in the former year.

satellite-derived patterns suggest a high degree of regional and seasonal interdependence of ice – cyclone feedbacks. A major problem in relating sea-ice conditions with the atmospheric circulation pattern on the hemispheric scale involves the pronounced asymmetry in ice extent across the Arctic basin. However, based on the observation that there was a greater area of sea ice in January 1979 compared with January 1978, and in April 1977 compared with April 1978, an attempt was made to examine the assertion of previous workers that cyclones are displaced further south in years of more extensive Arctic sea ice. By computing zonal (all-longitude) totals of satellite-derived $\Sigma\overline{\Delta p}$ and $\Sigma\overline{\Delta\Phi}$ for each $10°$ latitude band, it was found that the more extensive sea ice in January 1979 was accompanied by cyclone tracks located further south, and with depressions of greater frequency, than in the milder January of 1978. A similar pattern was noted for the two April months studied (1977, 1978), where cyclone-related negative pressure and height anomalies were reduced at high latitudes ($71 – 90°N$) in April 1977 compared with April 1978, but greater further south at $51 – 70°N$, corresponding to the southward displacement of depression tracks.

8.6. The climatic significance of cryosphere – atmosphere interactions

In the context of global climate, the polar regions are dynamic heat sinks that maintain their climatic equilibrium temperature by energy advected from extrapolar latitudes (Polar Group, 1980). They are therefore crucial for the regulation of the general circulation, operating via various inter-related non-linear feedback processes. Since variations in snow and ice limits constitute an index of oceanic and atmospheric anomalies, a key research problem involves circulation teleconnections between the high and lower latitudes and the time-scales of their operation. Of particular interest is the extent to which a circulation anomaly, once established, may be augmented by ice – albedo feedback in the climatically sensitive zone at about $60 – 70°$ latitude. Ice – albedo feedback is likely to be particularly critical at the pack ice – ocean boundary. Recent numerical modelling experiments indicate that the ice – albedo feedback may enhance a global climate sensitivity by up to $25 – 30\%$. Various studies document some of these large-scale anomaly teleconnections (for reviews see Polar Group, 1980; and Carleton, 1983). For example, the well-known oscillation in winter temperatures between northern Europe and Greenland observed in certain years is expressed as distinct regional ensembles of ice – ocean – atmosphere interaction involving characteristic variations in intensity of the Icelandic and Aleutian 'mean' lows and the upper-tropospheric long waves. Tropical as well as mid-latitude tele-connections have been identified in association with this phenomenon.

With respect to snowcover, connections have been made between the augmentation of high-latitude atmospheric blocking sequences initiated in middle latitudes and anomalous snowcover extent and even the depth of winter snow in Tibet and the onset of the Indian summer monsoon. As a general 'model' of snowcover–atmosphere interaction it is postulated that anomalous snowcover extent serves to chill further the overlying atmosphere, decreasing the 1000–500 mb thicknesses in the lower troposphere, and thereby augmenting lower-level anticyclonic circulations over the snow while enhancing storm development near the snowline, especially in coastal zones. Frequencies of cyclogenesis are then increased, which serve to augment the established anomaly by disturbing the 'normal' upper-air pattern. The observational results given in this chapter seem to confirm the essential features of this model, but further studies are required. The problem of disentangling the relative importances and dominant directions of feedback is of current and crucial interest.

There also appears to be a seasonal dependence to the leads and lags of ice–atmosphere interaction. Statistical analysis of pressure/height fields over the Arctic indicates that in autumn there is a two-month atmospheric response to the ice advance that is slightly greater than the response of the ice to atmospheric forcing. In the other seasons, especially in spring, the atmosphere leads the ice by up to four months. However, the synoptic scale implications of these feedbacks are not well understood. Many studies show the effects of anomalies in the lower tropospheric pressure, wind and air temperature fields in generating regional sea-ice anomalies. More controversial (and harder to detect) is the extent to which the sea-ice variations forced by the atmospheric anomalies may subsequently modify the circulation, especially on shorter synoptic time-scales. An attempt has been made in this chapter to present evidence for such forcing on the basis of satellite image interpretation.

This chapter has shown the value of using satellite cloud data in the investigation of cryosphere–atmosphere interactions. The major emphasis has been on identifying organized cloud systems (cyclones) appearing on visible and infra-red imagery and relating their frequencies of occurrence, stages of development and statistical intensities to the location and movement of the climatically sensitive snow and ice boundaries. However, this is only one aspect of cryosphere–cloud interactions. Very little is known about the relationships between cryosphere boundaries and mean (climatological) cloud parameters, such as cloud types, amounts and characteristic heights, and their mutual variations. This has been shown to be mainly the result of an acute lack of concurrent ice/cloud information in the largely inaccessible polar regions. The advent of new satellite sensors promises availability of simultaneous cloud–cryosphere data without the problems involved in comparing different types of data averaged over relatively long (monthly, seasonal) time-scales. Only from an analysis of the short-term surface – atmosphere interaction in high latitudes, arising from development of a detailed cloud climatology, is a more precise appraisal of the feedback processes important to global weather and climate and its variations possible.

Acknowledgements

I am grateful to Dr R. G. Barry for his comments on an initial draft of this chapter.

References

Afanas'eva, V. B., Esakova, N. P. and Klimentova, R. V., 1979, Relation of the planetary upper-air frontal zone to the position of the snow limit during the fall and spring. *Soviet Meteorology and Hydrology* **9**, 87 – 89.

Carleton, A. M., 1981*a*, Ice – ocean – atmosphere interactions at high southern latitudes in winter from satellite observation. *Australian Meteorological Magazine* **29**, 183 – 95.

Carleton, A. M., 1981*b*, Monthly variability of satellite-derived cyclonic activity for the Southern Hemisphere winter *Journal of Climatology* **1**, 21 – 38.

Carleton, A. M., 1983, Variations in Antarctic sea ice conditions and relationships with Southern Hemisphere cyclonic activity, winters 1973 – 77. *Archives for Meteorology, Geophysics, and Bioclimatology* **B32**, 1 – 22.

Carsey, F. D., 1982, Arctic sea ice distribution at end of summer 1973 – 1976 from satellite microwave data. *Journal of Geophysical Research* **87**, 5809 – 35.

Comiso, J. C. and Zwally, H. J., 1982, Antarctic sea ice concentrations inferred from Nimbus 5 ESMR and Landsat imagery. *Journal of Geophysical Research* **87**, 5836 – 44.

Gloersen, P. and Salomonson, V. V., 1975, Satellites — new global observing techniques for ice and snow. *Journal of Glaciology* **15**, 373 – 89.

Polar Group, 1980, Polar atmosphere – ice – ocean processes: a review of polar problems in climate research. *Reviews of Geophysics and Space Physics* **18**, 525 – 43.

Streten, N. A., 1974, A satellite view of weather systems over the North American Arctic. *Weather* **29**, 369 – 80.

Troy, B. E., Hollinger, J. P., Lerner, R. M. and Wisler, M. M., 1981, Measurement of the microwave properties of sea ice at 90 GHz and lower frequencies. *Journal of Geophysical Research* **86**, 4283 – 89.

Vowinckel, E. and Orvig, S., 1970, The climate of the North Polar Basin. In *Climates of the Polar Regions*, World Survey of Climatology, Vol. 14, edited by S. Orvig. (Amsterdam: Elsevier), pp. 129 – 252.

Wendler, G., 1973, Sea ice observation by means of satellite. *Journal of Geophysical Research* **78**, 1427 – 48.

Epilogue

Over the 4·5 billion year history of our planet, water has played a major environmental role by modifying the chemical composition of the atmosphere, controlling the climate and forming surface features. Water molecules originally entrapped in the planet during its formation were released to the atmosphere via volcanic outgassing early in the Earth's history. As a result of the temperature and pressure distribution of the early atmosphere, the outgassed water vapour rapidly condensed on to the surface, forming the oceans and giving rise to the hydrological cycle. Gases in the early atmosphere energized by lightning and ultraviolet radiation from the Sun formed complex organic molecules: the precursors of living systems.

Viewed from space today, the Earth's colours of blue and white are the result of the liquid and solid phases of water on the surface and in the atmosphere. The Earth's surface, atmosphere and hydrosphere can only be understood fully by continuous and continual development of research and operational satellite systems. The research community, aerospace industry and technical applications contractors are capable of initiating new developments and promoting rapid evolution of research tools. The impressive development over the last two decades of sensor and satellite capabilities has led to a much wider application of remote sensing information than might have been anticipated in the 1950s. In this text we have sought to draw together fundamental information relating to remote sensing of the cloudy planet—Earth.

Index

329